LABORATORY MANUAL
FOR INTRODUCTORY GEOLOGY

ALLAN LUDMAN Queens College, City University of New York

STEPHEN MARSHAK University of Illinois, Urbana-Champaign

W. W. Norton & Company • New York • London

W. W. Norton & Company has been independent since its founding in 1923, when William Warder Norton and Mary D. Herter Norton first published lectures delivered at the People's Institute, the adult education division of New York City's Cooper Union. The firm soon expanded its program beyond the Institute, publishing books by celebrated academics from America and abroad. By mid-century, the two major pillars of Norton's publishing program—trade books and college texts—were firmly established. In the 1950s, the Norton family transferred control of the company to its employees, and today—with a staff of four hundred and a comparable number of trade, college, and professional titles published each year—W. W. Norton & Company stands as the largest and oldest publishing house owned wholly by its employees.

Composition by TEXTECH
Manufacturing by Worldcolor, Inc.
Illustrations by Precision Graphics

Editor: Jack Repcheck
Senior project editor: Thomas Foley
Production manager: Christopher Granville
Book Design: Brian Salisbury
Art Direction: Hope Miller Goodell
Design Director: Rubina Yeh
Copy editor: JoAnn Simony
eMedia editor: Rob Bellinger
Photography editor: Junenoire Mitchell
Editorial assistant: Jason Spears
Managing editor, college: Marian Johnson

ISBN: 978–0–393–92814–3

W. W. Norton & Company, Inc., 500 Fifth Avenue, New York, N.Y. 10110
www.wwnorton.com

W. W. Norton & Company Ltd., Castle House, 75/76 Wells Street, London W1T 3QT

1 2 3 4 5 6 7 8 9 0

CONTENTS

PREFACE

This laboratory manual is based on our collective 60+ years of teaching and coordinating introductory Geology courses, experience that has helped us understand how students best learn geologic principles and which strategies help instructors arouse their interest and enhance the learning process. Based on that experience, this manual (1) provides up-to-date, comprehensive background focused on the hands-on tasks at the core of introductory Geology labs; (2) offers clearer, more patient step-by-step explanations than can be accommodated in textbooks; (3) engages students in thinking like geologists to solve real-life problems important to society; and (4) communicates the passion and excitement that we still feel after decades as geologists and teachers.

Students often ask us how we maintain this excitement. They begin to understand when we share with them both the joys and the frustrations of facing and solving real-world geologic problems. You will find many of those types of problems in the following pages—modified for the introductory nature of the course, but still reflecting their challenges and the rewards of solving them. If we are successful, this manual will *engage* students in their geologic learning experience by *explaining* concepts clearly and providing avenues for further *exploration*.

UNIQUE ELEMENTS

As you read through this manual, you will find several elements that distinguish it from others, including:

Hands-on, inquiry-based pedagogy

We believe that students learn science best by doing science, not just by memorizing facts. Beginning in Chapter 1 (on page 2) and continuing in each subsequent chapter, students are guided through real geologic puzzles so they understand concepts more deeply and learn to think like a geologist. In Chapter 9, for example, students work out the rules of contour lines by comparing a topographic map with a digital elevation model of the same area. In Chapter 5, they reason out the cooling rates for plutons of different sizes and shapes, and so on.

Innovative exercises that engage students

Scaffolded exercises are carefully integrated into the text, guiding students to understand concepts for themselves and then helping them to use what they just learned to take the next step. These unique exercises, based on real-life situations, engage students because they get to see how geologic principles are important in our everyday activities. Some examples include using map-reading skills to survive on an island after a plane crash (Ch. 9); using contouring skills to identify groundwater contaminant plumes (Ch. 12) or to cope with sea-level rise (Ch. 14); using profiling skills to site a microwave communication tower (Ch. 9); and using knowledge of faults to explore for coal deposits (Ch. 15).

There are more exercises in most chapters than can be completed in a single lab session. This was done intentionally, in order to provide options for instructors in class and as potential out-of-class assignments. The complexity and rigor of the exercises increase with each subsequent chapter, enabling instructors to use the manual for both non-majors and potential majors by assigning only those exercises appropriate for the class.

Superb illustration program

Readers expect a superior illustration program in Norton geology texts, and the extensive photos, line drawings, maps, and DEMs continue the tradition of Marshak's *Earth: Portrait of a Planet* and *Essentials of Geology*. Hundreds of new figures closely integrated with the text explain concepts and sequential processes with exceptional clarity.

Reader-friendly language and layout

Our decades of teaching introductory Geology help us identify the concepts most difficult for students to understand. A conversational style addresses these issues clearly and uses analogies with everyday experiences to heighten understanding. The crisp, open layout makes the book more attractive and reader-friendly than other laboratory manuals, which are crammed with pages of multiple-column text.

Unique mineral and rock labs

Students learn the difference between minerals and rocks by classifying Earth materials in a simple exercise (Ch. 3) that leads to the importance of physical properties and a logical system for identifying minerals. In Chapter 4, students make intrusive and extrusive igneous "rocks," clastic and chemical sedimentary rocks, and a foliated metamorphic rock to understand how the rock-forming processes are indelibly recorded in a rock's texture. The goal in studying rocks in Chapters 5–7 is to interpret the processes and conditions by which they formed, not just to find their right names.

Self-contained topographic maps

Custom-designed topographic maps created with *National Geographic's TOPO4!™* software clearly illustrate landscapes and classic landform features, using shaded relief to enhance student understanding of the features. Inset maps show the larger context for many individual quadrangle segments.

Digital elevation models (DEMs)

Digital elevation models are used to enhance understanding of contour lines and to build map-reading skills. Online instructions show how to make a DEM for any area in the United States, so instructors can design exercises based on landscapes familiar to their students.

Unmatched online support for instructors

The Norton Resource Library for instructors offers not only PowerPoint slides for every photograph, line drawing, topographic map, and digital elevation model in the Manual, but also the GeoTours, animations, and videoclips that users of Marshak's *Essentials* and *Earth* have found so valuable. Instructors can download lab syllabi and outlines for individual laboratory sessions that have worked for our students, as well as answers for all of the exercises.

Exceptional online support for students

Norton's Student StudySpace also provides an introduction to *Microdem* (GIS freeware designed by Dr. Peter Guth of the U.S. Naval Academy). Online tutorials show

how easily *Microdem* can be used for standard map analyses such as topographic profiling, measuring stream gradient and sinuosity, calculating areas of drainage basins, and the like. For the first time, students can create slope analysis maps and use them to interpret potential landslide hazards. A library of digital elevation models for every topographic map in the Manual makes it easy for students to do map exercises digitally using any computer with access to the Internet. *Students can download Microdem and do sophisticated landscape analyses, extending the laboratory to their homes or dorms.*

ACKNOWLEDGMENTS

We are indebted to the talented team at W. W. Norton & Company whose zealous quest for excellence is matched by its ingenuity in solving layout problems, finding that special photograph, and keeping the project on schedule.

We are also grateful to the following expert reviewers for their input and expertise in making this Lab Manual the best it can be:

Theodore Bornhorst, *Michigan Tech University*
Lee Anne Burrough, *Prairie State College*
John Dassinger, *Chandler-Gilbert Community College*
Mark Evans, *Central Connecticut State University*
Lisa Hammersley, *Sacramento State University*
Bernie Housen, *Western Washington University*
Ryan Kerrigan, *University of Maryland*
Alfred Pekarek, *St. Cloud State University*
Ray Russo, *University of Florida*
Lori Tapanila, *Idaho State University*
JoAnn Thissen, *Nassau Community College*
Peter Wallace, *Dalhousie University*
Victor Zabielski, *Northern Virginia Community College*

CHAPTER 1

SETTING THE STAGE FOR LEARNING ABOUT THE EARTH

PURPOSE

- Introduce the challenges geologists face when studying a body as large and complex as the Earth.
- Use concepts of dimension, scale, and order of magnitude to describe the Earth.
- Review the materials and forces you will encounter when studying the Earth.
- Learn how geologists discuss ages of geologic materials and events, and the rates at which geologic processes take place.
- Help you become familiar with the basic types of diagrams and images used by geologists.

MATERIALS NEEDED

- Triple-beam or electronic balance
- 500-ml graduated cylinder
- Metric ruler
- Calculator
- Compass

1.1 The Challenge of Studying a Planet

Learning about the Earth is like training to become a detective. Both geologists and detectives need keen powers of observation, curiosity about slight differences, broad scientific understanding, and instruments to analyze samples. And both ask the same questions: What happened? How? When? Although much of the logical thinking is the same, there are big differences between the work of a detective and a geologist. A detective's "cold" case may be 30 years old, but "old" to a geologist means hundreds of millions or billions of years. To a detective, a "body" is a human body, but to a geologist a body may be a mountain range or a continent. Eyewitnesses can help detectives, but for most of Earth's history there weren't any humans to act as eyewitnesses for geologists. To study the Earth, geologists must therefore develop different strategies from other kinds of investigators. The overall goal of this manual is to help you look at the Earth and think about its mysteries like a geologist.

To illustrate geologic thinking, let's start with a typical geologic mystery. Almost 300 years ago, settlers along the coast of Maine built piers to load and unload ships. Some of these piers are now submerged to a depth of 1 meter (39 inches) below sea level (**Fig. 1.1**). Tourists might not think twice about this before heading for a lobster dinner at the local restaurant, but a geologist would want to know how rapidly the pier was submerged and what caused the submergence. How would a geologist go about tackling this problem?

FIGURE 1.1 Subsidence along the coast of Maine.

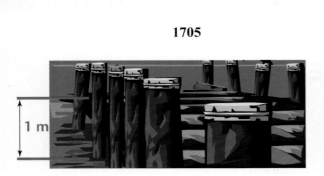

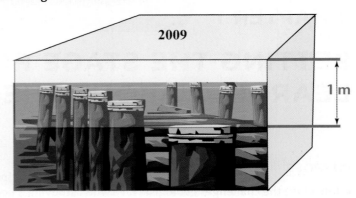

Let's first calculate the submergence rate. Since we haven't been watching the pier constantly for 300 years, we have to make some assumptions—geologists often do this to make estimates. Assume that the piers were originally 1 m *above* sea level, as many are built today, and that submergence occurred at a constant rate. With these assumptions, calculating the rate of submergence for the past 300 years becomes simple arithmetic.

EXERCISE 1.1   **Submergence Rate Along the Maine Coast**

Name: _____   Section: _____

Course: _____   Date: _____

The rate of submergence is the total change in elevation of the pier (_____ m) divided by the total amount of time involved (_____ years), and is therefore _____cm/yr. (Remember, 1 m = 100 cm.)

A geologist would also want to explain *why* the piers were submerged. When faced with such a problem, geologists typically try to come up with as many explanations as possible. Which of the following explanations could explain the submergence? Check all that might apply.

☐ The sea level has risen.

☐ The land has sunk.

☐ Both the sea level and land have risen, but sea level has gone up more.

☐ Both the sea level and land have sunk, but the land has sunk more.

If you checked all four choices (correctly!), you realize that explaining submergence along the Maine coast may be more complicated than it seemed at first. To find the answer, you need more **data**—more observations and/or measurements. One way to obtain more data would be to see if submergence is restricted to Maine or the east coast of North America, or if it is perhaps worldwide. As it turns out, submergence is observed worldwide, suggesting that the first choice above (sea level rise) is the most probable explanation. The next question is why, but that's another story.

With even more data, we could answer questions like, When did submergence happen? and Was the sea level rise constant? Maybe all submergence occurred in the first 100 years and then stopped. Or perhaps it began slowly and then accelerated. Unfortunately, we may not be able to answer all of these questions because, unlike television detectives who always get the bad guys, geologists don't always have enough data and must often live with uncertainty. We still do not have the answers to many questions about the Earth.

Problems like submergence along the Maine coast pose challenges to geologists and geology students who are trying to learn about the Earth. These challenges require us to:

- understand the many kinds of materials that make up the Earth and how they behave.
- be aware of how energy causes changes at Earth's surface and beneath it.
- consider features at a wide range of sizes and scales—from the atoms that make up rocks and minerals to the planet as a whole.
- think in *four* dimensions, because geology involves not just the three dimensions of space but also an enormous span of *time*.
- realize that some geologic processes occur in seconds but others take millions or billions of years and are so slow that we can only detect them with very sensitive instruments.

The rest of this chapter examines these challenges and how geologists cope with them. You will learn basic geologic terminology and tools of observation and measurement that will be useful throughout your geologic studies. Some terms will probably be familiar to you from previous science classes.

1.2 Studying Matter and Energy

Earth is a dynamic planet. Unlike the airless Moon, which has remained virtually unchanged for billions of years, Earth's gases, liquids, and solids constantly move from one place to another. They also change from one state to another through the effects of heat, gravity, other kinds of energy, and living organisms. We refer to all of Earth's varied materials and the processes that affect them as the **Earth System**, and the first step in understanding the Earth System is to understand the nature of matter and energy and how they interact with one another.

1.2.1 The Nature of Matter

Matter is the "stuff" of which the Universe is made. Geologists, chemists, and physicists have shown that matter consists of ninety-two naturally occurring elements and that some of these elements are much more abundant than others. Keep the following definitions in mind as you read further (Table 1.1).

TABLE 1.1 Basic definitions of matter.

- An **element** is a substance that cannot be broken down chemically into other substances.
- The smallest piece of an element that still has all the properties of that element is an **atom**.
- Atoms combine with one another chemically to form **compounds**; the smallest possible piece of a compound is called a **molecule**.
- Atoms in compounds are held together by **chemical bonds**.
- A simple **chemical formula** describes the combination of atoms in a compound. For example, the formula H_2O shows that a molecule of water contains two atoms of hydrogen and one of oxygen.

Matter occurs on Earth in three states—solid, liquid, or gas. Atoms in *solids*, like minerals and rocks, are held in place by strong chemical bonds. As a result, solids retain their shape over long periods. Bonds in *liquids* are so weak that atoms or molecules move easily, and as a result, liquids adopt the shape of their containers. Atoms or molecules in *gases* are barely held together at all, so a gas expands to fill whatever container it is placed in. Matter changes from one state to another in many geologic processes, as when the Sun evaporates water to produce water vapor, or when water freezes to form ice, or when lava freezes to become solid rock.

We describe the amount of matter by indicating its **mass**, and the amount of space it occupies by specifying its **volume**. The more mass packed into a given volume of matter, the greater the **density** of the matter. You notice density differences every day: it's easier to lift a large box of popcorn than a block of rock of the same size because the rock is much denser; it has much more mass packed into the same volume and therefore weighs much more.

1.2.2 Distribution of Matter in the Earth System

Matter is stored in the Earth System in five major **reservoirs** (Fig. 1.2). Most gas is in the **atmosphere**, a semitransparent blanket composed of about 78% nitrogen (N_2) and 21% oxygen (O_2), with minor amounts of carbon dioxide (CO_2), water vapor (H_2O), ozone (O_3), and methane (CH_4). Nearly all liquid occurs as water in the **hydrosphere**—Earth's oceans, rivers, and lakes, and the groundwater found in cracks and pores beneath the surface. Frozen water makes up the **cryosphere**, including snow, thin layers of ice on the surface of lakes or oceans, and huge masses of ice in glaciers and the polar ice caps.

The solid Earth is called the **geosphere**, which geologists divide into concentric layers like those in a hard-boiled egg (Fig. 1.3). The outer layer, the **crust**, is relatively thin, like an eggshell, and consists *mostly* of rock. Below the crust is the **mantle**, which also consists *mostly* of different kinds of rock and, like the white of an egg, contains most of Earth's volume. We say *mostly* because about 2% of the crust and mantle has melted to produce liquid material called **magma** (known as **lava** when it erupts on the surface). The central part of the Earth, comparable to the egg yolk, is the **core**. The outer core consists mostly of a liquid alloy of iron and nickel, and the inner core is a solid iron-nickel alloy.

FIGURE 1.2 Earth's major reservoirs of matter.

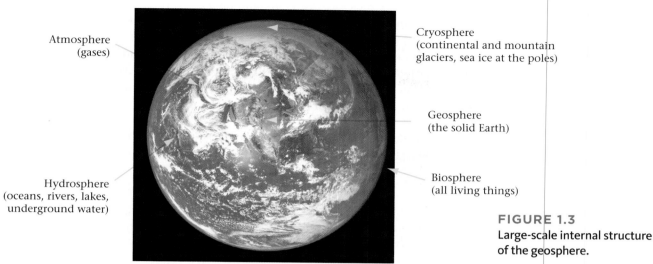

Atmosphere (gases)

Cryosphere (continental and mountain glaciers, sea ice at the poles)

Geosphere (the solid Earth)

Hydrosphere (oceans, rivers, lakes, underground water)

Biosphere (all living things)

FIGURE 1.3
Large-scale internal structure of the geosphere.

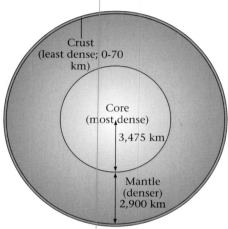

Crust (least dense; 0-70 km)

Core (most dense) 3,475 km

Mantle (denser) 2,900 km

Continents make up about 30% of the crust and are composed of relatively low-density rocks. The remaining 70% of the crust is covered by the oceans. Oceanic crust is both thinner and denser than the crust under the continents. Three types of solid materials are found at the surface: **bedrock**, a solid aggregate of minerals and rocks attached to those below; **sediment**, unattached boulders, sand, and clay; and **soil**, sediment and rock modified by interactions with the atmosphere, hydrosphere, and organisms so that it can support plant life.

The **biosphere** is the realm of living organisms, extending from a few kilometers below Earth's surface to a few kilometers above. Geologists have learned that organisms—from bacteria to mammals—are important

EXERCISE 1.2   **Reservoirs in the Earth System**

Name: _____   Section: _____
Course: _____   Date: _____

What Earth materials did you encounter in the last 24 hours? List them in the following table without worrying about the correct geologic terms (for example, "dirt" is okay *for now*). Place each Earth material in its appropriate reservoir and indicate whether it is a solid (S), liquid (L), or gas (G).

Atmosphere	Hydrosphere	Geosphere	Cryosphere	Biosphere

parts of the Earth System. They exchange gases with the atmosphere, absorb and release water, break rock into sediment, and help convert sediment and rock to soil.

The movement of materials from one reservoir to another is called a **flux** and happens in many geologic processes. For example, rain is a flux in which water moves from the atmosphere to the hydrosphere. Rates of flux depend on the materials, the reservoirs, and the processes involved. In some cases, a material moves among several reservoirs but eventually returns to the first. We call such a path a **geologic cycle**. You will learn about several cycles in your geology class, such as the rock cycle (the

---

EXERCISE 1.3   **Fluxes in the Earth System**

Name: _____   **Section:** _____

Course: _____   **Date:** _____

Even without a geology course, you already have a sense of how water moves from one reservoir to another in the Earth System. Based on your experience with natural phenomena on Earth, complete the following table that describes fluxes associated with the *hydrologic cycle*.

Process	What happens?	Movement from _____ to _____
Sublimation	*Solid ice becomes water vapor.*	cryosphere to atmosphere
Ice melting		
Evaporation of a puddle		
Water freezing		
Plants absorbing water		
Raindrop formation		
Cloud formation		
Steam erupting from a volcano		

movement of atoms from one rock type to another) and the hydrologic cycle (the movement of water from the hydrosphere to and from the other reservoirs).

1.2.3 Energy in the Earth System

Natural disasters in the headlines remind us of how dynamic the Earth is: rivers flood cities and fields, lava and volcanic ash bury villages, earthquakes topple buildings, and hurricanes ravage coastal regions. However, many geologic processes are much slower and less dangerous, like the movement of ocean currents and the almost undetectable creep of soil downhill. All are caused by energy, which acts on matter to change its character, move it, or split it apart.

Energy for the Earth System comes from (1) Earth's *internal* heat, which melts rock, causes earthquakes, and builds mountains (some of this heat is left over from the formation of the Earth, but some is being produced today by radioactive decay); (2) *external* energy from the Sun, which warms air, rocks, and water on the Earth's surface; and (3) the pull of Earth's gravity. Heat and gravity, working independently or in combination, drive most geologic processes.

Heat energy is a measure of the degree to which atoms or molecules move about (vibrate) in matter—including in solids. When you heat something in an oven, for

---

EXERCISE 1.4   **Sources of Heat for Earth Processes**

Name: _____   **Section:** _____
Course: _____   **Date:** _____

Some of the heat that affects geologic processes comes from the Sun and some comes from inside the Earth. What role does each of these heat sources play in Earth processes?

(a) If you take off your shoes on a beach and walk on the bare sand on a hot, sunny day, is the sand hot or cold? Why?

_____

_____

• Now, dig down in the sand just a few inches. What do you feel now, and why?

_____

_____

• What does this suggest about the depth to which heat from the Sun can penetrate the Earth?

_____

_____

• Based on this conclusion, is the Sun's energy or Earth's internal heat the cause of melting rock within the Earth? Explain.

_____

_____

_____

(b) The deeper down one goes into mines or drill holes, the hotter it gets. This temperature increase is called the **geothermal gradient**. Does this phenomenon support or contradict your conclusion above? Explain.

_____

_____

_____

continued

Name: _____   **Section:** _____
Course: _____   **Date:** _____

• In the upper 10 km of the crust, the geothermal gradient is typically about 25°C per km, but can range from 15°C/km to 50°C/km. In the diagram below, plot these three geothermal gradients for the upper 10 km of the Earth.

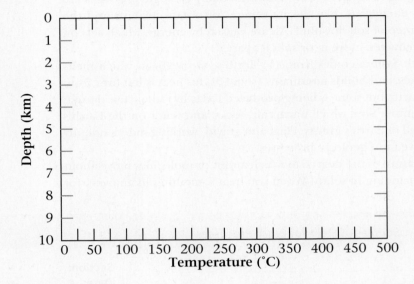

• The deepest mine on Earth penetrates to a depth of about 2 km. How hot is it in the bottom of this mine? What assumptions did you make to come up with this answer?

example, the atoms in the material vibrate faster and move farther apart. Heat energy drives the flux of material from one state of matter to another or from one reservoir of the Earth System to another. For example, heating ice causes **melting** (solid → liquid; crysosphere → hydrosphere) and heating water causes **evaporation** (liquid → gas; hydrosphere → atmosphere). Cooling slows the motion, causing **condensation** (gas → liquid, atmosphere → hydrosphere) or **freezing** (liquid → solid, hydrosphere → cryosphere).

Gravity, as Isaac Newton showed over three centuries ago, is the force of attraction that every object exerts on other objects. The strength of this force depends on the amount of mass in each object and how close the objects are to one another. The greater the mass and the closer the objects are, the stronger the gravitational attraction. The smaller the mass and the farther apart the objects are, the weaker the attraction. The Sun's enormous mass produces a force of gravity sufficient to hold Earth and the other planets in their orbits. Earth's gravitational force is far less than the Sun's, but is strong enough to hold the Moon in orbit, hold you on its surface, cause rain or volcanic ash to fall, and enable rivers and glaciers to flow.

1.3 Units for Scientific Measurement

Before we begin to examine components of the Earth System scientifically, we must first consider its dimensions and the units used to measure them. We can then examine the challenges scale geologists face when studying Earth and the atoms of which it is made.

1.3.1 Units of Length and Distance

If you described this book to a friend as being "big," would your friend have a clear picture of its size? Is it big compared to a quarter, or to a car? Without providing a frame of reference for the word "big," your friend wouldn't have enough information to visualize the book. A *scientific* description would much more accurately give the book's dimensions of length, width, and thickness using units of distance.

People have struggled for thousands of years to describe size in a precise way with standard units of measurement. Scientists everywhere and people in all countries except the United States use the **metric system** to measure length and distance. The largest metric unit of length is the kilometer (km), which is divided into smaller units: 1 km = 1,000 meters (m); 1 m = 100 centimeters (cm); 1 cm = 10 millimeters (mm). Metric units differ from each other by a factor of ten, making it very easy to convert one unit into another. For example, 5 km = 5,000 m = 500,000 cm = 5,000,000 mm. Similarly, 5 mm = 0.5 cm = 0.005 m = 0.000005 km.

The United States uses the **statute system** to describe distance. Distances are given in miles (mi), yards (yds), feet (ft), and inches (in), where 1 mi = 5,280 ft; 1 yd = 3 ft; and 1 ft = 12 in. As scientists, we use metric units in this book, but statute equivalents are also given (in parentheses).

Appendix 1.1, at the end of this chapter, provides basic conversions between English and metric units.

1.3.2 Other Dimensions, Other Units

Distance is just one of the dimensions of the Earth that you will examine during this course. We still need other units to describe other aspects of the Earth, its processes, and its history: units of time, velocity, temperature, mass, and density.

Time is usually measured in seconds (s), minutes (min), hours (h), days (d), years (yrs), centuries (hundreds of years), and millennia (thousands of years). A year is the amount of time it takes for the Earth to complete one orbit around the Sun. Because the Earth is very old, geologists also have to use much larger units of time: thousand years ago (abbreviated **Ka**, for "kilo-annum"), million years ago (**Ma**, for "mega-annum"), and billion years ago (**Ga**, for "giga-annum"). The 4,570,000,000-year age of the Earth can thus be expressed as 4.57 Ga or 4,570 Ma.

Velocity (speed) is described by units of distance divided by units of time, such as meters per second (m/s), feet per second (ft/s), kilometers per hour (km/h), or miles per hour (mph). You will learn later that the velocity at which geologic materials move ranges from extremely slow (mm/yr) to extremely fast (km/s).

Temperature is a measure of how hot an object is relative to a standard. It is measured in degrees Celsius (°C) in the metric system and degrees Fahrenheit (°F) in the English system. The reference standards are the freezing and boiling points of water: 0°C and 100°C or 32°F and 212°F, respectively. Note that there are 180°F between freezing and boiling but only 100°C. A change of 1°C is thus 1.8 times larger than a change of 1°F (180°/100°). To convert Fahrenheit to Celsius, or vice versa, see Appendix 1.1.

Mass refers to the amount of matter in an object, and **weight** refers to the force with which one object is attracted to another. The weight of an object on Earth therefore depends not only on its mass, but also on the strength of Earth's gravitational force. Objects that have more mass than others also weigh more on Earth because of the force of Earth's gravity. While the mass of an object remains the same whether it is on the Earth or on the Moon, the object *weighs* less on the Moon because of the Moon's weaker gravity.

Grams and kilograms (1 kg = 1,000 g) are the units of mass in the metric system; the English system uses pounds and ounces (1 lb = 16 oz). For those who don't read the metric equivalents on food packages, 1 kg = 2.2046 lb and 1 g = 0.0352 oz.

We saw earlier that the **density** (δ) of a material is a measure of how much mass is packed into each unit of volume, generally expressed as units of g/cm³ (**Fig. 1.4**). We instinctively distinguish unusually low-density materials like Styrofoam and feathers from low-density materials like water (δ = 1 g/cm³) and high-density materials like steel (δ = ~7 g/cm³) because the former feel very light *for their sizes* and the latter feel unusually heavy *for their sizes*.

To measure the density of a material, we need to know its mass and volume. We measure mass on a balance, and can use simple mathematical formulas to determine the volumes of cubes, bricks, spheres, or cylinders (**Fig. 1.5a**). But for an irregular chunk of rock, it is easiest to submerge the rock in a graduated cylinder partially filled with water. Measure the volume of water before the rock is added, and then with the rock in the cylinder. The rock displaces a volume of water equivalent to its own volume, so simply subtract the volume of the water before the rock was added from that of the water plus rock to obtain the volume of the rock (**Fig. 1.5b**). The density of a rock can be calculated simply from the definition of density: density = mass ÷ volume.

Thousands of years ago, the Greek scientist Archimedes recognized that density determines whether an object floats or sinks when placed in a liquid. For example,

FIGURE 1.4 Weight of materials with different densities.

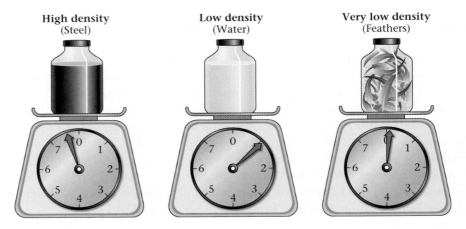

FIGURE 1.5 Measuring the volume of materials.

(a) For a rectangular solid, volume = length × width × height.

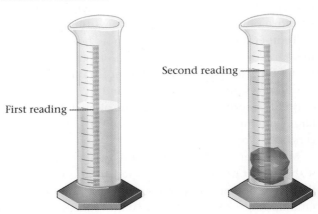

(b) For an irregular solid, the volume of the rock is the volume of the displaced liquid, which equals the difference between the first and second readings.

icebergs float in the ocean because ice is less dense than water. Nearly all rocks sink when placed in water because the density of most rocks is at least 2.5 times greater than the density of water.

EXERCISE 1.5　　　　**Measuring the Density of Earth Materials**

Name: _____　　Section: _____

Course: _____　　Date: _____

(a) Determine the density of water, given a graduated cylinder, a balance, and a container of water. _____ g/cm³

(b) Your instructor will provide you with two samples of rock: *granite*, a light-colored rock that makes up a large amount of the continental crust, and *basalt*, a dark-colored rock that makes up most of the oceanic crust and the lower part of the continental crust.

 • What is the density of granite?　　　　　　　　　　　_____ g/cm³

 • What is the density of basalt?　　　　　　　　　　　_____ g/cm³

 • What is the ratio of the density of granite to the density of basalt?　　_____

(c) Most modern ships are made of steel, which has a density of about 7.85 g/cm³. This is much greater than the density of water, so how can ships float?

_____

_____

1.4 Expressing Earth's "Vital Statistics" with Appropriate Units

Now that you are familiar with some of the units used to measure the Earth, we can look at some of the planet's "vital statistics."

 • *The Solar System:* Earth is the third planet from the Sun in a Solar System of eight planets (Pluto has been reclassified as a "dwarf planet"). It is one of four terrestrial planets, along with Mercury, Venus, and Mars, whose surfaces are made of rock rather than of frozen gases. Earth is, on average, 150,000,000 km (93,000,000 mi) from the Sun.

 • *Earth's orbit:* It takes Earth 1 year (365.25 days, or 3.15×10^7 seconds) to complete one slightly elliptical orbit around the Sun.

 • *Earth's rotation:* It takes Earth 1 day (24 hours, or 86,400 seconds) to rotate once on its axis. The two points where the axis intersects the surface of the planet are the north and south **geographic poles**. At present, Earth's axis is tilted at about 23.5° to the plane in which it orbits the Sun (the **plane of the ecliptic**), but Earth wobbles a bit so this tilt changes over time.

 • *Shape:* Our planet is almost, but not quite, a sphere. Centrifugal force caused by its rotation causes Earth to bulge just a bit at the equator. The equatorial radius of 6,400 km (~4,000 mi) is 21 km (~15 mi) longer than its polar radius.

 • *Temperature:* Earth's surface temperature ranges from 58°C (136°F) in deserts near the equator to −89°C (−129°F) near the South Pole. *Average* daily temperature in New York City ranges from 0°C (32°F) in the winter to 24°C (76°F) in the summer. The geothermal gradient is about 15°C to 30°C per km in the upper crust, and temperatures may reach 5,000°C at the center of the Earth.

- *Some additional dimensions:* The highest mountain on Earth is Mt. Everest, at 8,850 m (29,035 ft) above sea level. The average depth of the world's oceans is about 4,500 m (14,700 ft), and the deepest point on the ocean floor—the bottom of the Mariana Trench in the Pacific Ocean—is 11,033 m (35,198 ft) below the surface. The *relief*, the vertical distance between the highest and lowest points, is thus just a little less than 20 km (12 mi). It would only take about 10 minutes to drive this distance at highway speed! This relief is extremely small compared to the overall size of the planet.

1.5 The Challenge of Scale when Studying a Planet

Geologists deal routinely with objects as incredibly small as atoms and others as incredibly large as the Appalachian Mountains or the Pacific Ocean. Sometimes we have to look at a feature at different scales, as in **Figure 1.6**, to fully understand it.

The challenge of scale is often a matter of perspective: to a flea, the dog on which it lives is its entire world, but to a parasite inside the flea, the flea is *its* entire world. For most of our history, humans have had a flea's-eye view of the world, unable to recognize its dimensions or shape from our vantage point on its surface. As a result, we once thought Earth was flat and at the center of the Solar System. It's easy to laugh at such misconceptions now, but Exercise 1.6 gives a feeling for the challenges of scale that still exist.

1.5.1 Scientific Notation and Orders of Magnitude

Geologists must cope with the enormous ranges in the scale of distance (atoms to sand grains to planets), temperature (below 0°C in the cryosphere and upper atmosphere to more than 1,000°C in some lavas, to millions of degrees Celsius in the Sun), and velocity (continents moving at 2 cm/yr to light moving at 299,792 km/s). We sometimes describe scale in approximate terms, and sometimes more precisely. For example, the terms "mega-scale," "meso-scale," and "micro-scale" denote enormous, moderate, and tiny features, respectively, but don't tell exactly how large or small they are, because they depend on a scientist's frame of reference. For example, mega-scale to an astronomer

FIGURE 1.6 The white cliffs of Dover seen at two different scales.

(a) The chalk cliffs of Dover, England; the person in the foreground gives an idea of the size of the cliffs.

(b) Microscopic view of the chalk (plankton shells) of which the cliffs are made. The eye of a needle gives an idea of the miniscule sizes of the shells that make up the cliffs.

might mean intergalactic distances, but to a geologist it may mean the size of a continent or a mountain range. Micro-scale could refer to a sand grain, a bacterium, or an atom. Geologists commonly approximate scale in terms that clearly specify the frame of reference. For example, your instructor may use "outcrop-scale" for a feature in a single exposure of rock or "hand specimen–scale" for a rock the size of your fist.

Geologists use a system based on powers of ten to describe things spanning the entire range of scales that we study, sometimes using the phrase **orders of magnitude** to indicate differences in scale. A feature that is an order of magnitude larger than another is 10 times larger; a feature one-tenth the size of another is an order of

---

EXERCISE 1.6   **The Challenge of Perspective and Visualizing Scale**

Name: _____   Section: _____
Course: _____   Date: _____

The enormous difference in size between ourselves and our planet gives us a limited perspective on large-scale features and makes understanding major Earth processes challenging. To appreciate this challenge, consider the relative sizes of familiar objects (**Fig. 1.7a, b**).

FIGURE 1.7 A matter of perspective.

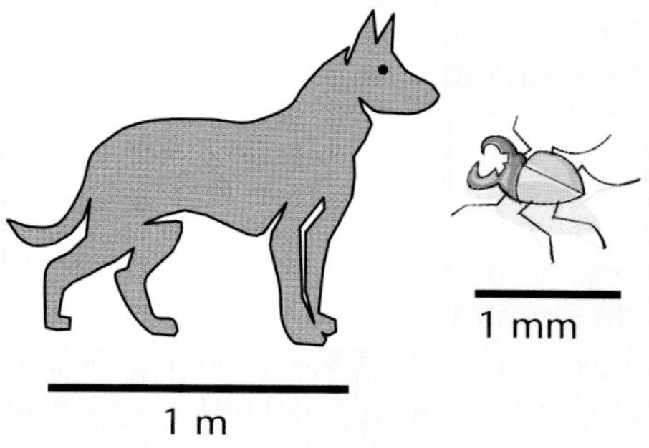

1 mm

1 m

(a) Relative sizes of a dog and a flea.

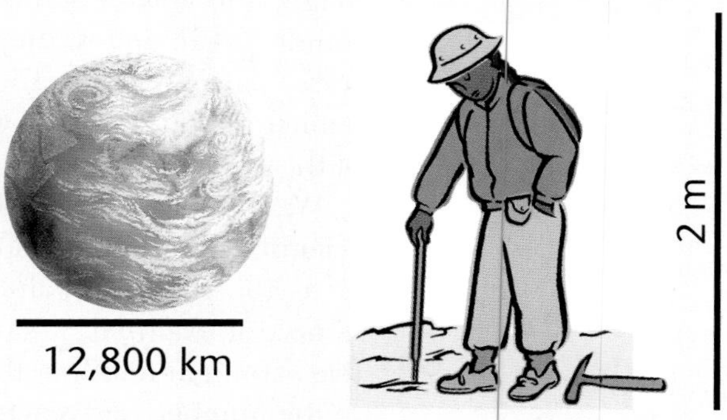

12,800 km

2 m

(b) Relative sizes of the Earth and a tall human being.

(a) Assume that the flea is 1 mm long and that a dog is 1 m long (Fig. 1.7a).

- To relate this to our English system of measurement, how long is this flea in inches? _____ in
- How many times larger is the dog than the flea? _____
- How long is the bar scale representing the flea? _____
- How long would the bar representing the dog have to be, if the dog and the flea were shown at the same scale? _____

(b) Now, think about the relative dimensions of a geologist and the Earth (Fig. 1.7b).

- How many times larger than the geologist is the Earth? _____
- How large would the drawing of the Earth have to be, if it were drawn at the same scale as the geologist? Give your answer in kilometers and in miles. _____ km _____ mi
- Based on the relative sizes of flea and dog versus human and Earth, does a flea have a better understanding of a dog than a human has of the Earth? Or vice versa? Explain.

_____

_____

_____

Table 1.2 Orders of magnitude defining lengths in the Universe (in meters).

~2 × 10²⁶	Radius of the observable Universe
2.1 × 10²²	Distance to the nearest galaxy (Andromeda)
9.0 × 10²⁰	Diameter of the Milky Way galaxy
1.5 × 10¹¹	Diameter of Earth's orbit
6.4 × 10⁶	Radius of the Earth
5.1 × 10⁶	East-west length of the United States
8.8 × 10³	Height of Mt. Everest above sea level (Earth's tallest mountain)
1.7 × 10⁰	Average height of an adult human
1.0 × 10⁻³	Diameter of a pinhead
6.0 × 10⁻⁴	Diameter of a living cell
2.0 × 10⁻⁶	Diameter of a virus
1.0 × 10⁻¹⁰	Diameter of a hydrogen atom (the smallest atom)
1.1 × 10⁻¹⁴	Diameter of a uranium atom's nucleus
1.6 × 10⁻¹⁵	Diameter of a proton (a building block of an atomic nucleus)
~1 × 10⁻¹⁸	Diameter of an electron (a smaller building block of an atom)

magnitude *smaller*. Something 100 times the size of another is two orders of magnitude larger, and so on. **Table 1.2** shows that the range of dimensions in our Universe spans an almost incomprehensible forty-four orders of magnitude, from the diameter of the particles that make up an atom—about 10^{-18} m across—to the radius of the observable Universe—about 10^{26} m.

We simplify the numbers that describe such a wide range with a method called **scientific notation** based on powers of 10. In scientific notation, 1 is written as 10^0, 10 as 10^1, 100 as 10^2, and so on. Numbers less than 1 are shown by negative exponents: for example, $\frac{1}{10} = 0.1 = 10^{-1}$; $\frac{1}{100} = 0.01 = 10^{-2}$; and $\frac{1}{1000} = 0.001 = 10^{-3}$. A positive exponent tells how many places the decimal point has to be moved to the right of the number, and a negative exponent how many places to the left.

In scientific notation, the 150,000,000-km (93,000,000-mi) distance from the Earth to the Sun is written as 1.5×10^8 km (9.3×10^7 mi)—start with 1.5 and move the decimal eight places to the *right*, adding zeroes as needed. Similarly, a small number such as 0.0000034 would be written as 3.4×10^{-6} (start with 3.4 and move the decimal six places to the *left*, adding zeroes as needed).

We use scientific notation in the following exercises and throughout the rest of this manual. Exercise 1.7 shows how our limited perspective challenges us to recognize that even the basic processes of Earth's rotation and orbit around the Sun are taking place.

EXERCISE 1.7   Moving Along

Name: _____   Section: _____

Course: _____   Date: _____

Complete the following calculations to get a sense of the distances that the Earth travels and the speed at which it moves. **Figure 1.8** shows the two types of motion involved.

(a) For simplicity, we picture the Earth's orbit around the Sun as a circle (Fig. 1.8a), with a diameter of 150,000,000 km (93 million miles). It takes 1 year for the Earth to orbit the Sun (365.25 days, _____hours).

continued

Name: _____   **Section:** _____

Course: _____   **Date:** _____

FIGURE 1.8 Earth's orbit and rotation.

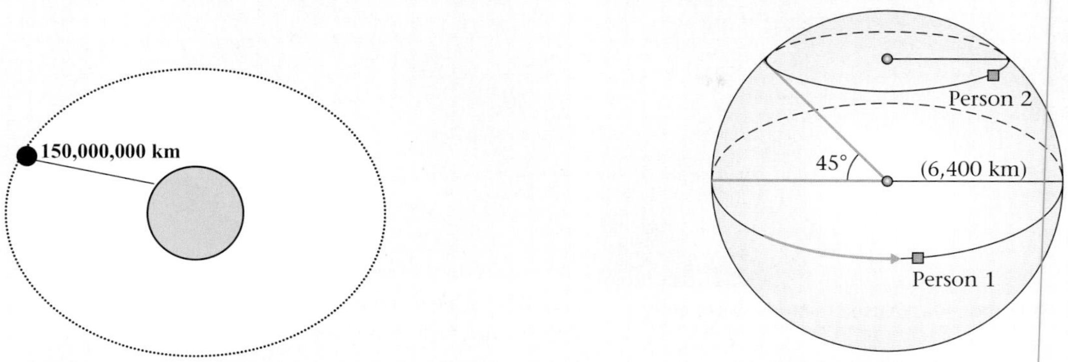

(a) Earth orbits the Sun in an elliptical path. To simplify calculation, here we assume a circular orbit with a radius of 150,000,000 km (93,000,000 mi).

(b) Earth rotates on its axis every 24 hours.

- What distance does the Earth travel during a single complete orbit? Remember that the circumference of a circle is $2\pi r$, and that $\pi \approx 3.14$. Orbital distance = _____ km (_____ mi)
- How many hours does it take Earth to complete a single orbit? _____ hours
- What is the velocity of the Earth as it orbits the Sun? Give your answer in both kilometers per hour and miles per hour. _____ km/h _____ mph
- Express the above results using scientific notation. _____ km/h _____ mph

(b) Earth rotates on its axis in 1 day. At what velocity is a person standing at the equator (Person 1 in Fig. 1.8b) moving due to our planet's rotation? _____ km/h _____ mph

(c) Now consider a person standing on Earth's surface at a point halfway between the equator and the geographic pole (Person 2 in Fig. 1.8b). Is this person moving faster or slower than the person standing at the equator? Explain your answer.

_____

_____

(d) Many components of the Earth System are too *small* for us to comprehend. For example, atoms are too small to be seen even with the most powerful optical microscope. A sodium atom has an approximate diameter of 2×10^{-8} cm. The diameter of the period at the end of this sentence is approximately 4×10^{-4} cm.

- If you lined up sodium atoms one next to the other, how many sodium atoms would it take to span the diameter of a period? _____ Express this number in scientific notation: _____
- How many orders of magnitude larger is the period than the atom? _____

1.5.2 Coping with Scale Problems: Maps, Diagrams, and Models

Figure 1.6 showed how we use microscopes to help see very *small* features. One way to cope with the challenge of very *large* geologic features is to construct scale models like that in **Figure 1.9a**. Using these models, we can recreate the

(a) A scale model of St. Paul's harbor, Alaska used to study water and sediment movement.

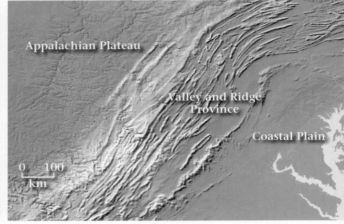

(b) Maps scale down portions of the Earth to make them easier to visualize and understand.

EXERCISE 1.8   Scaling Down Features to Visualize Them

Name: _____   **Section:** _____
Course: _____   **Date:** _____

(a) The United Kingdom is about 900 km long from north to south. By how many orders of magnitude would you have to scale down a map of the United Kingdom so it would fit on a sheet of paper 20 cm long? _____

(b) Earth's radius is about 6,500 km. Using a compass and ruler, draw a cross section of the Earth in the left-hand box below, using Point E to locate the center of the Earth. *Indicate the scale that you used in the lower left corner of the box.*

(c) The Moon has a radius of 1,750 km. At the same scale that you used in (a), draw a cross section of the Moon in the right-hand box below, using Point M to locate the center of the Moon.

(d) Using the data in Figure 1.3, compare the size of the Moon with that of Earth's core. Which is larger?

• E   • M

■■■■■■ _____ km

conditions of floods or the ground shaking during earthquakes to better understand these processes. Scaled-down diagrams like the one of the geothermal gradient in Exercise 1.4 help show relationships within the Earth, and we use maps, aerial photographs, and satellite images to visualize the Earth's surface (**Fig. 1.9b**). Depending on the degree to which we scale down the area, these images can show the entire Earth on a single sheet of paper or just the part of it we wish to study. If you have a good high-speed Internet connection, you can download *Google Earth*™ or *NASA World Wind* to see spectacular images of any part of the world at almost any scale. Exercise 1.8 gives simple examples of how scaled-down maps and diagrams are made.

1.6 The Challenge of Working with Geologic Time

Geologists have amassed a large body of evidence showing that Earth is about 4.57 billion years old (4,570,000,000 or 4.57×10^9 years). This enormous span is referred to as **geologic time** and understanding its vast scope is nearly impossible for humans, who live less than 100 years. Our usual frame of reference for time is based on human lifetimes: a war fought two centuries ago happened two lifetimes ago, and 3,000-year-old monuments were built about thirty lifetimes ago. To help visualize geologic time, we can compare its scale to more familiar dimensions. For example, if geologic time (4.57 billion years) was represented by a 1-km-long field, each millimeter would represent 45,700 years, each centimeter 457,000 years, each meter 45,750,000 years. If all of geologic time was condensed to a single year, each second would represent almost 53,000 years.

EXERCISE 1.9   **Picturing Geologic Time**

Name: _____   **Section:** _____
Course: _____   **Date:** _____

Complete the following calculations to appreciate the duration of geologic time and how slow processes can lead to major changes.

(a) If you want to represent the 4.57 billion years of geologic time with a roll of adding machine tape so that 1 inch represents 1 million years, how long would the tape have to be? *Answer using scientific notation.*
_____ in _____ mi _____ km

(b) If you want to represent geologic time with a 24-hour clock, how many years would be represented by a single second?
_____ a minute? _____ an hour? _____

(c) If you saved a penny a day, how long would it take you to save $1 million? _____ days

   • If you collected a penny a year for a hundred years, how much money would you have? $ _____

   • How much would you have in a million years? $ _____ a billion years? $ _____

FIGURE 1.10 The principle of uniformitarianism.

(a) Ripple marks in 145 million-year-old sandstone, at Dinosaur Ridge, Colorado.

(b) Ripple marks in modern sand on the shore of Cape Cod, Massachusetts.

1.6.1 Uniformitarianism: The Key to Understanding the Geologic Past

The challenge in studying Earth history is that many of the processes that made the rocks, mountains, and oceans we see today occurred millions or billions of years ago. How can we figure out what those processes were if they happened so long ago? The answer came nearly 300 years ago when a Scottish geologist, James Hutton, noted that some features observed in rocks resembled those forming in modern environments (**Fig. 1.10a, b**). For example, the sand on a beach or a stream's floor commonly forms a series of ridges called ripple marks. You can see identical ripple marks preserved in ancient layers of sandstone, a rock made of sand grains.

Based on his observations, Hutton proposed the **principle of uniformitarianism** as a basis for interpreting Earth history. According to uniformitarianism, most ancient geologic features formed by the same processes as modern ones. The more we know about the physical and chemical processes by which modern features form, the more we can say about the ancient processes. This principle is often stated more succinctly as "the present is the key to the past."

1.6.2 Rates of Geologic Processes

Many geologic processes occur so slowly that it's difficult to know they are happening. For example, mountains rise and are worn away (**eroded**) at rates of about 1 to 1.5 *mm*/yr, and continents move about one to two orders of magnitude faster—1 to 15 *cm*/yr. Some geologic processes are much faster. A landslide or volcanic eruption can happen in minutes; an earthquake may be over in seconds; and a meteorite impact takes just a fraction of a second.

Before the significance of uniformitarianism became clear, scholars thought that Earth was about 6,000 years old. Uniformitarianism made scientists rethink this estimate and consider the possibility that the planet is much older. When geologists observed that it takes a very long time for even a thin layer of sand to be deposited or for erosion to deepen a modern river valley, they realized that it must have taken much more than all of human history to deposit the layers of rock in the Grand Canyon. Indeed, the submergence of coastal Maine discussed at the opening of this chapter turns out to be one of the *faster* geologic processes. Understanding rates of geologic processes thus helped us recognize Earth's vast age hundreds of years before the discovery of radioactivity enabled us to measure the age of a rock in years.

Small changes can have big results when they are repeated over enormous spans of time, as shown by Exercise 1.9. And at the very slow rates at which some geologic processes take place, familiar materials may behave in unfamiliar ways. For example, if you drop an ice cube on your kitchen floor, it behaves brittlely and shatters into pieces. But given time and the weight resulting from centuries of accumulation, ice in a glacier can flow plastically at tens of meters a year (**Fig. 1.11a**). Under geologic conditions and over long enough periods, even layers of solid rock can be bent into folds like those seen in **Figure 1.11b**.

FIGURE 1.11 Solid Earth materials change shape over long periods under appropriate conditions.

(a) Athabasca Glacier, Alberta, Canada.

(b) Folded sedimentary rocks in eastern Ireland.

1.6.3 "Life Spans" of Mountains and Oceans

You will learn later that mountains and oceans are not permanent landscape features. Mountains form by uplift or intense folding, but as soon as land rises above the sea, running water, ice, and wind begin to erode it away. When the forces that cause the uplift cease, the mountains are gradually leveled by the forces of erosion. Oceans are also temporary features. They form when continents split and the pieces move apart from one another, and they disappear when the continents on their margins collide. Exercise 1.10 explores the rates at which mountains and oceans form and are then destroyed.

Name: _____   Section: _____
Course: _____   Date: _____

(a) Rates of uplift and erosion. The following questions give you a sense of the rates at which uplift and erosion take place. We will assume that uplift and erosion do not occur at the same time—that the mountains are first uplifted and only then does erosion begin—whereas the two processes actually operate simultaneously.

- If mountains rose by 1 mm/yr, how high would they be (in meters) after 1,000 years? _____ m 10,000,000 years? _____ m   50 million years? _____ m
- The Himalayas now reach an elevation of 8.8 km, and radiometric dating suggests that their uplift began about 45 Ma. Assuming a constant rate of uplift, how fast did the Himalayas rise? _____ km/yr _____ m/yr _____ mm/yr
- Evidence shows that there were once Himalayan-scale mountains in northern Canada, an area now eroded nearly flat. If Earth were only 6,000 years old as was once believed, how fast would the rate of erosion have had to be for these mountains to be eroded to sea level in 6,000 years? _____ m/yr _____ mm/yr
- Observations of modern mountain belts suggest that ranges erode at rates of 2 mm per 10 years. At this rate, how long would it take to erode the Himalayas down to sea level? _____ years

(b) Rates of sea-floor spreading. Today the Atlantic Ocean is about 5,700 km wide at the latitude of Boston. At one time, however, there was no Atlantic Ocean because the east coast of the United States and the northwest coast of Africa were joined in a huge supercontinent. The Atlantic Ocean started to form "only" 185,000,000 years ago, as modern North America split from Africa and the two slowly drifted apart in a process called *sea-floor spreading*.

- Assuming that the rate of sea-floor spreading has been constant, at what rate has North America been moving away from Africa? _____ per year _____ per million years

1.7 Applying the Basics to Interpreting the Earth

The concepts presented in this chapter are the foundations for understanding the topics that will be covered throughout this course. This section shows how you can apply these concepts with a little geologic reasoning to arrive at significant conclusions about the Earth.

1.7.1 Pressure in the Earth

The pull of Earth's gravity produces **pressure**. For example, the pull of gravity on the atmosphere creates a pressure of 1.03 kg/cm² (14.7 lbs/in²) at sea level. This means that every square centimeter of the ocean, the land, or your body is affected by a weight of 1.03 kg. We call this amount of pressure 1 **atmosphere** (atm). Scientists commonly specify pressures using a unit called the *bar* (from the Greek *barros*, meaning weight), where 1 bar ≈1 atm. Two kinds of pressure play important roles in the hydrosphere and geosphere. **Hydrostatic pressure**, due to the weight of water, increases as you descend into the ocean and can crush a submarine in the deep ocean. **Lithostatic pressure**, due to the weight of overlying rock, increases as you go deeper in the geosphere and is great enough in the upper mantle to change the graphite in a pencil into diamond. The rate at which lithostatic pressure increases is called the **geobaric gradient**, which along with the geothermal gradient plays a major role in determining what materials can exist at depth. Lithostatic pressures are so great that we must measure them in **kilobars** (Kb), where 1 Kb = 1,000 bars. Exercise 1.11 shows how pressure varies in the Earth.

EXERCISE 1.11   **Thinking about Pressure in the Earth**

Name: _____   **Section:** _____
Course: _____   **Date:** _____

Hydrostatic and lithostatic pressures are caused by the weight of the overlying water and rock, respectively, and these pressures, in turn, depend on the density of the overlying material. You calculated earlier that water has a density of $1 \, g/cm^3$, whereas rock in the crust beneath continents has an average density of about $2.8 \, g/cm^3$.

(a) What is the pressure on the sea floor at a depth of 5 km? (*Hint:* Consider the mass of an overlying column of water $1 \, cm \times 1 \, cm \times 5 \, km$ high.) Give your answer in _____ g/cm^2 _____ bars

(b) What is the pressure at a depth of 5 km below the surface of a continent? _____ g/cm^2 _____ Kb
(*Hint:* Consider the mass of an overlying column of rock $1 \, cm \times 1 \, cm \times 5 \, km$ high.)

(c) Why does lithostatic pressure increase more rapidly with depth than hydrostatic pressure?

_____

_____

_____

(d) The geobaric gradient beneath continents is about 1 Kb for every 3 km of depth.

- On the top of the diagram used in Exercise 1.4 for the *geothermal* gradient, construct a scale showing pressure from 1 to 5 kilobars and draw the *geobaric* gradient in a different color.
- What is the lithostatic pressure at a depth of 8 km below the surface of a continent? _____ Kb
- What are the temperature and pressure conditions in the Earth at a depth of 3.5 km?
  _____ °C   at 5.6 km? _____ °C
  _____ Kb   _____ Kb

METRIC-ENGLISH CONVERSION CHART

To convert English units to metric units	To convert metric units to English untis
Length or distance inches $\times$ 2.54 = centimeters feet $\times$ 0.3048 = meters yards $\times$ 0.9144 = meters miles $\times$ 1.6093 = kilometers	centimeters $\times$ 0.3937 = inches meters $\times$ 3.2808 = feet meters $\times$ 1.0936 = yards kilometers $\times$ 0.6214 = miles
Area $in^2 \times 6.452 = cm^2$ $ft^2 \times 0.929 = m^2$ $mi^2 \times 2.590 = km^2$	$cm^2 \times 0.1550 = in^2$ $m^2 \times 10.764 = ft^2$ $km^2 \times 0.3861 = mi^2$
Volume $in^3 \times 16.3872 = cm^3$ $ft^3 \times 0.02832 = m^3$ U.S. gallons $\times$ 3.7853 = liters	$cm^3 \times 0.0610 = in^3$ $m^3 \times 35.314 = ft^3$ liters $\times$ 0.2642 = U.S. gallons
Mass ounces $\times$ 28.3495 = grams pounds $\times$ 0.45359 = kilograms	grams $\times$ 0.03527 = ounces kilograms $\times$ 2.20462 = pounds
Density $lb/ft^3 \times 0.01602$ g/cm^3	$g/cm^3 \times 62.4280 = lb/ft^3$
Velocity ft/s $\times$ 0.3048 = m/s mph $\times$ 1.6093 = km/h	m/s $\times$ 3.2804 = ft/s km/h $\times$ 0.6214 = mph
Temperature $(1.8 \times °F) + 32 = °C$	$0.55 \times (°C - 32) = °F$
Pressure $lb/in^2 \times 0.0703 = kg/cm^2$	$kg/cm^2 \times 14.2233 = lb/in^2$

CHAPTER 2

THE WAY THE EARTH WORKS: EXAMINING PLATE TECTONICS

PURPOSE

- Examine the evidence that led to the development of the plate tectonic theory.
- Use major Earth features to understand the processes involved in plate tectonics.
- Learn how geologists estimate the direction and rate of plate motion.

MATERIALS NEEDED

- Tracing paper
- Colored pencils
- Ruler with divisions in tenths of an inch or millimeters
- Protractor
- Calculator, or pencil and paper for simple arithmetic

2.1 Introduction

Earthquakes and volcanic eruptions show that Earth is a dynamic planet with internal processes strong enough to affect the surface. In the 1960s and 1970s, geologists developed the **theory of plate tectonics** that recognizes these events as parts of a global process that splits continents and moves the pieces across the globe, opens and closes oceans, and creates mountains. Plate tectonics is a unifying theory for geology because it answers many questions that had puzzled geologists for years. This chapter explores the observations and reasoning that led geologists to propose the plate tectonics theory.

Before going any farther, it is important to understand what scientists mean by the word *theory*. A **theory** is an idea that successfully explains many observations, can be used to make predictions that can be tested by experiments, and has not been proven wrong by conflicting observations after many years of testing. Plate tectonics continues to meet all these criteria after nearly 50 years of rigorous testing, and therefore joins the ranks of other important theories, such as the germ theory of disease, the theory of relativity, and the theory of evolution.

2.2 The Plate Tectonics Theory

Plate tectonics is based on many aspects of the Earth that you will examine throughout this course. The basic concepts of the theory include the following:

- Earth's crust and 100 to 150 km of the mantle form a relatively rigid outermost layer called the **lithosphere**. The lithosphere is not a single shell, but consists of several large pieces called **lithosphere plates** or, simply, **plates** (Fig. 2.1). There are about twelve major plates that are thousands of kilometers wide and several minor plates that are hundreds of kilometers wide.

FIGURE 2.1 Earth's major lithosphere plates.

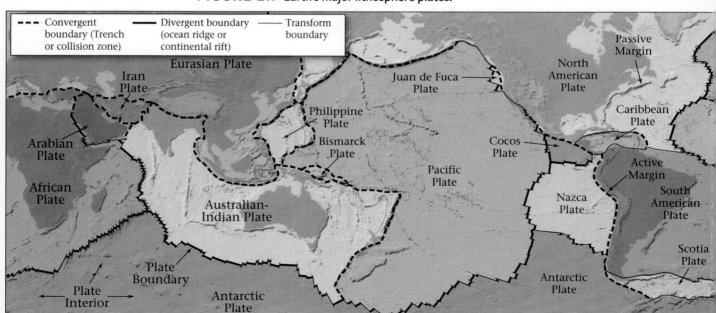

- Continental lithosphere is thicker than oceanic lithosphere because continental crust alone (without the mantle component) is 25 to 70 km thick, whereas the entire oceanic lithosphere is only about 7 km thick.

- The plates rest on the **asthenosphere**, a zone in the upper mantle (**Fig. 2.2**) that, although solid, has such low rigidity that it can flow like soft plastic. The asthenosphere acts as a lubricant, permitting the plates above it to move. Plates move relative to one another at 1 to 15 cm/yr, roughly the rate at which fingernails grow.

- A place where two plates make contact is called a **plate boundary**, and in some places *three* plates come together at a **triple junction**. There are three different kinds of plate boundaries defined by the relative motions of the adjacent plates (Figs. 2.1 and 2.2).

1. At a **divergent boundary**, plates move away from one another at the axis of submarine mountain ranges called **mid-ocean ridges**, or oceanic ridges. Molten material rises from the asthenosphere to form new oceanic lithosphere at the ridge axis. The ocean grows wider through this process, called **sea-floor spreading**, during which the new lithosphere moves outward from the axis to the flanks of the ridge.

2. At a **convergent boundary**, two plates move toward each other and the oceanic lithosphere of one plate sinks (the subducted plate) into the mantle below the other (the overriding plate) in **subduction zones**. The lithosphere of the overriding plate may be oceanic or continental. The boundary between the two plates is a deep-ocean **trench**. At a depth of about 150 km, gases released from the heated subducted plate rise and partially melt the overlying asthenosphere. The resulting magma erupts near the edge of the overriding plate, forming a **volcanic island arc** where the overriding plate is made of oceanic lithosphere, or a **continental arc** (like the Andes Mountains) where the overriding plate is made of continental lithosphere.

3. At a **transform boundary** two plates slide past one another along a vertical zone of fracturing called a **transform fault**. Most transform faults break ocean ridges into segments and are also called *oceanic fracture zones*. A few, however, such as the San Andreas Fault in California, the Alpine Fault in New Zealand, and the Great Anatolian Fault in Turkey, cut through continental plates.

FIGURE 2.2 Cross section showing activity at convergent, divergent, and transform plate boundaries.

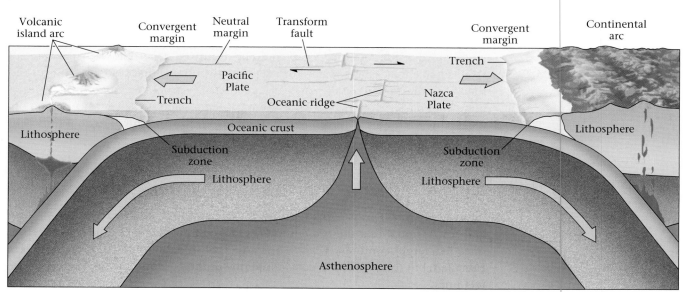

- Continental crust cannot be subducted because it is too buoyant. When subduction completely consumes an oceanic plate between two continents, **continental collision** occurs, forming a **collisional mountain belt** like the Himalayas, Alps, or Appalachians. Folding during the collision thickens the crust to the extent that the thickest continental crust is found in these mountains.

- **Continental rifts** are places where continental lithosphere is stretched in the process of breaking apart at a new divergent margin. If rifting is "successful," a continent splits into two pieces separated by a new oceanic plate, which gradually widens by sea-floor spreading.

- In the **tectonic cycle**, new oceanic lithosphere is created at the oceanic ridges, moves away from the ridges during sea-floor spreading, and returns to the mantle in subduction zones. Oceanic lithosphere is neither created nor destroyed at transform faults, where movement is almost entirely horizontal.

EXERCISE 2.1   **Recognizing Plates and Plate Boundaries**

Name: _____   Section: _____
Course: _____   Date: _____

(a) What is the name of the plate on which the United States resides? _____

(b) Does this plate consist of continental lithosphere? Oceanic lithosphere? Or both? _____

(c) Where does the lithosphere of the Atlantic Ocean form? _____

(d) What kind of plate boundary occurs along the west coast of South America? _____

(e) Is the west coast of Africa a plate boundary? _____

2.3 Early Evidence for Plate Tectonics

Because of a simple scale problem, it wasn't until the late 1960s that geologists discovered plate tectonics. Lithosphere plates are so large and move so slowly that we didn't realize they were moving. It was also difficult to believe that North America could move or the Atlantic Ocean could be destroyed. We will look first at some of the evidence that led geologists to accept the hypothesis that plates can move, and then see how we deduce the rates and geometry of processes at the three types of plate boundaries. Today, global positioning satellites and sensitive instruments confirm many of those early hypotheses by directly measuring the directions and rates of plate movement.

2.3.1 Evidence from the Fit of the Continents

As far back as 500 years ago, mapmakers drawing the coastlines of South America and Africa noted that the two continents looked as if they might have fit together once in a larger continent (**Fig. 2.3**). Those foolish enough to say it out loud were ridiculed, but today this fit is considered one of the best lines of evidence for plate tectonics.

Name: _____   **Section:** _____
Course: _____   **Date:** _____

(a) Sketch the shorelines of South America and Africa in Figure 2.3 on separate pieces of tracing paper. Rearrange the continents so that they fit snugly without overlapping or leaving large gaps. How well do the continents fit? Where are the problem areas?

_____

_____

_____

(b) If Africa and South America split apart tens of millions of years ago, would their shorelines be the best indicators of the shapes of the original pieces, or the true extent of either continent? What factors other than rifting and sea-floor spreading could have modified the shape of the current shorelines?

_____

_____

_____

FIGURE 2.3 Physiographic map of the South Atlantic Ocean floor and adjacent continents.

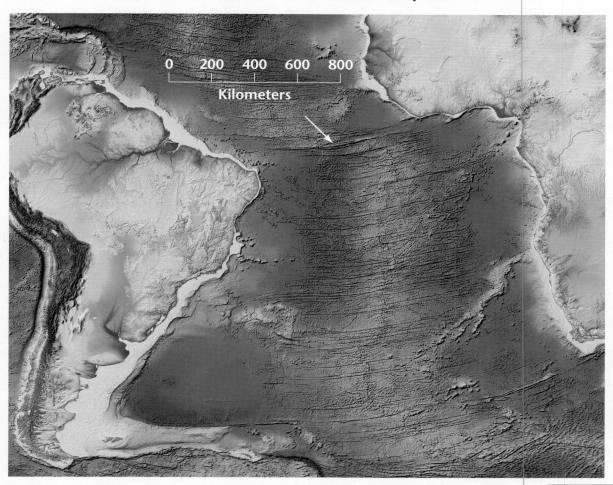

continued

Name: _____ Section: _____
Course: _____ Date: _____

(c) Trace the outlines of the continents again, this time using the edges of their continental shelves (the shallow, flat areas adjacent to the land) rather than the shoreline, and attempt to join them. Which reconstruction produces the best fit? In what ways is it better than the other?

_____

_____

_____

(d) Based on this evidence, what is the true edge of a continent?

_____

(e) When you fit the continents together, you could rotate them however you wished. However, Figure 2.3 also contains clues that show exactly how Africa and South America spread apart. Place the *best-fit* tracings of South America and Africa over those continents on Figure 2.3. Now bring them closer to one another until they join, *using the oceanic fracture zones to guide the direction in which you move the two plates.* Do the continents fit well when moved this way?

_____

(f) What does this suggest about the age and origin of the fracture zones?

_____

_____

_____

FIGURE 2.4
Earth's major climate zones.

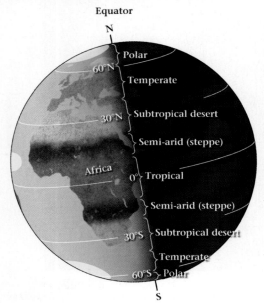

2.3.2 Evidence from Reconstructing Paleoclimate Belts

Earth's climate zones are distributed symmetrically about the equator today (**Fig. 2.4**), and the tropical, temperate, and polar zones support animals and plants unique to each environment. For example, walruses live in Alaska but palm trees don't, and large coral reefs grow in the Caribbean Sea but not in the Arctic Ocean. Climate zones also produce distinctive rock types: regional sand dunes and salt deposits form in arid regions, rocks made of debris deposited by continental glaciers accumulate in polar regions, and coal-producing plants grow in temperate and tropical forests.

However, geologists have found many rocks and fossils far outside the modern climate belts where we expect them to form. For example, 390-million-year-old (390-Ma) limestone containing reef-building organisms crops out along the entire Appalachian Mountain chain, some far north of where reefs exist today; 420-Ma salt underlies humid, temperate Michigan, Ohio, and New York; and deposits of 260- to 280-Ma continental glaciers are found in near-equatorial Africa, South America, India, and Madagascar.

According to the principle of uniformitarianism, these ancient salt deposits, reefs, and glacial deposits should have formed in locations similar to those where they form today. Either uniformitarianism doesn't apply to these phenomena or the landmasses have moved from their original climate zone into another. Observations like these led Alfred Wegener, a German meteorologist, to suggest in the 1920s that the continents had moved—a process he called *continental drift*. Modern geologists interpret these anomalies as the result of plate motion—when continents change position on the globe by moving apart during sea-floor spreading or coming together as subduction closes an ocean.

EXERCISE 2.3   Paleoclimate Evidence That Plates Move

Name: _____   Section: _____
Course: _____   Date: _____

Figure 2.5 shows simplified paleoclimate zones for 280 Ma as revealed by rocks in North America, South America, Greenland, Africa, Europe, and Turkey. Note that these do not fit Earth's modern climate belts. Copy these landmasses and their climate zones on tracing paper and cut each into a separate piece. Based on the zonation shown in Figure 2.4 and the shapes of the landmasses, restore them to their ancient positions on the globe in **Figure 2.6** that makes good climatic sense.

FIGURE 2.5 Fossils and rock types indicate 280-Ma climate zones.

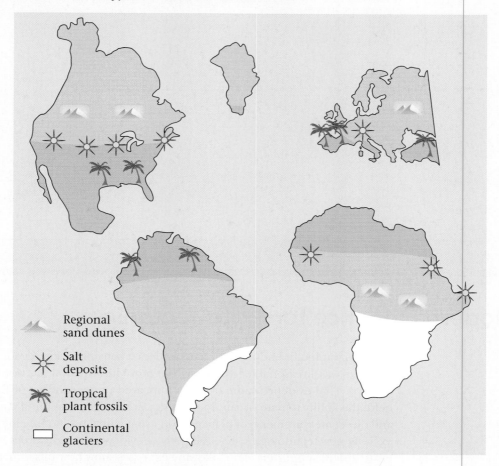

Regional sand dunes

Salt deposits

Tropical plant fossils

Continental glaciers

Note: Colors indicate same climate zones shown in Figure 2.4.

continued

EXERCISE 2.3 continued

Name: ______________________ **Section:** ____________
Course: ______________________ **Date:** ____________

FIGURE 2.6 Blank map for continents in Figure 2.5.

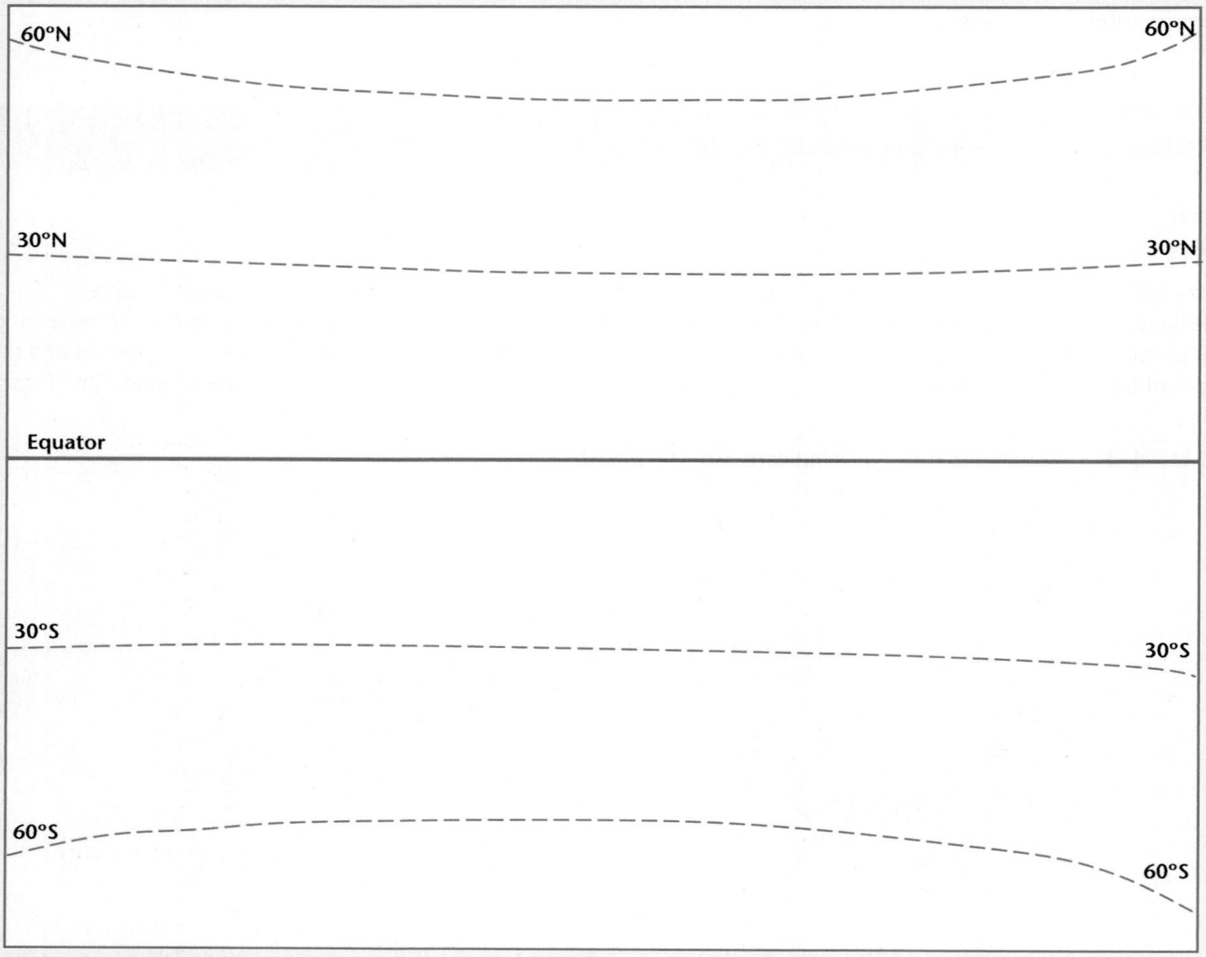

2.4 Modern Evidence for Plate Tectonics

The geographic fit and paleoclimate evidence convinced some geologists that plate tectonics was a reasonable hypothesis, but more information was needed to convince the rest. That evidence came from an improved understanding of Earth's magnetic field, the ability to date ocean floor rocks, careful examination of earthquake waves, and direct measurements of plate motion using global positioning satellites and other exciting new technologies. The full body of evidence has converted nearly all doubters to ardent supporters. The following exercises show how this evidence is used.

2.4.1 Evidence for Sea-Floor Spreading: Oceanic Magnetic Anomalies

Earth's magnetic field can be thought of as having "north" and "south" poles like a bar magnet. Geologists have learned that some rocks preserve a weak record of Earth's ancient magnetic field, called **paleomagnetism**. We have also learned that the magnetic field sporadically reverses polarity so that what is now the north magnetic pole becomes the south magnetic pole and vice versa. During periods of **normal polarity**, the field is the same as it is today, but during periods of **reversed polarity**, a compass needle that points to today's North Pole would swing around and point south. Geologists are able to accurately date the timing of magnetic reversals back to 4.5 Ma by determining the age of rocks with different polarities throughout the world (**Fig. 2.7**). The pattern of reversals has been found at every oceanic ridge, proving that the reversals are truly worldwide events.

Ocean ridge basalt records the magnetic field polarity in effect at the time it freezes from lava, and this paleomagnetism is convincing evidence for sea-floor spreading. The measured strength of the magnetic field in the oceans is the result of two components: (1) Earth's modern magnetic field strength and (2) the paleomagnetism (remanent magnetic field) of the oceanic crust. If the paleomagnetic polarity of a rock is the same as today's magnetic field, the rock's weak paleomagnetism *adds* to the modern field strength, resulting in an observed magnetic field *stronger* than today's field. If the paleomagnetic polarity is reversed, the rock's paleomagnetism *subtracts* from the modern field and the result is a measurement *weaker* than the average modern magnetic field (**Fig. 2.8**).

Measurements of the magnetic field in the oceans reveal linear belts (informally called **magnetic stripes**) in which the field alternates between being anomalously stronger (a positive anomaly) or weaker (a negative anomaly) than the average magnetic field (**Fig. 2.9**). Let's see how geologists use these magnetic stripes to prove that sea-floor spreading takes place.

FIGURE 2.7
Radiometric dating of lava flows indicates the magnetic reversals over the past 4 Ma.
Major intervals of positive or negative polarity are called *chrons* and are named after scientists who contributed to the understanding of the magnetic field. Short-duration reversals are called *subchrons*.

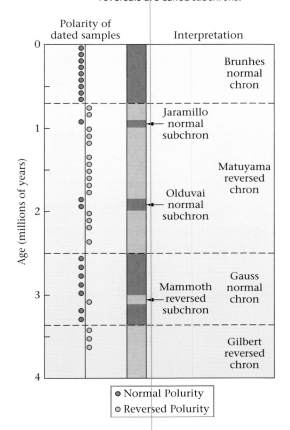

FIGURE 2.8 Components of Earth's magnetic field strength in the oceans.

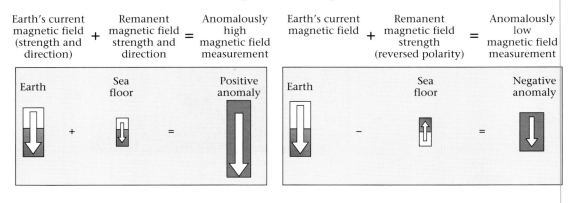

Name: _____   **Section:** _____

Course: _____   **Date:** _____

Examine the magnetic anomaly stripes in the North Atlantic and North Pacific oceans (Fig. 2.9).

(a) Compare the orientation of the magnetic stripes with those of the two ridge axes. Are they parallel? Perpendicular? At some other angle?

_____

(b) Describe the relationship between the ages of magnetic stripes and their distance from the ridge crests.

_____

_____

_____

(c) Explain how the process of sea-floor spreading can produce these orientations and relationships.

_____

_____

_____

(d) Some magnetic stripes are wider than others. Knowing what you do about sea-floor spreading and magnetic reversals, suggest an explanation.

_____

_____

_____

FIGURE 2.9   **Magnetic anomaly stripes.**

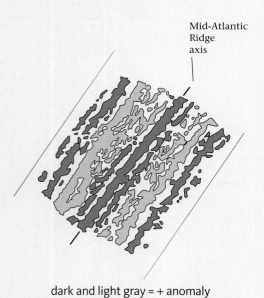

dark and light gray = + anomaly
beige = – anomaly

(a) The Mid-Atlantic Ridge southwest of Iceland.

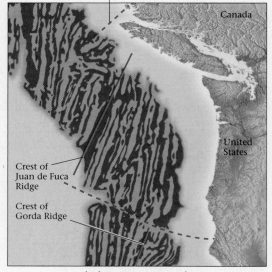

dark gray = + anomaly
light gray = – anomaly

(b) The Juan de Fuca and Gorda ridges in the North Pacific off Washington and British Columbia.

2.5 Plate Tectonic Processes Revealed by Earth Features

The next few exercises examine the three kinds of plate boundaries and show how geologists deduce details of their geometry, the rates of plate motion involved, and their histories. Let's start with information that we can gather about sea-floor spreading.

2.5.1 Sea-Floor Spreading

The South Atlantic Ocean formed by sea-floor spreading at the Mid-Atlantic Ridge. Geologists can get a rough estimate of the spreading rate (i.e., the relative motion of South America with respect to Africa) by measuring the distance between the two continents in a direction parallel to the fracture zones and determining the time over which the spreading occurred.

(a) Measure the distance between South America and Africa along the fracture zone indicated by the arrow in Figure 2.3.
_____ km

The oldest rocks in the South Atlantic Ocean, immediately adjacent to the African and South American continental shelves, are 120,000,000 years old.

(b) Calculate the average rate of sea-floor spreading for the South Atlantic Ocean over its entire existence. Express your answer in _____ km/million years = _____ km/yr = _____ cm/yr = _____ mm/yr

(c) Assuming someone born today lives to the age of 100, how much wider will the Atlantic Ocean become during his or her lifetime? _____ cm

Magnetic reversals are found worldwide, so magnetic stripes should be the same width in every ocean *if the rate of sea-floor spreading is the same at all ridges*. If a particular anomaly is wider in one ocean than another, however, it must result from faster spreading. **Figure 2.10** shows simplified magnetic stripes from the South Atlantic and Pacific oceans, the ages of the rocks, and the distance from the spreading center. Note that only half of each ocean basin is shown; a symmetrical series of stripes exists on the other side of each ridge as in Figure 2.9. For simplicity, only the most recent 80 million years of data are shown for the two oceans.

continued

Name: _____   Section: _____
Course: _____   Date: _____

FIGURE 2.10 Magnetic anomalies associated with different ocean ridges.

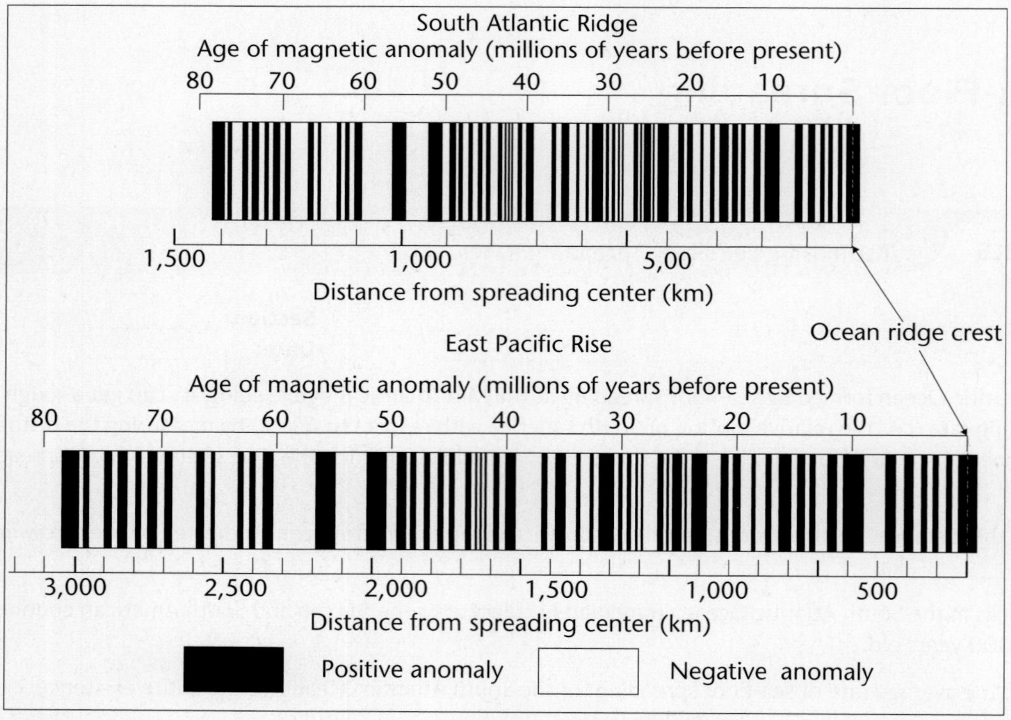

South Atlantic Ridge
Age of magnetic anomaly (millions of years before present)

80   70   60   50   40   30   20   10

1,500   1,000   5,00
Distance from spreading center (km)

Ocean ridge crest

East Pacific Rise
Age of magnetic anomaly (millions of years before present)

80   70   60   50   40   30   20   10

3,000   2,500   2,000   1,500   1,000   500
Distance from spreading center (km)

Positive anomaly   Negative anomaly

(a) Measure the distance from the spreading center to the farthest magnetic anomaly stripe in each ocean and determine the age of that anomaly.
South Atlantic _____ km   age: _____
North Pacific _____ km   age: _____

(b) Divide the distance by the age to determine the average rate of spreading. [*Then double this rate, because the data are for only one side of the oceanic ridge.*]
South Atlantic _____ mm/yr   North Pacific _____ mm/yr

(c) Have these two ocean ridges spread at the same rate?

_____

(d) Compare the results of this exercise with those from Exercise 1.10b. Are the two estimates for the opening of the South Atlantic Ocean the same? If not, what factors could explain the difference?

_____

_____

_____

(e) Why are the anomalies in the North Pacific wider than those in the South Atlantic?

_____

_____

_____

Name: _____   **Section:** _____

Course: _____   **Date:** _____

Figure 2.11 shows bathymetric maps of parts of the Mid-Atlantic Ridge and East Pacific Rise at the same scale.

FIGURE 2.11 **Bathymetric maps of the East Pacific Rise and Mid-Atlantic Ridge.**

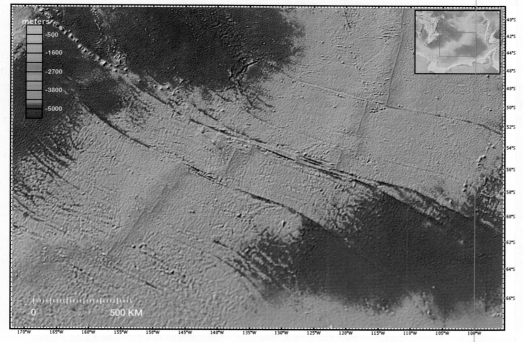

(a) The East Pacific Rise in the South Pacific Ocean.

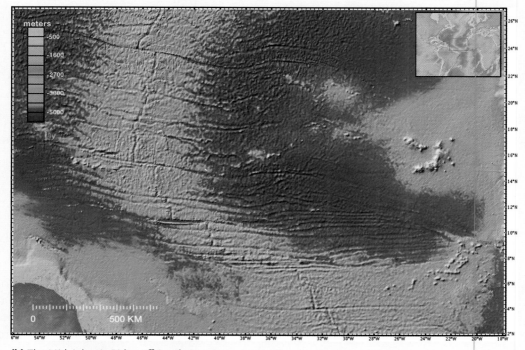

(b) The Mid-Atlantic Ridge off South America.

continued

Name: _____   **Section:** _____

Course: _____   **Date:** _____

(a) Ocean ridges typically have a rift valley at their axes. Which of the two ridges in Figure 2.11 has the best developed rift valley?

_____

   Do all ocean ridges have the same shape? Most are similar, but different spreading rates cause variations in shape be-cause of the way the cooling lithosphere behaves. Lithosphere near the ridge axis is young, thin, still hot, and therefore has a lower density than older, colder lithosphere far from the axis. As a result, the ridge axis area floats relatively high on the underlying asthenosphere and the water above it is relatively shallow. As sea-floor spreading moves the oceanic lithosphere away from the ridge axis, the rocks cool and get thicker and denser. The farther it is from the ridge axis, the lower the oceanic lithosphere sits on the asthenosphere and the deeper the water will be. This concept is known as the **age versus depth relationship**.

(b) Keeping in mind the age versus depth relationship, why is the belt of shallow sea wider over the East Pacific Rise than over the Mid-Atlantic Ridge?

_____

_____

(c) On the graph provided, plot depth (on the vertical axis) against distance from the ridge axis (on the horizontal axis) for both the East Pacific Rise and the Mid-Atlantic Ridge. Use five to ten points for each ridge. Connect the dots for the East Pacific Rise with red pencil and those for the Mid-Atlantic Ridge with green pencil to make cross sections of each ridge.

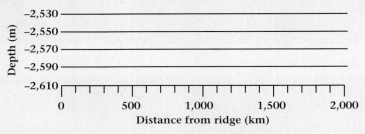

(d) Does the rate at which the depth with distance from the ridge stay the same over time, decrease over time, or increase over time?

_____

2.5.2 Hot-Spot Volcanic Island Chains: Evidence for the Direction and Rate of Plate Motion

Paleoclimate anomalies show that plates have moved and magnetic stripes help measure the rate of sea-floor spreading, but how can we determine the direction and rate at which a plate has moved (its absolute motion)? Do plates always move in a single direction or can they change direction? The answers come from the study of hot-spot volcanic island chains.

FIGURE 2.12 Origin of hot-spot island chains and seamounts.

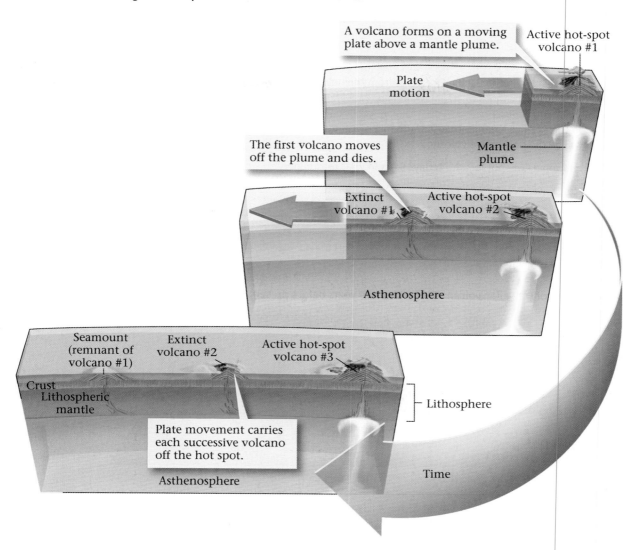

A volcano forms on a moving plate above a mantle plume.

Active hot-spot volcano #1

Plate motion

Mantle plume

The first volcano moves off the plume and dies.

Extinct volcano #1

Active hot-spot volcano #2

Asthenosphere

Seamount (remnant of volcano #1)

Extinct volcano #2

Active hot-spot volcano #3

Crust
Lithospheric mantle

Plate movement carries each successive volcano off the hot spot.

Lithosphere

Asthenosphere

Time

Figure 2.12 shows how hot-spot island chains form. Each volcano in the chain forms at a **hot spot**—an area of unusual volcanic activity not associated with processes at plate boundaries. The cause is still controversial, but many geologists propose that hot spots form above a narrowly focused source of heat called a **mantle plume**—a column of very hot rock that rises by slow plastic flow from deep in the mantle. When the plume reaches the base of the lithosphere, it melts the lithosphere rock and produces magma that rises to the surface, erupts, and builds a volcano. The plume is thought to be relatively motionless. If the plate above it moves, the volcano is carried away from its magma source and becomes extinct. A new volcano then forms above the hot spot until it, too, is carried away from the hot spot.

Over millions of years, a chain of volcanic islands forms, the youngest at the hot spot, the oldest farthest from it. As the volcanoes cool, they become denser, subside, and are eroded by streams and ocean waves. Eventually, old volcanoes sink below the ocean surface, forming **seamounts**. The chain of islands and seamounts traces plate motion above the hot spot, just as footprints track the movement of animals.

Name: _____   **Section:** _____

Course: _____   **Date:** _____

The Hawaiian Islands, located in the Pacific Ocean far from the nearest oceanic ridge, are an excellent example of hot-spot volcanic islands (**Fig. 2.13**). Volcanoes on Kauai, Oahu, and Maui haven't erupted for millions of years, but the island of Hawaii hosts five huge volcanoes, one of which (Kilauea) has been active for the past 20 years. In addition, a new volcano, Loihi, is growing on the Pacific Ocean floor just southeast of Kilauea. As the Pacific Plate moves, Kilauea will become extinct and Loihi will be the only active volcano.

(a) Where is the Hawaiian hot spot located today? Explain your reasoning.

_____

_____

(b) Draw a line connecting the volcanic centers (highlighted in red) on Hawaii and Maui. Do the same for Maui to Molokai, Molokai to Oahu, and Oahu to Kauai.

FIGURE 2.13 Ages of Hawaiian volcanoes in millions of years before present.

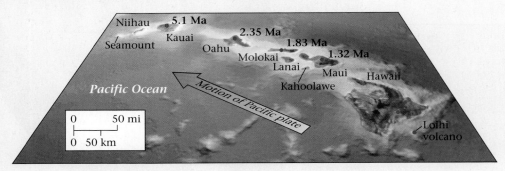

(c) Measure the direction and distance between these centers using a ruler, a protractor, and the map scale. Calculate the rate of plate motion (distance between volcanoes divided by the time interval between eruption ages) and fill in **Table 2.1**. Express the rates in millimeters per year. Use the protractor in your toolkit to measure direction using the azimuth system (see below).

TABLE 2.1 Movement of the Pacific Plate over the Hawaiian hot spot.

	Distance between islands (km)	Number of years of plate motion	Rate of plate motion (mm/yr)	Azimuth direction of plate motion (e.g., 325°)
Hawaii to Maui				
Maui to Molokai				
Molokai to Oahu				
Oahu to Kauai				

Geologists use the *azimuth system* to describe direction precisely. In the azimuth system, north is 0°, east is 090°, south is 180°, and west is 270°, as shown in **Figure 2.14**. The green arrow points to 045° (northeast). To practice using the azimuth system, estimate and then measure the directions shown by arrows A through D.

continued

Name: _____   Section: _____
Course: _____   Date: _____

FIGURE 2.14 The azimuth system.

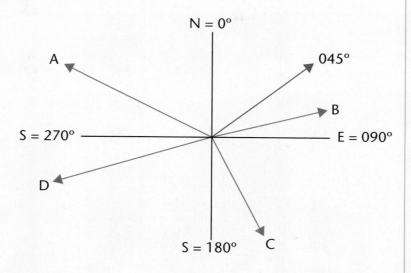

The Hawaiian Islands are the youngest volcanoes in the Hawaiian-Emperor seamount chain (**Fig. 2.15**). Most of the older volcanoes are seamounts detected by underwater oceanographic surveys. The Hawaiian-Emperor seamount chain tracks the motion of the Pacific Plate and lets us interpret Pacific Plate motion for a longer time span than that recorded by the Hawaiian Islands alone.

EXERCISE **2.9**   **Long-Term Movement of the Pacific Plate**

Name: _____   Section: _____
Course: _____   Date: _____

The oldest volcano of the Hawaiian-Emperor seamount chain was once directly above the hot spot, but is now in the northern Pacific, thousands of kilometers away (Fig. 2.15). Seamount ages show that the hot spot has been active for a long time and reveal the direction and rate at which the Pacific Plate has moved.

(a) What evidence is there that the Pacific Plate has not always moved in the same direction?

_____

_____

_____

continued

Name: _____   Section: _____

Course: _____   Date: _____

FIGURE 2.15 Physiography of the Pacific Ocean floor showing ages of volcanoes in the Hawaiian-Emperor seamount chain.

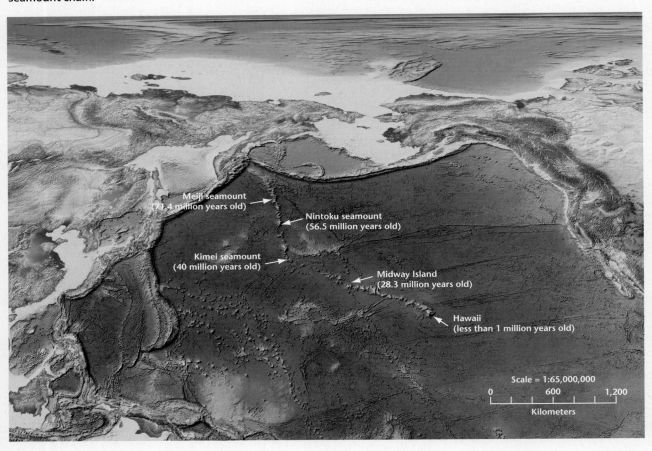

(b) How many years ago did it change direction? Explain your reasoning.

_____

_____

_____

(c) Based on the information in Figure 2.15, in what direction did the Pacific Plate move originally?

_____

(d) How far has the Meiji seamount moved from the hot spot? Explain your reasoning.

_____

_____

_____

continued

Name: _____   Section: _____
Course: _____   Date: _____

(e) At what rate has the Pacific Plate moved
　　(i) based on data from the Hawaiian Islands alone? _____
　　(ii) based on data from the Hawaii-Midway segment? _____
　　(iii) based on data from the Hawaii-Kimei segment? _____
　　(iv) based on data from the Kimei-Meiji segment? _____

(f) What is the best set of data to use for estimating the rate of motion of the Pacific Plate over the full 71-million-year span? Explain.

_____

_____

_____

(g) Has the Meiji seamount moved at a constant rate? Explain your reasoning.

_____

_____

_____

(h) In what direction is the Pacific Plate moving today? Explain your reasoning.

_____

_____

_____

(i) Assuming that the current direction of motion continues, what will be the eventual fate of the Meiji seamount? Explain in as much detail as possible.

_____

_____

_____

_____

_____

_____

2.5.3 Continental Rifting

A new ocean forms when rifting takes place beneath a continent. The continental crust first thins, then breaks into two pieces separated by an oceanic ridge. As seafloor spreading proceeds, an ocean basin grows between the fragments of the original continent. This process is in an early stage today in eastern Africa.

Name: _____   Section: _____

Course: _____   Date: _____

Examine the bathymetric/physiographic map of the region that includes eastern Africa, the Red Sea, and the western Indian Ocean (Fig. 2.16). Eastern Africa is beginning to break away from the western part. The split is occurring along the East African Rift, a deep valley locally filled with lakes and bordered by steeply uplifted land.

FIGURE 2.16   Red Sea rift zone.

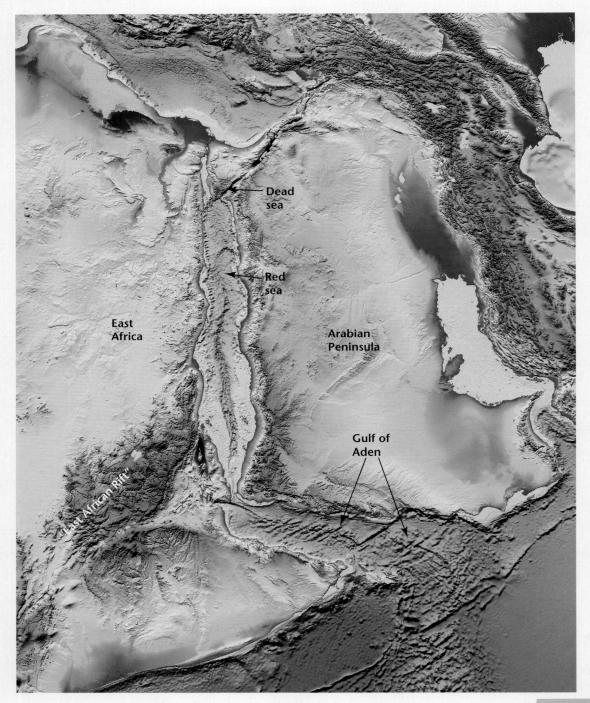

continued

Name: _____   Section: _____

Course: _____   Date: _____

Rifting has split Africa from the Arabian Peninsula to form the Red Sea. A new ocean ridge and ocean lithosphere have formed in the southern two-thirds of the Red Sea. A narrow belt of deeper water defines the trace of this ridge. At the northern end of the Red Sea, the ridge/rift axis is cut by a transform fault that runs along the eastern side of the Sinai Peninsula and through the Dead Sea.

(a) Use a red line to show the trace of the Red Sea rift/ridge axis. Use a purple line to show the trace of the Dead Sea transform fault.

(b) The narrow ocean bordering the southeast edge of the Arabian Peninsula is the Gulf of Aden. Use red and purple lines to trace the ridge segments and transform fault in this narrow sea.

(c) Based on the geometry of the ridges and transform faults in the Red Sea and Gulf of Aden, draw an arrow showing the motion of the Arabian Peninsula (the Arabian Plate) relative to Africa.

2.5.4 Subduction Zones: Deducing the Steepness of Subduction

When two oceanic plates collide head-on, one is subducted beneath the other and re-turns to the mantle, completing the tectonic cycle begun when the ocean crust initially erupted at a mid-ocean ridge. Melting occurs above the subducted plate when it reaches a depth of about 150 km. As a result, the volcanic arc forms at some distance from the ocean trench. The area between the volcanic arc and trench typically contains an ac-cretionary prism, composed of highly deformed sediment scraped off the sinking plate, and a fore-arc basin in which debris eroded from the arc accumulates (**Fig. 2.17**).

FIGURE 2.17 Anatomy of an island arc-trench system.

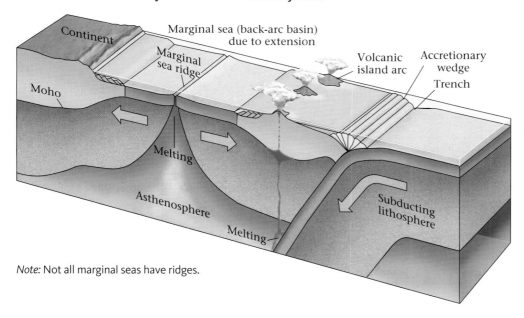

Note: Not all marginal seas have ridges.

No two subduction zones are identical. Differences can include the width of the accretionary prism, the rate of subduction, and the steepness of the subducted plate as it moves into the mantle. Earthquakes occur in the subducted plate as it moves, concentrated in what geologists call a *Wadati-Benioff zone*. We can track the plate as

Name: _____   **Section:** _____
Course: _____   **Date:** _____

(a) Based on island arc-trench geometry illustrated in Figure 2.17, what factors determine the width of the arc-trench gap? Explain.

_____

_____

_____

(b) Based on your answer to (a), sketch two island arc-trench systems, one with a wider arc-trench gap than the other.

Figure 2.18 shows profiles across four segments of the Aleutian island arc, which extends westward from mainland Alaska. The positions of the volcanic arc and trench are indicated in each profile.

(c) In the profiles showing the earthquakes (on the next page), sketch the upper and lower boundaries of the subducted plate, assuming that the deepest part of the Aleutian Trench is at the middle of the trench. Explain why you drew the boundaries where you did.

_____

_____

_____

(d) At what depth does melting apparently begin beneath the Amchitka segment of the arc? Explain your reasoning.

_____

_____

_____

(e) *Assuming that melting occurs at the same depth everywhere* beneath the Aleutian arc, complete the other three profiles by drawing the subducted plates.

(f) From these profiles, estimate the arc-trench gap for the four segments of the Aleutian arc.

	Amchitka	Shumagin Islands	Cook Inlet	Skwentna
Arc-trench gap (km)				

continued

Name: _____   Section: _____

Course: _____   Date: _____

FIGURE 2.18 Profiles across the Aleutian island arc-trench system.
Orange and green dash/dot lines illustrate calculation of melting depth.

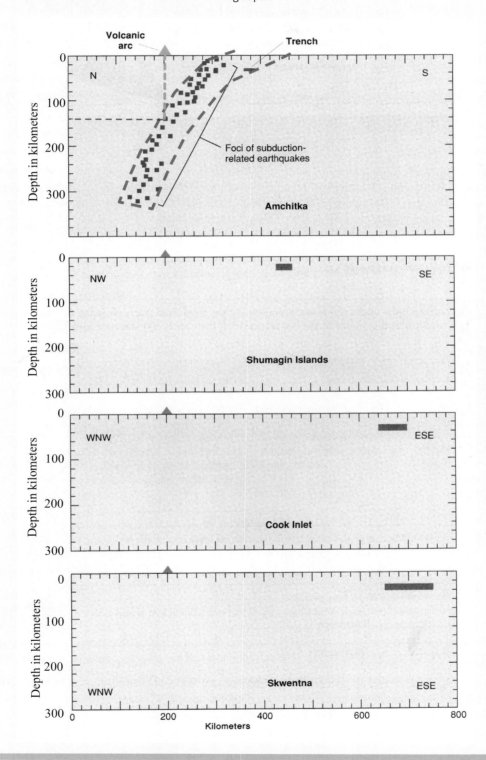

it is subducted by locating the depth and location of the earthquake foci (points where the energy is released). With a little geologic reasoning, you can also estimate the steepness of subduction from details of its arc-trench system.

2.5.5 Transform Faults

The transform faults that you have looked at thus far are in the ocean, but some also cut continental lithosphere. The most famous of these is the San Andreas Fault system of California, one that has caused major damage and loss of life over the past century. Exercise 2.12 explores the age, size, and effects of the San Andreas Fault system.

EXERCISE 2.12   **Movement on Continental Transform Faults**

Name: _____   Section: _____
Course: _____   Date: _____

The San Andreas Fault system is a continental transform that links the East Pacific Rise, an actively spreading oceanic ridge segment in the Gulf of California, with another ridge segment to the northwest in the Pacific Ocean (**Fig. 2.19a**). **Figure 2.19b** shows four groups of distinctive rocks separated by the San Andreas system. The bodies shown in green are thought to have been a single mass at one time and are now separated by offset movement along the San Andreas transform fault system. The same conclusion has been reached about the other pairs of similarly colored rocks.

FIGURE 2.19 San Andreas Fault system.

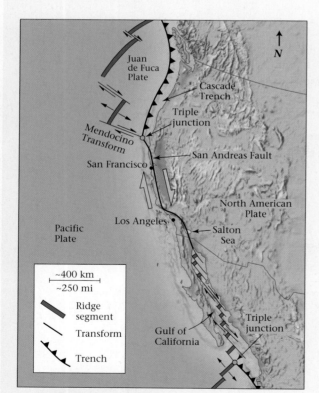

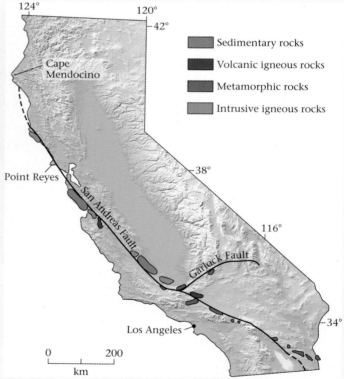

(a) Geologic setting of the San Andreas Fault system. At its northern end, the San Andreas links to the Cascade Trench and an oceanic tranform; at is southern end, it links to a mid-ocean ridge in the Gulf of California.

(b) Amount of fault movement indicated by offset bodies of identical rock.

continued

Name: _____   **Section:** _____

Course: _____   **Date:** _____

(a) Based on the offset of the rock bodies shown in Figure 2.19b, draw arrows to indicate the sense of slip along the San Andreas Fault system.

(b) Measure the amount of offset for each of the four rock types and record it below.

	Type of offset rock	**Measured offset (km)**	**Average rate of fault motion (mm/yr)**
Sedimentary			
Volcanic igneous			
Metamorphic			
Intrusive igneous			

(c) Motion along the San Andreas Fault system is thought to have begun around 20 Ma. Assuming that the four offset bodies of rock are older than 20 million years, calculate the average rate of fault motion for each.

(d) If an offset igneous body was dated at 10 million years old, how would its mapped displacement differ from that of these rocks? Why might the estimate of displacement rate be different?

_____

_____

_____

2.6 Active versus Passive Continental Margins

Earthquakes and volcanic eruptions are common on the west coast of North America, but there are no active volcanoes on the east coast and earthquakes are rare. Geologists call the west coast a (tectonically) **active continental margin** and the east coast a **passive continental margin**. These differences can be found on several continents. Most passive continental margins have broad continental shelves, whereas active continental margins typically have narrow continental shelves.

EXERCISE 2.13 Why Are There More Earthquakes and Volcanoes on One Side of Some Continents than on the Other?

Name: ______________________ **Section:** __________

Course: ______________________ **Date:** __________

Try to explain why some continental margins are active and others passive based on what you know about plate tectonics.

(a) Are all continental coastlines plate boundaries? *Explain.*

____

____

____

(b) Compare the west coast of South America with the west coast of Africa. Which has a broad continental shelf? A narrow shelf? Which is close to an ocean trench? Which coast would you expect to have the most earthquakes? Explain.

____

____

____

(c) On the plate tectonic map of the world (Fig. 2.1), label active continental margins with a red letter A and passive margins with a blue letter P.

CHAPTER 3

MINERALS

PURPOSE

- Practice scientific classification and learn how minerals are different from other materials.
- Develop the skills involved in identifying minerals.
- Become familiar with common rock-forming minerals.
- Learn what minerals can reveal about Earth processes and Earth history.

MATERIALS NEEDED

- Sets of mineral and rock specimens
- Hand lens, streak plate, glass plate, knife or steel nail, and a penny to determine the physical properties of minerals
- Dilute hydrochloric acid (HCl) for a simple chemical test

3.1 Introduction

This chapter begins our study of Earth materials. It starts by examining the different kinds of materials in the geosphere and then focuses on minerals, the basic building blocks of most of the Earth. You will learn how minerals are different from other substances, how to study their physical properties, and how to use those properties to identify common minerals.

3.2 Classifying Earth Materials

Imagine that an octopus is swimming in the ocean when a container falls off a freighter overhead. The container breaks up and spills its entire cargo of sneakers, sandals, flip-flops, shoes, moccasins, and boots into the sea. The octopus is curious about these objects and wants to learn about them. How would it classify them? Remember, an octopus doesn't have heels or toes, has eight legs, doesn't understand "left" and "right," and doesn't wear clothes. One system might be to separate items that are mostly enclosed (shoes, boots, sneakers, moccasins) from those that are open (sandals, flip-flops). Another might be to separate objects made of leather from those made of cloth; or brown objects from black ones; or big ones from small ones. There are many ways to classify footwear, some of which might lead our octopus to a deeper understanding of the reasons for these differences.

Early geologists faced a similar task in the seventeenth century when they began to study Earth materials systematically by classifying them. Why classify things? Because classification shows us relationships between things that lead to understanding them and the processes by which they were made. Biologists classify organisms, art historians classify paintings, and geologists classify Earth materials. Exercise 3.1 introduces the thought processes involved in developing a classification scheme.

EXERCISE 3.1   **Classifying Earth Materials**

Name: _____   Section: _____
Course: _____   Date: _____

(a) Examine the specimens of Earth materials provided by your instructor. Group them into categories you believe are justified by your observations, and explain the criteria by which you set up the groups.

Group	Defining criteria for each group	Specimens in group

continued

Name: _____   **Section:** _____

Course: _____   **Date:** _____

(b) Compare your results with others in the class. Did you all use the same criteria? Are your classmates' specimens in the same groups as yours?

_____

_____

_____

(c) What does your comparison tell you about the process of classification?

_____

_____

_____

_____

3.3 What Is a Mineral and What Isn't?

Most people know that the Earth is made of minerals and rocks but don't know the difference between them. The words *mineral* and *rock* have very specific meanings to geologists, often much more precise than those used in everyday language. For example, what a dietitian calls a mineral is not a mineral to a geologist. Geologically, a **mineral** is *a naturally occurring, homogeneous, inorganic solid with an ordered internal arrangement of atoms and a distinctive chemical composition.* Let's look at this definition more closely.

- *Naturally occurring* means that a mineral forms by natural Earth processes. Thus, man-made materials like steel and plastic are not minerals.

- *Homogeneous* means that a piece of a mineral contains the same pure material throughout.

- *Inorganic* means that things formed by the life processes of animals and plants cannot be minerals. This rules out sugar, pearls, amber, bones, and wood.

- *Solid* means that minerals retain their shape indefinitely under normal conditions. Therefore, liquids like oil and water and gases like air and propane cannot be minerals.

- An *ordered internal arrangement of atoms* is an important characteristic that separates minerals from substances that may fit all other parts of the definition. Atoms in minerals occupy positions in a grid called a *crystalline structure*. Solids in which atoms occur in random clusters rather than in a crystalline structure are called *glasses*.

- *Distinctive chemical composition* means that the elements present in a mineral and the proportions of their atoms can be expressed by a simple formula—for example, quartz is SiO_2 and calcite is $CaCO_3$—or by one that is more complex—the mineral muscovite is $KAl_2(AlSi_3O_{10})(OH)_2$.

When a mineral grows without interference from other minerals, it develops smooth faces and a symmetrical geometric shape that we call a **crystal**. When a mineral forms in an environment where other minerals interfere with its growth, it has an irregular shape but still has the appropriate crystalline structure for that mineral, as would a piece broken off a crystal during erosion. An irregular piece of mineral is a **grain**, and a single piece of a mineral, either crystal or grain, is called a *specimen*.

In the geosphere, most minerals occur as parts of rocks. It is important to know the difference between a mineral specimen and a rock. A **rock** is *a naturally occurring, inorganic solid consisting of an aggregate of mineral grains, pieces of older rocks, or a mass of natural glass.* Some rocks, like granite, contain grains of several different minerals, and some, like rock salt, are made of many grains of a single mineral. Others are made of fragments of previously existing rock that are cemented together. And a few kinds of rock are natural glasses, cooled so rapidly from a molten state that their atoms did not have time to form the grid-like crystalline structures required for minerals.

EXERCISE 3.2   **Is It a Mineral or a Rock?**

Name: _____   Section: _____
Course: _____   Date: _____

(a) Based on the definitions of *mineral* and *rock*, determine which specimens used in Exercise 3.1 are minerals and which are rocks.

Minerals	Rocks	Other

(b) Look carefully at one of the *rock* specimens. How many different minerals are in this rock? _____

(c) How do you know? What visual or other clues did you use to determine how each mineral is different from its neighbors?

_____

_____

_____

(d) Describe each of the minerals in this rock in your own words.

Mineral #1

_____

_____

_____

Mineral #2

_____

_____

_____

continued

Name: _____　　**Section:** _____
Course: _____　　**Date:** _____

Mineral #3

_____

_____

_____

Mineral #4

_____

_____

_____

3.4 Physical Properties of Minerals

Mineralogists (geologists who specialize in the study of minerals) have named more than 4,000 minerals that differ from one another in composition and crystalline structure. These characteristics, in turn, determine a mineral's **physical properties**, which include how it looks (color and luster), breaks, feels, smells, and even tastes. Some minerals are colorless and nearly transparent; others are opaque, dark colored, and shiny. Some are hard, others soft. Some form long, needle-like crystals, others blocky cubes. You probably instinctively used some of these physical properties in Exercise 3.2 to decide how many minerals were in your rocks and then to describe them. We discuss the major physical properties of minerals below so you can use them to identify common minerals—in class, at home, or while on vacation.

3.4.1 Diagnostic versus Ambiguous Properties

Geologists use physical properties to identify minerals much as detectives use physical descriptions to identify suspects. And, as with people, some physical properties are **diagnostic properties**—they immediately help identify an unknown mineral or rule it out as a possibility. Other properties are **ambiguous properties** because they may vary in different specimens of the same mineral. For example, color is a notoriously ambiguous property in many minerals (Fig. 3.1). Size doesn't really matter either; a large specimen of quartz has the same properties as a small one. Exercise 3.3 shows how diagnostic and ambiguous properties affect everyday life.

3.4.2 Luster

One of the first things we notice when we pick up a mineral is its luster—the way light interacts with its surface. For mineral identification, we distinguish minerals that have a metallic luster from those that are nonmetallic. Something with a *metallic* luster is shiny and opaque like an untarnished piece of metal. Materials with a *nonmetallic* luster look earthy (dull and powdery like dirt), glassy (vitreous), waxy, silky, or pearly—terms relating luster to familiar materials. Luster is a diagnostic property for many minerals, but be careful: some minerals may tarnish and their metallic luster may be dulled.

EXERCISE 3.3 People Have Diagnostic Properties Too

Name: ______________________ **Section:** __________
Course: ______________________ **Date:** __________

Your father has asked you to pick up his old college roommate at the airport. You've never met him, but your father gave you a yearbook photo and described what he looked like 30 years ago: height, weight, hair color, beard, eye color. Which of these features would still be diagnostic today? Which, considering the passage of time, might be ambiguous? What other properties might also be diagnostic?

Property	Diagnostic (explain)	Ambiguous (explain)
Height		
Weight		
Hair color		
Beard		
Eye color		
Others		

3.4.3 Color

The color we see when we look at a mineral is controlled by how the different wavelengths of visible light are absorbed or reflected by the mineral's atoms. Color is generally a diagnostic property for minerals with a metallic luster and for *some* with a nonmetallic luster. Specimens of some nonmetallic minerals, like the fluorite in **Figure 3.1**, have such a wide range of colors that they were once thought to be different minerals. We now know that the colors are caused by impurities. For example, rose quartz contains a very small amount of titanium.

3.4.4 Streak

The streak of a mineral is the color of its powder, which we can determine by rubbing a mineral against an unglazed porcelain plate. Streak and color are the same for most minerals, but for some they are different. In these cases, the difference between streak and color is an important diagnostic property. A mineral's color may vary widely, as in Figure 3.1, but its streak is generally similar for all specimens regardless of their color as seen in **Figure 3.2**.

FIGURE 3.2 Color and streak.

(a) Red hematite has a reddish-brown streak.

(b) But this dark, metallic-looking specular hematite also has a brownish streak.

3.4.5 Hardness

The hardness of a mineral is a measure of the ease with which it can scratch or be scratched by other substances. A nineteenth-century mineralogist, Fredrick Mohs, created a mineral hardness scale that we still use today, using ten familiar minerals. He assigned a hardness of 10 to the hardest mineral, and a hardness of 1 to the softest (Fig. 3.3). This is a **relative scale**, meaning that a mineral can scratch those lower in the scale but cannot scratch those that are higher. It is not an *absolute scale* in which diamond would be 10 times harder than talc and corundum would be 3 times harder than calcite. The hardness of common materials can also be described using the Mohs hardness scale, such as the testing materials listed in Figure 3.3. To determine the hardness of a mineral, see which of these materials it can scratch and which can scratch it.

FIGURE 3.3 Mohs hardness scale and its relationship to common testing materials.

Mineral	Rank on Mohs hardness scale	Testing material
Diamond	10	
Corundum	9	
Topaz or beryl	8	
Quartz	7 ←	Streak plate (H = 6.5–7)
Orthoclase	6 ←	Window glass, steel-cut nail (H = 5.5)
Apatite	5 ←	Common nail, knife (H = 5–5.5)
Fluorite	4	
Calcite	3 ←	U.S. penny (H = 3)
	←	Fingernail (H = 2.5)
Gypsum	2	
Talc	1	

EXERCISE 3.4   **Constructing and Using a Relative Hardness Scale**

Name: _____   Section: _____
Course: _____   Date: _____

Your instructor will tell you which specimens from your mineral set to use for this exercise. Arrange them in order of increasing hardness by seeing which can scratch the others and which are most easily scratched.

Softest ⟶ *Hardest*

Specimen #:   _____   _____   _____   _____   _____

Now use the testing materials listed in Figure 3.3 to determine the Mohs hardness of minerals in your set.

Mohs hardness: _____   _____   _____   _____   _____

3.4.6 Crystal Habit

Crystal shapes found in the mineral kingdom range from simple cubes with six faces to complex twelve-, twenty-four-, or forty-eight-sided (or more) crystals (Fig. 3.4a-i). Some crystals are flat like a knife blade, others are needle-like. Each mineral has its own diagnostic **crystal habit**, a preferred crystal shape that forms when it grows unimpeded by other grains. For example, the habit of halite is a cube, that of garnet is an equant twelve-sided crystal. The habit of quartz is elongate hexagonal crystals topped by a six-sided pyramid. *Remember:* Crystal growth requires very special conditions. As a result, most mineral specimens are irregular grains; very few display their characteristic crystal habit.

FIGURE 3.4 Crystal habits of some common minerals.

(a) Cubes of pyrite

(b) 12-Sided garnet crystals

(c) Slender tourmaline crystals

(d) Prismatic quartz crystals

(e) Potassic feldspar

FIGURE 3.4 Crystal habits of some common minerals (*con't.*).

(f) Bladed kyanite crystals

(g) Calcite scalenohedra

(h) Needle-like crystals of natrolite

3.4.7 Breakage

Some minerals break along one or more smooth planes, others along curved surfaces, still others in irregular shapes. The way a mineral breaks is controlled by whether or not there are zones of weak bonds in its structure. Instead of breaking the minerals in your sets with a hammer, we examine specimens to see how they have already broken—using a microscope or magnifying glass to see more clearly. Two kinds of breakage are important: fracture and cleavage.

Fracture occurs when there are no zones of particularly weak bonding within a mineral. When such a mineral breaks, either irregular (**irregular fracture**) or curved surfaces (**conchoidal fracture**; **Fig. 3.5**) form. Conchoidal fracture surfaces are common in thick glass and in minerals in which bond strength is nearly equal in all directions (e.g., quartz, garnet).

Cleavage occurs when bonds holding atoms together are weaker in some directions than in others. The mineral breaks along these zones of weakness, producing flat, smooth surfaces. Some minerals have a single zone of weakness, but others may have two, three, four, or six (**Fig. 3.6**). If there is more than one zone of weakness, a mineral cleaves in more than one direction. It is important to note *how many* directions there are and *the angles between those directions*. Two different minerals might each have two directions of cleavage, but those directions might be at 90° in one mineral but not in the other (see Fig. 3.6). For example, amphiboles and pyroxenes (two important groups of minerals) are similar in most other properties and have two directions of cleavage, but amphiboles cleave at 56° and 124°, whereas pyroxenes cleave at 90°.

Note that in Figure 3.6 there may be many cleavage *surfaces*, but several of those surfaces are parallel to one another, as shown for halite. All of these parallel surfaces define a single *cleavage direction*. To help observe a mineral's cleavage, hold it up to the light and rotate it. Parallel cleavage surfaces reflect light at the same time, making different cleavage directions easy to see.

FIGURE 3.5 Conchoidal fracture in quartz.

FIGURE 3.6 Types of cleavage commonly observed in minerals.

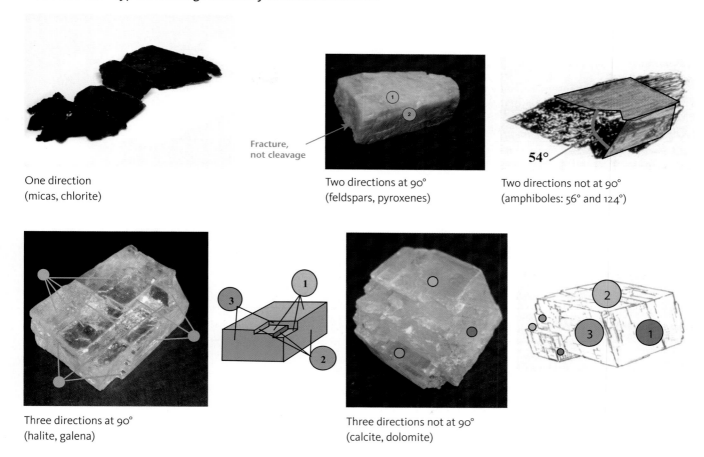

One direction
(micas, chlorite)

Fracture,
not cleavage

Two directions at 90°
(feldspars, pyroxenes)

54°

Two directions not at 90°
(amphiboles: 56° and 124°)

Three directions at 90°
(halite, galena)

Three directions not at 90°
(calcite, dolomite)

Both crystal faces and cleavage surfaces are smooth, flat planes and might be mistaken for one another. If you can see many small, parallel faces, these are cleavage faces because crystal faces are not repeated. In addition, breakage occurs after a crystal has grown, thus cleavage or fracture surfaces generally look less tarnished or altered than crystal faces.

3.4.8 Specific Gravity

The specific gravity (**SpG**) of a mineral is a comparison of its density with the density of water. The density of pure water is 1 g/cm³, so if a mineral has a density of 4.68 g/cm³, its specific gravity is 4.68.

$\dfrac{4.68 \text{ g/cm}^3}{1.00 \text{ g/cm}^3}$ The units cancel, so **SpG = 4.68**.
          This means that the mineral is 4.68 times denser than water.

You can measure specific gravity by calculating the density of a specimen (density = mass ÷ volume). But geologists generally estimate specific gravity by *hefting* a specimen and determining if it seems heavy or light. To compare the specific gravities of two minerals, pick up similar-sized specimens to get a general feeling for their densities. You will feel the difference—just as you would feel the difference between a box of Styrofoam packing material and a box of marbles.

Name: _____   Section: _____
Course: _____   Date: _____

Examine the specimens indicated by your instructor. Which have cleaved and which have fractured? For those with cleavage, indicate the number of directions and the angles between them.

_____

_____

_____

_____

_____

_____

_____

_____

_____

_____

_____

_____

EXERCISE 3.6   Heft and Specific Gravity

Name: _____   Section: _____
Course: _____   Date: _____

Separate the minerals provided by your instructor into those with relatively high specific gravity and those with relatively low specific gravity.

(a) What luster do most of the minerals in the high specific gravity group have? _____ This is not a coincidence. In general, minerals with this luster have higher specific gravities than minerals with other lusters.

(b) To become familiar with the range of specific gravity in common minerals, select the most dense and least dense specimens and measure their specific gravities as shown above.

High specific gravity _____   Low specific gravity _____

continued

Name: _____   **Section:** _____
Course: _____   **Date:** _____

(c) Do not try this procedure with halite (rock salt). Why wouldn't it work?

_____

_____

_____

3.4.9 Magnetism

A few minerals are attracted to a magnet or act like a magnet and attract metallic objects like nails or paper clips. The most common example is, appropriately, called *magnetite*. Because so few minerals are magnetic, this is a diagnostic property.

3.4.10 Feel

Some minerals feel greasy or slippery when you rub your fingers over them. They are greasy because their chemical bonds are so weak in one direction that the pressure of your fingers is enough to break them and to slide planes of atoms past one another. Talc and graphite are common examples.

3.4.11 Taste

Yes, we sometimes taste minerals! Taste is a *chemical* property, determined by the presence of certain elements. The most common example is halite (common salt), which tastes salty because of the chloride ion (Cl^-). **Do not taste minerals in your set unless instructed to do so!** We taste minerals only *after* we have narrowed the possibilities down to a few for which taste would be the diagnostic property. Why not taste every mineral? Because some taste bitter (like sylvite—KCl), some are poisonous, and you don't want to taste other students' germs!

3.4.12 Odor

As geologists we use all of our senses to identify minerals. A few minerals, and the streak of a few others, have a distinctive odor. For example, the streak of minerals containing sulfur smells like rotten eggs, and the streak of some arsenic minerals smells like garlic.

3.4.13 Reaction with Dilute Hydrochloric Acid

Many minerals containing the carbonate anion (CO_3^{2-}) effervesce (fizz) when dilute hydrochloric acid is dropped on them. The acid frees carbon dioxide from the mineral and the bubbles of gas escaping through the acid produce the fizz.

3.4.14 Tenacity

Tenacity refers to the way in which materials respond to being pushed, pulled, bent, or sheared. Most adjectives used to describe tenacity are probably familiar: *malleable* materials can be bent or hammered into a new shape; *ductile* materials can be pulled into wires; *brittle* materials shatter when hit hard; and *flexible* materials can bend. Flexibility is a diagnostic property for some minerals. After being bent, thin sheets of **elastic** minerals return to their original, unbent shape, but sheets of **flexible** minerals retain the new shape.

3.5 Identifying Mineral Specimens

You are now ready to use these physical properties to identify minerals. Although there are more than 4,000 minerals, only about 30 occur commonly and an even smaller number make up most of the Earth's crust—the part of the geosphere with which we are most familiar. Identification is easier if you follow the systematic approach used by geologists:

STEP 1 Assemble the equipment available in most geology classrooms to study minerals.
  • A glass plate, penny, and knife or steel nail to test hardness
  • A ceramic streak plate to test streak
  • A magnifying glass, hand lens, or microscope to help determine cleavage
  • Dilute hydrochloric acid to identify carbonate minerals
  • A magnet to identify magnetic minerals

STEP 2 Observe or measure the specimen's physical properties. Profile the properties on a standardized data sheet like the one at the end of this chapter, as shown in the example given in **Table 3.1**.

TABLE 3.1 Profile of a mineral's physical properties.

Specimen #	Luster	Color	Hardness	Breakage	Other diagnostic properties
1	Metallic	Dark gray	Less than a fingernail	Excellent cleavage in one direction	Leaves a mark on a sheet of paper

STEP 3 Eliminate from consideration all minerals that do not have the properties you have recorded. This can be done systematically using a **flow chart** (**Fig. 3.7**) that asks key questions in a logical sequence so that each answer eliminates entire groups of minerals until only a few remain (one, if you're lucky). Or you can use a **determinative table** in which each column answers the same questions as at the branches of a flow chart. The appendices at the end of this chapter provide flow charts and determinative tables. Experiment to find out which works best for you.

FIGURE 3.7 Flow chart used to identify minerals.

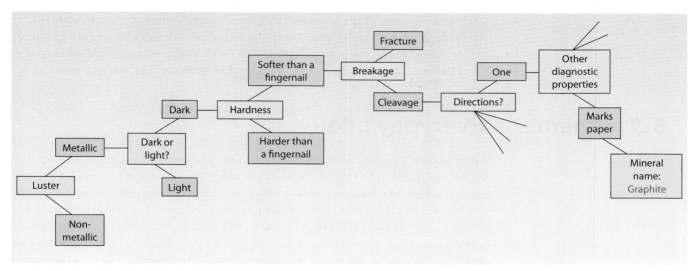

Note: For simplicity, only the path for the unknown mineral is shown.

EXERCISE 3.7   **Identifying Minerals**

Name: _____   **Section:** _____
Course: _____   **Date:** _____

Your instructor will provide a set of minerals to identify. Record the profile of physical properties for each specimen on the data sheets at the end of this chapter. Then use either the flow charts (Appendix 3.1) or determinative tables (Appendix 3.2) to identify each mineral. If these lead to more than one possibility, look at Appendix 3.3 for additional information.

3.6 Mineral Classification

The minerals in your set were chosen because they are important rock-forming minerals that make up most of the geosphere, are economically valuable resources, or illustrate the physical properties used to study minerals. Geologists classify all minerals into a small number of groups based on their chemical composition. These groups include:

- *silicates,* such as quartz, feldspars, amphiboles, and pyroxenes, which contain silicon and oxygen. Silicates are divided into *ferromagnesian* minerals, which contain iron and magnesium, and *nonferromagnesian* minerals, which do not contain those elements.
- *oxides,* such as magnetite (Fe_3O_4) and hematite (Fe_2O_3), in which a cation is bonded to oxygen anions.
- *sulfides,* such as pyrite (FeS_2) and sphalerite (ZnS), in which a cation is bonded to sulfur anions.
- *sulfates,* such as gypsum ($CaSO_4 \cdot 2H_2O$), in which cations are bonded to the sulfate complex (SO_4^{2-}).
- *halides,* such as halite (NaCl) and fluorite (CaF_2), in which cations are bonded to halogen anions (elements in the second column from the right in the periodic table).

- *carbonates*, such as calcite ($CaCO_3$) and dolomite ($CaMg[CO_3]_2$), containing the carbonate complex (CO_3)$^{2-}$.
- *native elements*, which are minerals that consist of atoms of a single element. The native elements most likely to be found in your mineral sets are graphite (carbon) and copper. It is unlikely that you would find more valuable native elements like gold, silver, and diamond (which is carbon, just like graphite).

3.7 Minerals in Everyday Life

When people think of minerals, many picture brilliant gemstones. Most minerals are less spectacular, but many are useful in modern society and extremely valuable. Indeed, stages of human evolution are named for the resources that our ancestors learned to obtain from rocks and minerals: the Stone Age (rocks), Bronze Age (copper and tin melted from minerals), and Iron Age (iron smelted from iron-bearing minerals). *Ore minerals*, commonly oxides and sulfides, are valuable because useful metals can be separated from them, usually by melting. Physical properties make other minerals valuable. For example, hard minerals are used as abrasives, soft and greasy minerals are used as lubricants, and minerals whose powder forms solid masses after becoming wet are the major component of wallboard and plaster of Paris. Exercise 3.8 explores everyday uses of common minerals.

EXERCISE 3.8   **Everyday Uses of Minerals**

Name: _____   Section: _____
Course: _____   Date: _____

The minerals listed below play important roles in everyday life. Some are sources of metals and others are used because of their physical properties. Match the mineral with its use and explain what makes it ideal for that purpose. Some minerals may be used for more than one purpose. You might want to check the chemical formulas of these minerals in the determinative table (Appendix 3.2) for clues.

garnet   halite   galena   quartz   magnetite   graphite   calcite   malachite   gypsum

Economic uses	Mineral	Why ideal for this use?
Melt ice on roads		
Very hard abrasive		
Make wallboard (Sheetrock)		
Major component of glass		
Decrease acidity of fields and gardens		
Source of lead		

Name: _____   Section: _____
Course: _____   Date: _____

Economic uses	Mineral	Why ideal for this use?
Source of iron		
Source of copper		
Lubricant for locks		
Writing material		
Navigation with a compass		

MINERAL IDENTIFICATION FLOW CHARTS

A. Minerals With Metallic Luster

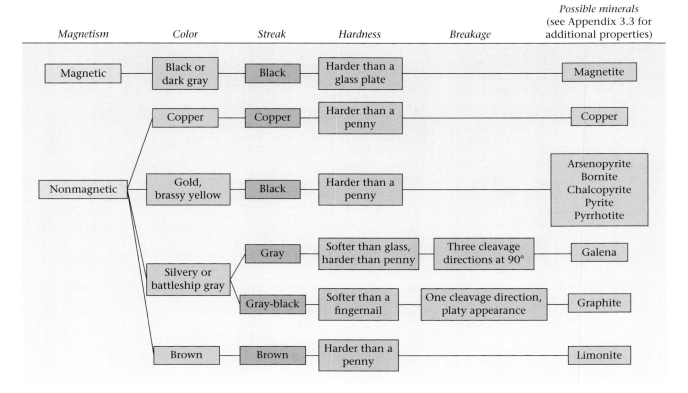

MINERAL IDENTIFICATION FLOW CHARTS

B. Minerals With Nonmetallic Luster, Dark Colored

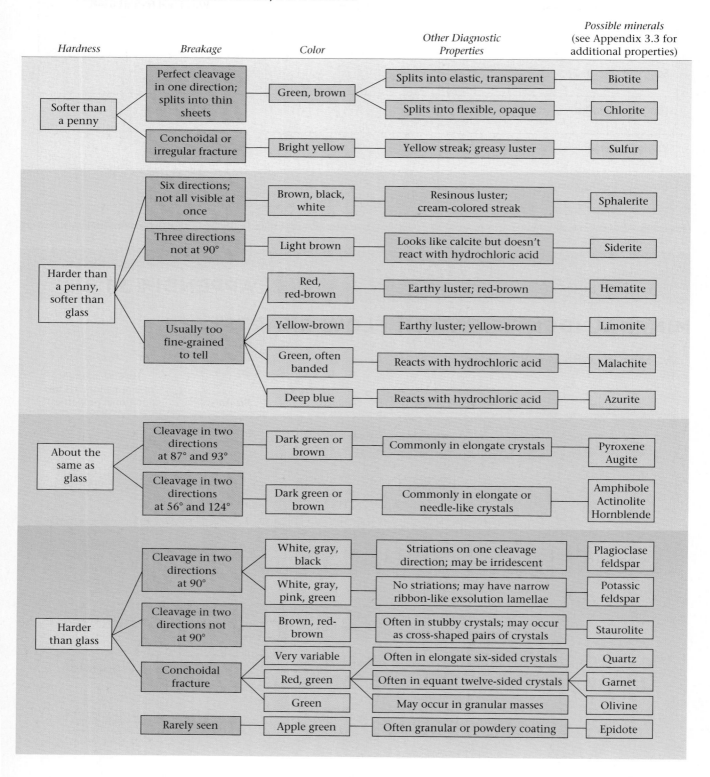

MINERAL IDENTIFICATION FLOW CHARTS

C. Minerals With Nonmetallic Luster, Light Colored

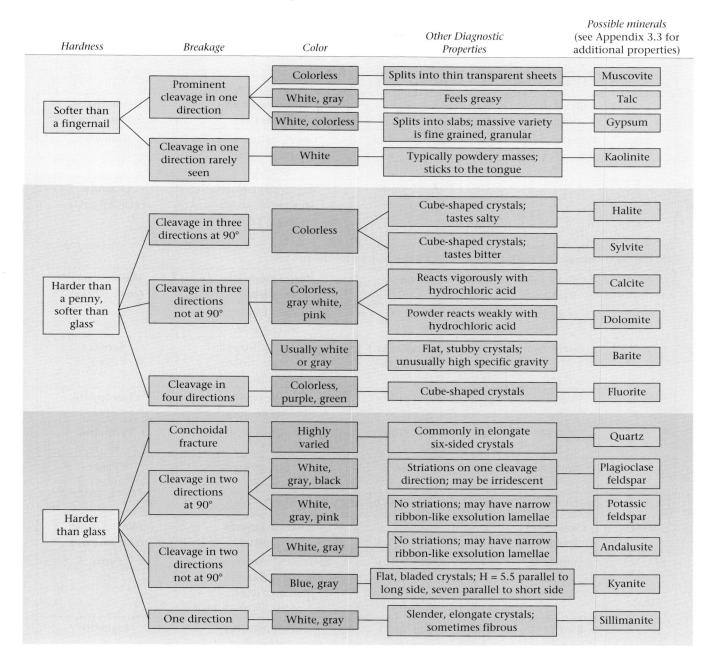

Hardness	Breakage	Color	Other Diagnostic Properties	Possible minerals (see Appendix 3.3 for additional properties)
Softer than a fingernail	Prominent cleavage in one direction	Colorless	Splits into thin transparent sheets	Muscovite
		White, gray	Feels greasy	Talc
		White, colorless	Splits into slabs; massive variety is fine grained, granular	Gypsum
	Cleavage in one direction rarely seen	White	Typically powdery masses; sticks to the tongue	Kaolinite
Harder than a penny, softer than glass	Cleavage in three directions at 90°	Colorless	Cube-shaped crystals; tastes salty	Halite
			Cube-shaped crystals; tastes bitter	Sylvite
	Cleavage in three directions not at 90°	Colorless, gray white, pink	Reacts vigorously with hydrochloric acid	Calcite
			Powder reacts weakly with hydrochloric acid	Dolomite
		Usually white or gray	Flat, stubby crystals; unusually high specific gravity	Barite
	Cleavage in four directions	Colorless, purple, green	Cube-shaped crystals	Fluorite
Harder than glass	Conchoidal fracture	Highly varied	Commonly in elongate six-sided crystals	Quartz
	Cleavage in two directions at 90°	White, gray, black	Striations on one cleavage direction; may be irridescent	Plagioclase feldspar
		White, gray, pink	No striations; may have narrow ribbon-like exsolution lamellae	Potassic feldspar
	Cleavage in two directions not at 90°	White, gray	No striations; may have narrow ribbon-like exsolution lamellae	Andalusite
		Blue, gray	Flat, bladed crystals; H = 5.5 parallel to long side, seven parallel to short side	Kyanite
	One direction	White, gray	Slender, elongate crystals; sometimes fibrous	Sillimanite

DETERMINATIVE TABLES FOR SYSTEMATIC MINERAL IDENTIFICATION

Sequence of questions: luster? Approximate hardness? Streak? Breakage? Hardness? Color? Other?

Table 1   Minerals with metallic luster.

(a) Hardness less than 2.5 (softer than a fingernail)

Streak	Cleavage or fracture	H	Color	Other diagnostic properties	Mineral name (composition)
Black	Perfect cleavage in one direction	1	Dark gray-black	Greasy feel; leaves a mark on paper; SpG = 2.23	**Graphite** C
Yellow-brown	Rarely seen	—	Yellow-brown	Very rarely in masses with meallic luster; more commonly dull, earthy; SpG = 3.6–4	**Limonite** $FeO(OH) \cdot nH_2O$

(b) Hardness between 2.5 and 5.5 (harder than a fingernail; softer than glass)

Streak	Cleavage or fracture	H	Color	Other diagnostic properties	Mineral name (composition)
Gray	Three directions at 90° angles	2.5	Lead gray	Commonly in cubic crystals; SpG = 7.4–7.6	**Galena** PbS
Black	Rarely seen	3	Bronze-brown when fresh	Commonly with purplish, iridescent tarnish; SpG = 5.06–5.08	**Bornite** Cu_5FeS_4
	Rarely seen	3.5–4	Brassy yellow	Often tarnished; similar to pyrite but not in cubes	**Chalcopyrite** $CuFeS_2$
	Rarely seen	4	Brown-bronze	Slightly magnetic; SpG = 4.62	**Pyrrhotite** $Fe_{1-x}S$
Copper-red	Rarely seen	2.5–3	Copper	Often in branching masses; SpG = 8.9	**Copper** (Cu)

(c) Hardness greater than 5.5 (harder than glass; cannot be scratched by a knife)

Streak	Cleavage or fracture	H	Color	Other diagnostic properties	Mineral name (composition)
Black	Conchoidal fracture	5	Brassy yellow	Commonly in twelve-sided crystals or cubes with striated faces; Sp.G. = 5.02	**Pyrite** FeS_2
	Rarely seen	6	Iron black	Strongly magnetic; SpG = 5.18	**Magnetite** Fe_3O_4
Black	Rarely seen	5.5–6	Silver white	Streak smells like garlic because of arsenic; SpG = 6.07	**Arsenopyrite** FeAsS
Red-brown	Rarely seen	5.5–6.5	Black, red	Black variety is metallic; red variety is more common and has nonmetallic, earthy luster.	**Hematite** Fe_2O_3

DETERMINATIVE TABLES FOR SYSTEMATIC MINERAL IDENTIFICATION

Table 2 **Minerals with non-metallic luster.**

(a) Hardness less than 2.5 (softer than a fingernail)					
Streak	**Cleavage or fracture**	**H**	**Color**	**Other diagnostic properties**	**Mineral name (composition)**
Yellow	Conchoidal or uneven fracture	1.5–2.5	Yellow	Resinous luster; SpG = 2.05–2.09	**Sulfur** S
White or colorless	Perfect cleavage in one direction	2–2.5	Colorless, light tan, yellow	Can be peeled into transparent, flexible sheets; SpG = 2.76–2.88	**Muscovite** $KAl_2(AlSi_3O_{10})(OH)_2$
	Perfect cleavage in one direction	1	Green, gray, white	Greasy feel; may occur in irregular masses (soapstone); SpG = 2.7–2.8	**Talc** $Mg_3Si_4O_{10}(OH)_2$
	Perfect cleavage in one direction; may show two other directions at 90°	2	Colorless, white, gray	Occurs in clear crystals or gray or white earthy masses (alabaster); SpG = 2.32	**Gypsum** $CaSO_4 \cdot 2H_2O$
	Three directions at 90°	2	Colorless, white	Cubic crystals like halite but has very bitter taste; SpG = 1.99	**Sylvite** KCl
	Perfect in one direction, but rarely seen	2–2.5	White	Usually in dull, powdery masses that stick to the tongue; SpG = 2.6	**Kaolinite** $Al_2Si_2O_5(OH)_4$
	Perfect in one direction, but not always visible	2–5	Green, white	Platy and fibrous (asbestos) varieties; greasy luster; SpG = 2.5–2.6	**Serpentine** $Mg_3Si_2O_5(OH)_4$
	—	—	White, brown, gray	Not really a mineral; rock often made of small spherical particles containing several clay minerals; SpG = 2–2.55	**Bauxite** Mixture of aluminum hydroxides
Green or brown	Perfect cleavage in one direction	2.5–3	Brown, black, green	Can be peeled into thin, flexible sheets; SpG = 2.8–3.2	**Biotite** $K(Fe,Mg)_2AlSi_3O_{10}(OH)_2$
	Perfect cleavage in direction	2–2.5	Green, dark green	A mica-like mineral, but sheets are not flexible; SpG = 2.6-3.3	**Chlorite** Complex Fe-Mg sheet silicate

(b) Hardness between 2.5 and 5.5 (harder than a fingernail; softer than glass)					
Streak	**Cleavage or fracture**	**H**	**Color**	**Other diagnostic properties**	**Mineral name (composition)**
Green	—	3.5–4	Bright green	Occurs in globular or elongate masses; reacts with HCl; SpG = 3.9–4.03	**Malachite** $Cu_2CO_3(OH)_2$

DETERMINATIVE TABLES FOR SYSTEMATIC MINERAL IDENTIFICATION

Table 2 Minerals with non-metallic luster (*continued*)

Streak	Cleavage or fracture	H	Color	Other diagnostic properties	Mineral name (composition)
Blue	—	3.5	Intense blue	Often in platy crystals or spherical masses; reacts with HCl; SpG = 3.77	**Azurite** $Cu_3(CO_3)_2(OH)_2$
Red-brown	—	—	Reddish brown	Usually in earthy masses; also occurs as black, metallic crystals; SpG = 5.5–6.5	**Hematite** Fe_2O_3
Yellow-brown	—	—	Brown, tan	Earthy, powdery masses and coatings on other minerals; SpG = 3.6–4	**Limonite** $FeO(OH)\cdot nH_2O$
	Three directions not at 90°	3.5–4	Light to dark brown	Often in rhombic crystals; reacts with hot HCl; SpG = 3.96	**Siderite** $FeCO_3$
	Six directions, few of which are usually visible	3.5	Brown, white, yellow, black, colorless	Resinous luster; SpG = 3.9–4.1	**Sphalerite** ZnS
White or colorless	Three directions at 90°	2.5	Colorless, white	Cubic crystals or massive (rock salt); salty taste; SpG = 2.5	**Halite** $NaCl$
	Three directions not at 90°	3	Varied; usually white or colorless	Rhombic or elongated crystals; reacts with HCl; SpG = 2.71	**Calcite** $CaCO_3$
	Three directions not at 90°	3.5–4	Varied; commonly white or pink	Rhombic crystals; *powder* reacts with HCl but crystals may not; SpG = 2.85	**Dolomite** $CaMg(CO_3)_2$
	Three directions at 90°	3–3.5	Colorless, white	SpG = 4.5 (unusually high for a nonmetallic mineral)	**Barite** $BaSO_4$
	Four directions	4	Colorless, purple, yellow, blue, green	Often in cubic crystals; SpG = 3.18	**Fluorite** CaF_2
	One direction, poor	5	Usually green or brown	Elongate six-sided crystals; may be purple, blue, colorless; SpG = 3.15–3.20	**Apatite** $Ca_5(PO_4)_3(OH,Cl,F)$

(c) Hardness between 5.5 and 8 (harder than glass or a knife; softer than a streak plate)

Streak	Cleavage or fracture	H	Color	Other diagnostic properties	Mineral name (composition)
	Conchoidal fracture	7	Red, green, brown, black, colorless, pink	Equant twelve-sided crystals; SpG = 3.5–4.3	**Garnet family** Complex Ca, Fe, Mg, Al, Cr, Mn silicate
	Two directions at 90°	6	Colorless, salmon, gray, green, white	Stubby prismatic crystals; three polymorphs: orthoclase, microcline, sanidine; may show exsolution lamellae; SpG = 2.54–2.62	**Potassic feldspar** $KAlSi_3O_8$

DETERMINATIVE TABLES FOR SYSTEMATIC MINERAL IDENTIFICATION

Table 2 Minerals with non-metallic luster (*continued*)

Streak	Cleavage or fracture	H	Color	Other diagnostic properties	Mineral name (composition)
White or colorless	Two directions at 90°	6	Colorless, white, gray, black	Striations on one of the two cleavage directions; solid solution between sodium (albite) and calcium (anorthite) plagioclase; SpG = 2.62–2.76	**Plagioclase feldspar** $CaAl_2Si_2O_8$ $NaAlSi_3O_8$
	Conchoidal fracture	7	Colorless, pink, purple, gray, black, green, yellow	Elongate six-sided crystals; SpG = 2.65	**Quartz** SiO_2
	Rarely seen	7.5	Gray, white, brown	Elongate four-sided crystals; SpG = 3.16–3.23	**Andalusite** Al_2SiO_5
	One direction	6–7	White, rarely green	Long, slender crystals, often fibrous; SpG = 3.23	**Sillimanite** Al_2SiO_5
	One direction	5 *and* 7	Blue-gray to white	Bladed crystals; *two hardnesses*: H = 5 parallel to long direction of crystal, H = 7 across the long direction	**Kyanite** Al_2SiO_5
Colorless to light green	Conchoidal fracture	6.5–7	Most commonly green	Stubby crystals and granular masses; solid solution between Fe (fayalite) and Mg (forsterite)	**Olivine family** Fe_2SiO_4 Mg_2SiO_4
	Two directions at 56° and 124°	5–6	Dark green to black	An amphibole with elongate crystals; SpG = 3–3.4	**Hornblende** Complex double-chain silicate with Ca, Na, Fe, Mg
	Two directions at 56° and 124°	5–6	Pale to dark green	An amphibole with elongate crystals; SpG = 3–3.3	**Actinolite** Double-chain silicate with Ca, Fe, Mg
	Two directions at 87° and 93°	5–6	Dark green to black	A pyroxene with elongate crystals; SpG = 3.2–3.3	**Augite** Single-chain silicate with Ca, Na, Mg, Fe, Al
	One perfect one poor; not at 90°	6–7	Apple green to black	Elongate crystals and fine-grained masses; SpG = 3.25–3.45	**Epidote** Complex twin silicate with Ca, Al, Fe, Mg
No streak; mineral scratches streak plate	One direction, imperfect	8	Blue-green, emerald green, yellow, pink, white	Six-sided crystals with flat ends; SpG = 2.65–2.8; gem variety: emerald (green)	**Beryl** $Be_3Al_2Si_6O_{18}$
	—	7–7.5	Red-brown	Stubby or cross-shaped crystals; SpG = 3.65–3.75	**Staurolite** Hydrous Fe, Al silicate

COMMON MINERALS AND THEIR PROPERTIES

Mineral	Additional diagnostic properties and occurrences
Actinolite	Elongate green crystals; cleavage at 56° and 124°; H = 5.5–6. An amphibole found in metamorphic rocks.
Amphibole*	Stubby rod-shaped crystals common in igneous rocks; slender crystals common in metamorphic rocks; two cleavage directions at 56° and 124°.
Andalusite	Elongate gray crystals with rectangular cross sections.
Apatite	H = 5; pale to dark green, brown, white; white streak; six-sided crystals.
Augite	H = 5.5–6; green to black rod-shaped crystals; cleavage at 87° and 93°. A pyroxene common in igneous rocks.
Azurite	Deep blue; reacts with HCl. Copper will plate out on a steel nail dipped into a drop of HCl and placed on this mineral.
Barite	H = 3–3.5; SpG is unusually high for nonmetallic mineral.
Bauxite	Gray-brown earthy *rock* commonly containing spherical masses of clay minerals and mineraloids.
Beryl	Six-sided crystals; H = 8.
Biotite	Dark-colored mica; one perfect cleavage into flexible sheets.
Bornite	High SpG; irridescent coating on surface gives it "peacock ore" nickname.
Calcite	Reacts with HCl. Produces double image from text viewed through transparent cleavage fragments.
Chalcopyrite	Similar to pyrite, but typically has iridescent tarnish.
Chlorite*	Similar to biotite, but does not break into thin, flexible sheets; forms in metamorphic rocks.
Copper	Copper-red color and high specific gravity are diagnostic.
Dolomite	Similar to calcite but only weak or no reaction with HCl placed on a grain of the mineral; *powder* reacts strongly. Slightly curved rhombohedral crystals.
Epidote	Small crystals and thin granular coatings form in some metamorphic rocks and [by alteration] of some igneous rocks.
Fluorite	Commonly in cube-shaped crystals with four cleavage directions cutting corners of the cubes.
Galena	Commonly in cube-shaped crystals with three perfect cleavages at 90°.
Garnet*	Most commonly dark red; twelve- or twenty-four-sided crystals in metamorphic rocks.
Graphite	Greasy feel; leaves a mark on paper.
Gypsum	Two varieties: *selenite* is colorless and nearly transparent with perfect cleavage; *alabaster* is a rock—an aggregate of grains with an earthy luster.
Halite	Cubic crystals and taste are diagnostic.
Hematite	Two varieties: most common is red-brown masses with earthy luster; rare variety is black crystals with metallic luster.
Hornblende	Dark green to black amphibole; two cleavages at 56° and 124°.
Kaolinite	Earthy, powdery white to gray masses; sticks to tongue.
Kyanite	Bladed blue-green crystals; H = 5 parallel to blade, H = 7 across blade.
Limonite	Earthy, yellow-brown masses, sometimes powdery; forms by the "rusting" (oxidation) of iron-bearing minerals.
Magnetite	Gray-black; H = 6; magnetic.
Malachite	Bright green. Copper will plate out on a steel nail dipped into a drop of HCl and placed on this mineral.
Muscovite	A colorless mica; one perfect cleavage; peels into flexible sheets.
Olivine	Commonly as aggregates of green granular crystals.

COMMON MINERALS AND THEIR PROPERTIES

Mineral	Additional diagnostic properties and occurrences
Plagioclase feldspar*	Play of colors and striations distinguish plagioclase feldspars from potassic feldspars, which do not show these properties.
Potassic feldspar*	
Pyrite	Gold color and black streak are diagnostic. Cubic crystals with striations on their faces or in twelve-sided crystals with five-sided faces.
Pyrrhotite	Brownish bronze color; black streak; may be slightly magnetic.
Pyroxene*	Two cleavages at 87° and 93°; major constituent of mafic and ultramafic igneous rocks.
Quartz	Wide range of colors; six-sided crystal shape, high hardness (7) and conchoidal fracture are diagnostic.
Serpentine	Dull white, gray, or green masses; sometimes fibrous (asbestos).
Siderite	Three cleavages not at 90°; looks like brown calcite; powder may react to HCl.
Sillimanite	Gray, white, brown; slender crystals, sometimes needle-like.
Sphalerite	Wide variety of colors (including colorless); distinctive cream-colored streak.
Staurolite	Red-brown stubby crystals in metamorphic rocks.
Sulfur	Bright yellow with yellow streak; greasy luster.
Sylvite	Looks like halite but has very bitter taste.
Talc	Greasy feel; H = 1.

*Indicates mineral family.

MINERAL PROFILE DATA SHEET

Sample	Luster	Hardness	Streak	Color	Cleavage or fracture (describe)	Other properties	Mineral name and composition

MINERAL PROFILE DATA SHEET

Sample	Luster	Hardness	Streak	Color	Cleavage or fracture (describe)	Other properties	Mineral name and composition

CHAPTER 4

MINERALS, ROCKS, AND THE ROCK CYCLE

PURPOSE

- Understand how relationships among grains differ in igneous, sedimentary, and metamorphic rocks because of the way in which the rocks form.
- Use textures and mineral identification to determine if a rock is igneous, sedimentary, or metamorphic.

MATERIALS NEEDED

- An assortment of igneous, sedimentary, and metamorphic rocks
- Supplies to make artificial igneous, sedimentary, and metamorphic rocks; tongs for moving hot petri dishes, hot plate, sugar, Thymol, Na-acetate in a dropper bottle, sand grains, glass petri dishes, $Ca_2(OH)_2$ solution, straws, *Play-doh*, plastic coffee stirring rods, plastic chips
- Magnifying glass or hand lens
- Mineral testing supplies: streak plate, steel nail, etc.

4.1 Introduction

A rock is an aggregate of mineral grains, fragments of previously existing rock, or a mass of natural glass. Rocks can form in a variety of ways—through freezing of a melt, by cementation of loose grains, by precipitation from water solutions, or from changes that happen in response to temperature and pressure underground. Geologists can identify rocks and interpret aspects of Earth's history by studying two principal characteristics of rocks: **rock composition** (the identity of minerals or glass that make up a rock) and **rock texture** (the dimensions and shape of grains, and the way in which grains are arranged, oriented, and held together). On our dynamic planet, rocks don't survive forever—nature recycles materials, using those in one rock to form new rocks through a series of steps called the *rock cycle.*

In this chapter, we first introduce the three basic classes of rocks. Then we describe how to observe the characteristics of rocks and how to use these characteristics to classify rocks. Then you will make your own "rocks" in the classroom and see how different processes produce different textures. By combining your mineral identification skills and knowledge of rock textures, you can face the challenge of determining how specimens of common rocks have formed.

4.2 The Three Classes of Rocks

Geologists struggled for a century with the question of how to classify rocks. They finally concluded that rocks can best be classified on the basis of how they formed. Rock composition and texture provide the basis for this classification, for these characteristics reflect the process of formation. In modern terminology, we distinguish three classes of rocks: igneous, sedimentary, and metamorphic.

• **Igneous rocks** form through the cooling and solidification of molten rock, which is created by the melting of preexisting rock in the mantle or lower crust. We refer to molten rock below Earth's surface as **magma**, and molten rock that has erupted onto the surface as **lava**. Some volcanoes erupt explosively, blasting rock fragments into the air, and when these fragments fall back to Earth, coalesce, and solidify, the resulting rock is also considered to be igneous.

• **Sedimentary rocks** form at or near the surface of the Earth in two basic ways: (1) when grains of preexisting rocks accumulate, are buried, and then are cemented together by minerals precipitating out of groundwater; and (2) when minerals precipitate out of water, either directly or through the life function of an organism, and either form a solid mass or are cemented together later. The grains that become incorporated in sedimentary rocks form when preexisting rocks are attacked and broken down by interactions with air, water, and living organisms. These interactions are called **weathering**. Some weathering simply involves the physical fragmentation of rock. Other kinds of weathering involve chemical reactions that produce new minerals, most notably clay. The products of weathering can be transported by water, ice, or wind to the site where they are deposited (settle out), buried, and transformed into new rock.

• **Metamorphic rocks** form when preexisting rocks are subjected to physical and chemical conditions within the Earth that are significantly different from those under which they first formed. For example, when buried very deeply, rock is warmed to high temperature and squeezed by high pressure. The texture and/or mineral content of the initial rock changes in response to the new conditions in *the solid state.*

Name: _____   **Section:** _____

Course: _____   **Date:** _____

Three settings are described below in which three different classes of rocks form. Fill in the blanks by indicating which class of rocks is the result.

(a) Along the coast, waves carry sand out into deeper, quieter water, where it accumulates, gradually becoming buried. A rock formed from sand grains cemented together is _____.

(b) Thick flows of glowing red lava engulf farms and villages on the flanks of Mt. Etna, a volcano in Sicily. The cold, black rock formed when these flows are cool enough to walk on is _____.

(c) The broad tundra plains of northern Canada expose gray, massive rock that formed many kilometers below a mountain belt. This rock, exposed only after the overlying rock was stripped away, is _____.

Metamorphic rock thus forms without the melting or weathering that make igneous and sedimentary rocks.

4.2.1 The Rock Cycle

An igneous rock that has been exposed at the surface of the Earth will not last forever. Minerals that crystallized from magma to form the rock can be broken apart at the surface by weathering, undergo **erosion** (the grinding effect of moving ice, water, or wind), be transported by streams, be deposited in the ocean, and be cemented together to form a sedimentary rock. These same minerals, now in a sedimentary rock, can later be buried so deeply beneath other sedimentary rocks that they are heated and squeezed to form a metamorphic rock with new minerals. Erosion may eventually expose the metamorphic rock at the Earth's surface, where it can be weathered to form new sediment and eventually a different sedimentary rock. Or the metamorphic rock may be buried so deeply that it melts to produce magma and eventually becomes a new igneous rock with still different minerals.

This flux (flow) of material from one rock type to another over geologic time is called the **rock cycle** (**Fig. 4.1**). The rock cycle involves the reuse of mineral grains or the breakdown of minerals into their constituent atoms and the reuse of those atoms to make other minerals. Rocks formed at each step of the rock cycle look very different from their predecessors because of the different processes by which they formed.

4.3 A Rock Is More than the Sum of Its Minerals

The first thing a geologist wants to determine about a rock is whether it is igneous, sedimentary, or metamorphic. Composition *alone* (i.e., the component minerals of the rock) is rarely enough to define a rock's origin, because some of the most common minerals (quartz, potassic feldspar, sodium-rich plagioclase feldspar) are found in all three rock classes. For example, a rock made of these minerals could be igneous (granite), sedimentary (sandstone), or metamorphic (gneiss). Texture *alone* can identify which type of process was involved for many rocks, but some textures can also develop in all three types of rock. The best clue to the origin of a rock is its unique combination of texture *and* mineralogy.

FIGURE 4.1 The rock cycle.

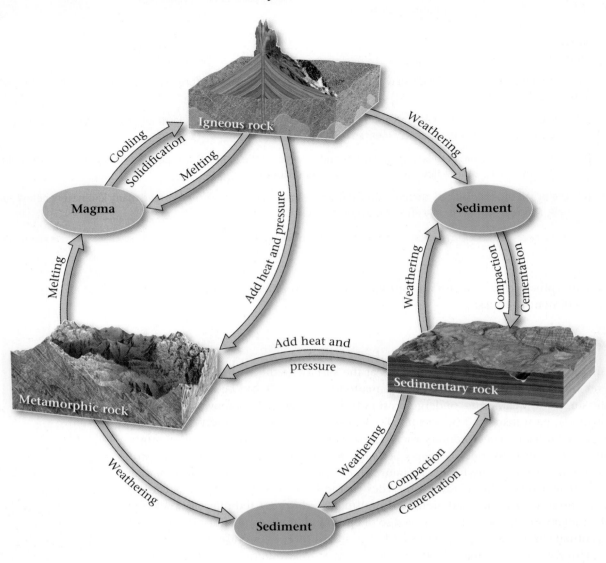

4.3.1 Describing Texture

We begin characterizing a rock's texture by asking, Does the rock consist of a mass of glass, or does it consist of grains? If it consists of glass, it will be shiny and develop conchoidal fractures (curving, ridged surfaces). Rocks composed of glass are relatively rare—they form only at certain kinds of volcanoes. In most cases, we jump to the next question: What are the grains like, and how are they connected? To answer this question, we first describe the size and shape of grains, then determine if the grains are interlocking, like pieces of a jigsaw puzzle, or if they are stuck together by cement (mineral "glue").

4.3.1a Grain Size A rock description should include the size(s) of grains and whether the rock has a single (homogeneous) grain size, a wide range of sizes, or two distinctly different sizes. Are the grains **coarse** enough to be identified easily; **medium** sized, so you can see that there are separate grains but can't easily identify them; **fine**, so that they can barely be recognized; or **very fine**, so small that they cannot be seen even with a magnifying glass? Special terms are used for grain size in igneous and sedimentary rocks, but here we use common words that work just as well (**Table 4.1** and **Fig. 4.2**).

TABLE 4.1 Common grain size terminology.

Grain size term	Practical definition	Approximate size range
Coarse grained	Grains are large enough so you can identify the minerals present.	Larger than 5 mm
Medium grained	Individual grains can be seen with the naked eye but are too small to be identified.	0.1–4.9 mm
Fine grained	Individual grains are too small to be seen with the naked eye or identified with a hand lens.	Smaller than 0.1 mm
Glassy	No grains at all; rock is a homogeneous mass of glass.	—

FIGURE 4.2 Grain sizes in sedimentary rocks.

(a) Very fine grained.

(b) Medium grained.

(c) Coarse grained.

(d) Very coarse grained.

Note that some rocks contain grains of different sizes, as in Figure 4.2d where very coarse grains are embedded in a very-fine-grained matrix. As we will see later, these mixed grain sizes give important information about the rock-forming processes.

4.3.1b Grain Shape When we describe the shape of grains, we ask, Are the grains **equant**, meaning they have the same dimension in all directions, or **inequant**, meaning that the dimension in different directions is different? Grains that resemble spheres or cubes are equant, whereas those that resemble rods, sheets, ovals, or have an irregular form are inequant. Next, we look at the grains to see if they are rounded, angular, or some combination of these shapes (**Fig. 4.3**). Finally we ask, Are *all* grains the same shape, or do they vary depending on mineral type or on position within the rock? Each of these characteristics tells us something about the formation of the rock.

FIGURE 4.3 Typical grain shapes.

(a) Rounded grains.

(b) Angular grains.

(c) Halite crystals in sedimentary rock.

(d) Garnet crystals in metamorphic rock.

4.3.2 Relationships among Grains

During most igneous- and metamorphic-rock-forming processes, and during some sedimentary-rock-forming processes, crystallizing minerals interfere with one another and they interlock like a three-dimensional jigsaw puzzle (**Fig. 4.4a, b**). We refer to the result as a **crystalline texture**. A very different texture results from sedimentary processes that deposit grains and then cement them together (**Fig. 4.4c, d**).

FIGURE 4.4 Grain relationships.

(a) Interlocking quartz and feldspar grains in an igneous rock.

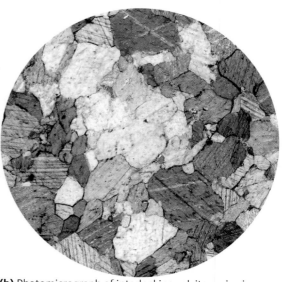

(b) Photomicrograph of interlocking calcite grains in a metamorphic rock.

(c) Cemented mineral grains in a sedimentary rock.

(d) Cemented fossils.

Such sedimentary rocks are said to have a **clastic texture**, and individual grains in these rocks are called **clasts**.

4.3.3 Grain Orientation and Alignment

The orientation of grains relative to one another often provides a key clue to the classification of a rock. For example, in most igneous rocks, the inequant grains are randomly oriented in that they point in a variety of directions. In many metamorphic rocks, however, inequant minerals, such as platy micas or elongate amphiboles, are oriented parallel to one another. Thus, when looking at a rock, you should ask, Are the inequant grains aligned parallel to one another or are they randomly oriented throughout the rock? If the grains are aligned parallel to one another, the rock develops a type of layering called **foliation**.

Now that we've developed a basic vocabulary for important characteristics of rocks, we can begin to describe a rock the way a geologist would.

Name: _____   Section: _____
Course: _____   Date: _____

Describe the *texture* of the rocks provided by your instructor. Be sure to consider the size and shape of grains, the relationships among grains, and whether grains are randomly oriented or aligned. Use your own words—the formal geologic terms for the textures will be introduced later.

(a) Rock 1:

_____

_____

_____

(b) Rock 2:

_____

_____

_____

(c) Rock 3:

_____

_____

_____

(d) What textural features are common to all three rock samples, and which are not?

_____

_____

_____

(e) Did all three rocks form by the same process? Explain your reasoning.

_____

_____

_____

4.4 The Processes That Produce Textures

The best way to understand how a rock's texture forms is to observe the rock-forming process in action. You can't go inside a lava flow to watch an igneous rock crystallize, deep below a mountain belt to see rock metamorphose, or under the ocean floor to watch sediment become a sedimentary rock—but you can *model* the formation of igneous, sedimentary, and metamorphic rocks well enough in the classroom to understand how textures form. You will use textural features more fully later in detailed interpretations of igneous, sedimentary, and metamorphic rocks, but the experiments in Exercise 4.3 will start you on the way.

Name: _____   Section: _____
Course: _____   Date: _____

(a) Crystalline igneous rock. Place a glass petri dish on a hot plate and add enough of the powder provided by your instructor to cover the bottom. Heat until the material melts completely. Carefully remove the dish from the hot plate using forceps and allow the liquid to cool. Observe the crystallization process closely. You may have to add a crystal seed if crystallization does not begin in a few minutes. Congratulations, you have just made a "magma" and then an "igneous rock" with an interlocking crystalline texture.

 • Describe the crystallization process. How and where did individual grains grow, and how did they eventually join with neighboring grains?

_____

_____

_____

Sketch and describe the texture of the cooled "rock."

Sketch: Describe:

This crystalline texture formed when crystals in the cooling "magma" grew until they interfered with one another and eventually interlocked to form a cohesive solid. Crystals only grow when the magma cools slowly enough for atoms to have time to fit into a crystalline lattice.

(b) Glassy igneous texture. Add sugar to another petri dish, melt it on a hot plate, and allow it to cool.

Describe the cooling process and sketch the resulting texture.

Sketch: Describe:

continued

This glassy texture is typical of igneous rocks that cool so quickly that atoms do not have time to arrange into crystal-line lattices. In some cooling lavas, the bubbles that you saw may be preserved as the lava freezes.

(c) Clastic sedimentary texture. Cover the bottom of another petri dish with a mixture of sand grains and small pebbles. Add a small amount of the liquid provided by your instructor and allow the mixture to sit until the liquid has evaporated. Now turn the dish upside down.

- Why don't the grains fall out of the dish?

_____

_____

Examine the "rock" with your hand lens. Sketch and describe the texture.

Sketch: Describe:

_____

_____

_____

_____

_____

_____

_____

_____

You have just made a clastic sedimentary rock composed of individual grains cemented to one another.

(d) Fine-grained chemical sedimentary texture. Using a straw, blow *gently* into the beaker of clear liquid provided by your instructor until you notice a change in the water. Continue blowing for another minute and then observe what happens.

- Describe the change that occurred in the liquid as you were blowing into it.

_____

_____

- Describe what happened to the liquid after you stopped blowing.

_____

_____

- Pour off the liquid and add a drop of dilute hydrochloric acid to the mineral that you formed. What is the mineral?

_____

You have just made a chemical sediment, one that forms when minerals precipitate (grow and settle out) from a solution. This sediment would be compacted and solidified to make a very-fine-grained sedimentary rock.

continued

Name: _____   Section: _____

Course: _____   Date: _____

(e) Crystalline chemical sedimentary texture. Suspend a piece of cotton twine in a supersaturated salt solution and allow the salt to evaporate over a few days.

  • Describe the texture of the salt crystals attached to the twine and compare it with the interlocking igneous texture from Exercise 4.3a.

_____

_____

_____

You've now seen that crystalline textures can form in both sedimentary and igneous rocks. How might you distinguish the rocks? (*Hint:* Think about whether the chemical composition of lava is the same as that of seawater.) By the way, crystalline textures also form in metamorphic rocks.

(f) Metamorphic foliation. Take a handful of small plastic chips and push them with random orientation into a mass of *Play-Doh*. Then flatten the *Play-Doh* with a book.

Sketch and describe the orientation of the plastic chips in the "rock." Are they randomly oriented as they were originally? How is their alignment related to the pressure you applied?

Sketch: Describe:

_____

_____

_____

_____

_____

_____

_____

_____

The parallel alignment of platy minerals is called *foliation* and is found in metamorphic rocks that have been strongly squeezed, as at a convergent boundary between colliding lithosphere plates.

  • What real minerals would you expect to behave the way the plastic chips did?

_____

_____

Name: _____   **Section:** _____

Course: _____   **Date:** _____

Now look at the rocks used in Exercise 4.2. Compare their textures with those that you just made, and suggest whether each is igneous, sedimentary, or metamorphic. Explain your reasoning.

Rock #1:

_____

_____

_____

Rock #2:

_____

_____

_____

Rock #3:

_____

_____

_____

4.5 Clues about a Rock's Origin from the Minerals It Contains

Some minerals can only form in one class of rocks, so their presence in a sample tells us immediately whether the rock is igneous, sedimentary, or metamorphic. For example, halite forms only by the evaporation of saltwater, and staurolite forms only by the metamorphism of aluminum-rich sedimentary rocks at high temperature. However, as mentioned earlier, several common minerals can form in more than one class of rocks. For example, quartz crystallizes in many magmas, but it also forms during metamorphic reactions and can precipitate from water to form a fine-grained chemical sedimentary rock or cement grains together in clastic sedimentary rocks.

Table 4.2 shows which common minerals indicate a unique rock-forming process and which can form in two or more ways. Remember the rock cycle: a grain of quartz that originally *formed* by cooling from magma may now be *found* in a sedimentary or metamorphic rock.

TABLE 4.2 Occurrence of common rock-forming minerals.

Mineral	Igneous	Sedimentary	Metamorphic
Plagioclase feldspar	▲	▼ ■	▲
Potassic feldspar	▲	▼ ■	▲
Quartz	▲	▼ ▲	▲
Hornblende	▲		▲
Actinolite			▲
Augite	▲		
Muscovite	▲	▼	▲
Biotite	▲		▲
Chlorite			▲
Olivine	▲		▲
Garnet	■	▼	▲
Andalusite	■		▲
Kyanite			▲
Sillimanite			▲
Epidote	■		▲
Halite		▲	
Calcite		▲	▲
Dolomite		▲	▲
Gypsum		▲	
Talc			▲
Serpentine	■		▲

▲ Commonly *forms* in these rocks   ■ Rarely *forms* in rocks   ▼ Commonly *found* in these rocks

4.6 Identifying Minerals in Rocks

The techniques you used to identify individual specimens of minerals are also used when minerals are combined with others in rocks, but the relationships among grains may make the process more difficult, especially when the grains are small. For example, it is usually easy to determine how many minerals are in coarse-grained rocks, because color and luster are as easy to determine in rocks as in minerals. But it is difficult to determine hardness and cleavage of small grains without interference from their neighbors. And when a rock contains many mineral grains cemented together, it may be difficult to distinguish between the hardness of individual minerals and the strength of the cement holding them together.

Extra care must be taken with rocks to be sure you are measuring the mineral properties you think you are measuring. In very-fine-grained rocks, geologists have to use microscopes and even more complex instruments (e.g., X-rays) to identify the minerals. Keep the following tips in mind as you determine the properties of minerals in rocks.

- *Color:* Whenever possible, look at a rock's weathered outer surface *and* a freshly broken surface, because weathering often produces a surface color different from the color of unaltered minerals. This difference can be helpful for identification. For example, regardless of whether plagioclase feldspar is dark gray, light gray, or colorless, weathering typically produces a very fine white coating of clay minerals. Weathering of a fine-grained rock can help to distinguish dark gray plagioclase (white weathering) from dark gray pyroxene (brown weathering) that might otherwise be hard to tell apart.

  A rock's color depends on its grain size as well as the color of its minerals. All other things being equal, a fine-grained rock appears darker than a coarse-grained rock made of the same minerals.

- *Luster:* Using a hand lens, rotate the rock in the light to determine how many different kinds of luster (and therefore how many different minerals) are present.

- *Hardness:* Use a steel safety pin or the tip of a knife blade and a magnifying glass to determine mineral hardness in fine-grained rocks. Be sure to scratch a single mineral grain or crystal, and when using a glass plate be sure a single grain is scratching it. Only then are you testing the mineral's hardness. Otherwise, you might be dislodging grains from the rock and demonstrating how strongly the rock is cemented together rather than measuring the hardness of any of its minerals.

- *Streak:* Similarly, be very careful when using a streak plate, because it may not be obvious which mineral in a fine-grained rock is leaving the streak. As a result, streak is generally not useful for identifying single grains in a rock that contains many minerals.

- *Crystal habit:* Crystal habit is valuable in identifying minerals in a rock when it has well-shaped crystals. But when many grains interfere with one another during growth, the result is an interlocking mass of irregular grains rather than crystals. Weathering and transportation break off corners of grains and round their edges, destroying whatever crystal forms were present.

- *Breakage:* Use a hand lens to observe cleavage and fracture. Rotate the rock in the light while looking at a single grain. Remember that multiple *parallel* shiny surfaces represent a single cleavage direction. Breakage can be a very valuable property in distinguishing light-colored feldspars (two directions of cleavage at 90°) from quartz (conchoidal fracture), and amphibole (two directions of cleavage not at 90°) from pyroxene (two directions of cleavage at 90°).

- *Specific gravity:* A mineral's specific gravity can only be measured from a pure sample. When two or more minerals are present in a rock, you are measuring the *rock's* specific gravity, and that includes contributions from all of the minerals present. However, the heft of a very-fine-grained rock can be very helpful in interpreting what combination of minerals might be in it.

- *Hydrochloric acid test for carbonates:* Put a *small* drop of acid on a fresh, unaltered surface. Be careful: a thin film of soil or weathering products may contain calcite and make a noncarbonate mineral appear to fizz. Use a hand lens to determine exactly where the carbon dioxide is coming from. Are all of the grains reacting with the acid, or is gas coming only from the cement that holds the grains together?

Name: _____   **Section:** _____
Course: _____   **Date:** _____

Your instructor will provide similar-sized samples of granite and basalt. Granite has grains that are large enough for you to identify. It forms by slow cooling of molten rock deep underground. Basalt also forms from cooling molten material, but it cools faster and has much smaller grains.

(a) Examine the granite. Does it have well-shaped crystals or irregular grains? Suggest an explanation for their shapes.

_____

_____

_____

(b) Identify three or four minerals in the granite. Use a hand lens or magnifying glass if necessary.

_____

_____

(c) Now examine the basalt. Identify the minerals present if you can, using a hand lens or magnifying glass. What problems did you have?

_____

_____

_____

(d) Heft the granite and basalt. Are their specific gravities similar, or is one denser than the other? If so, which is denser?

_____

_____

(e) Based on specific gravity, do you think that the most abundant minerals in the basalt are the same as those in the granite? Explain.

_____

_____

_____

(f) Your instructor will provide samples of sandstone (made mostly of quartz) and limestone (made mostly of calcite). Suggest two tests that will enable you to tell which sample is which. Explain.

_____

_____

_____

(g) Now, which is the limestone? Which is the sandstone?

_____

4.7 Interpreting the Origin of Rocks

You are now ready to begin your examination of rocks. The goal is to determine which of the three types of rock-forming processes was involved in the origin of each specimen. Later, you can examine each group separately and be more specific about details of those processes. **Figure 4.5** is a flow chart that combines textural and mineralogical features to make your task easier.

| EXERCISE 4.6 | Classifying Rocks: Igneous, Sedimentary, or Metamorphic? |

Name: _____   **Section:** _____
Course: _____   **Date:** _____

Use the flow chart in Figure 4.5 to separate the rocks provided by your instructor into igneous, sedimentary, and metamorphic piles based on their textures, mineral content, and the information in Table 4.2. A fourth category—Not sure yet—is reasonable in some cases, particularly for the finer-grained rocks. You will look at these in more detail in the next three chapters.

Igneous rocks

#	Reasons for classifying as igneous

Sedimentary rocks

#	Reasons for classifying as sedimentary

Metamorphic rocks

#	Reasons for classifying as metamorphic

continued

Name: _____   Section: _____

Course: _____   Date: _____

FIGURE 4.5 Flow chart for recognizing igneous, sedimentary, and metamorphic rocks.

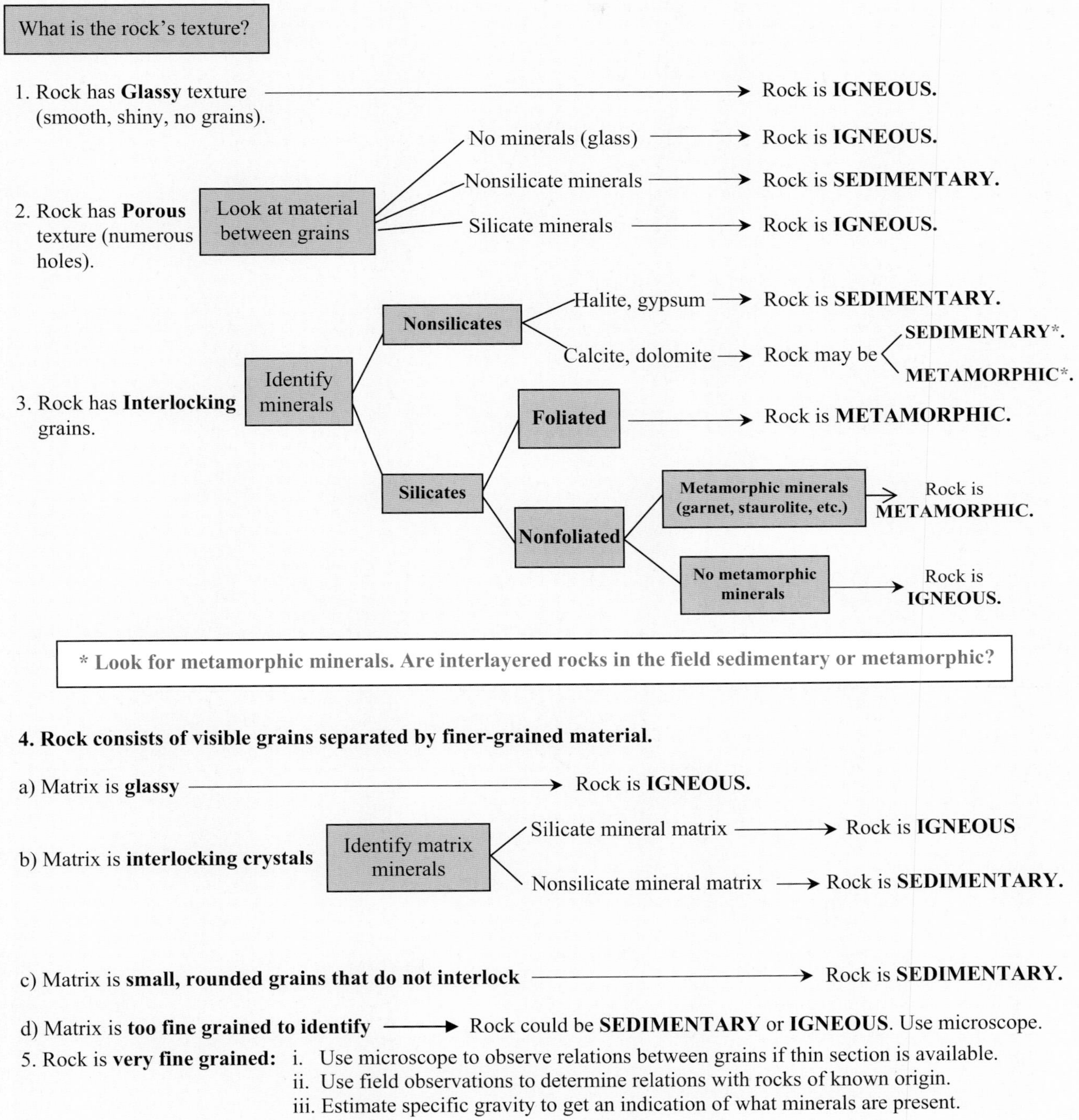

What is the rock's texture?

1. Rock has **Glassy** texture ——————————————→ Rock is **IGNEOUS.**
   (smooth, shiny, no grains).

                                        No minerals (glass) ————→ Rock is **IGNEOUS.**

                                        Nonsilicate minerals ————→ Rock is **SEDIMENTARY.**

2. Rock has **Porous**   Look at material
   texture (numerous   between grains
   holes).   Silicate minerals ————→ Rock is **IGNEOUS.**

                                        Halite, gypsum ——→ Rock is **SEDIMENTARY.**

                      Nonsilicates

                                        Calcite, dolomite ——→ Rock may be ⟨ **SEDIMENTARY*.**
                                                                            METAMORPHIC*.

3. Rock has **Interlocking**   Identify
   grains.   minerals   Foliated ————→ Rock is **METAMORPHIC.**

                                                            Metamorphic minerals   Rock is
                                                            (garnet, staurolite, etc.) ——→ **METAMORPHIC.**
                      Silicates
                                        Nonfoliated

                                                            No metamorphic   Rock is
                                                            minerals ——→ **IGNEOUS.**

> *** Look for metamorphic minerals. Are interlayered rocks in the field sedimentary or metamorphic?**

4. Rock consists of visible grains separated by finer-grained material.

a) Matrix is **glassy** ——————————————————→ Rock is **IGNEOUS.**

                                        Silicate mineral matrix ————→ Rock is **IGNEOUS**
b) Matrix is **interlocking crystals**   Identify matrix
                                         minerals
                                        Nonsilicate mineral matrix ——→ Rock is **SEDIMENTARY.**

c) Matrix is **small, rounded grains that do not interlock** ————————————→ Rock is **SEDIMENTARY.**

d) Matrix is **too fine grained to identify** ——→ Rock could be **SEDIMENTARY** or **IGNEOUS.** Use microscope.

5. Rock is **very fine grained:**   i.   Use microscope to observe relations between grains if thin section is available.
                                    ii. Use field observations to determine relations with rocks of known origin.
                                    iii. Estimate specific gravity to get an indication of what minerals are present.

## CHAPTER 5

USING IGNEOUS ROCKS TO INTERPRET EARTH HISTORY

PURPOSE

- Become familiar with igneous textures and mineral assemblages.
- Use texture and mineral content to interpret the history of igneous rocks.
- Learn to identify igneous rocks.

MATERIALS NEEDED

- Set of igneous rocks
- Magnifying glass or hand lens and, ideally, a microscope and thin sections of igneous rocks
- Standard supplies for identifying minerals (streak plate, glass plate, etc.)

5.1 Introduction

Every rock has a story to tell. The story of an igneous rock begins when rock in the lower crust or upper mantle melts to form molten material called **magma**, which rises up through the crust. Some magma flows or spatters out on the surface as **lava,** or explodes into the air as tiny particles of **volcanic ash** or larger blocks. Igneous rock that forms from solidified lava or ash is called **extrusive igneous rock** because it comes out (*extrudes*) onto the surface. Other magma never reaches the surface and solidifies underground to form **intrusive igneous rock**, so called because it squeezes into (*intrudes*) the surrounding rocks. Intrusions come in many shapes. Massive blobs are called **plutons**, and the largest of these—called **batholiths**—are generally composites of several plutons. Other intrusions that form thin sheets cutting across layering in the wall rock (the rock around the intrusion) are called **dikes**, and those that form thin sheets parallel to the layers of wall rock are called **sills** (**Fig. 5.1**).

When looking at an igneous rock, geologists want to know: Where did it cool (was it intrusive or extrusive)? Where in the Earth did the rock's parent magma form? In what tectonic setting—ocean ridge, mid-continent, subduction zone, hot spot—did it form? In this chapter, you will learn how to answer these questions and how to identify common types of igneous rock. Few new skills are needed—just observe carefully and apply some geologic reasoning. Exercise 5.1 shows how easy the process is.

Most of the questions geologists ask about an igneous rock can be answered with three simple observations: grain size, color, and specific gravity. The following sections provide the additional information you need to interpret the history of an igneous rock. Let's start with cooling history.

5.2 Interpreting the Cooling Histories of Igneous Rocks

Imagine that you are looking at an ancient igneous rock in an outcrop. That rock might have formed from volcanic debris blasted into the air, lava that frothed out of a volcano or flowed smoothly across the ground, magma that cooled just below the surface, or magma that solidified many kilometers below the surface. But millions of years have passed since the rock formed, and if a volcano had been involved it has long since been eroded away. If the rock was intrusive, kilometers of overlying

FIGURE 5.1 Intrusive and extrusive igneous rock bodies.

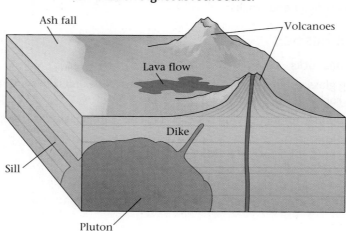

rock would have been removed to expose it at the surface. How can you determine which of these possibilities is the right one?

The key to understanding the cooling history of an igneous rock is its **texture**—the size, shape, and arrangement of its grains. The texture of an igneous rock formed by the settling of ash and other volcanic fragments looks very different from that of a rock formed by magma cooling deep underground or by lava cooling at the surface.

Specimens composed of interlocking grains—whether large enough to be identified or too small to be identified—are called **crystalline** rocks. Those that are shiny and contain no grains are called **glasses**; sponge-like masses are said to be **porous**; and those that appear to have pieces cemented together are said to be **fragmental**. Each of these textures indicates a unique cooling history—if you understand how to "read" the textural information. Your rock-reading lesson begins with grain size in crystalline igneous rock.

5.2.1 Grain Size in Crystalline Igneous Rock

Grain size is the key to understanding the cooling history of most igneous rocks that solidified underground or on the surface. When magma or lava begins to cool, small crystal seeds form (a process called *nucleation*), and crystals grow outward from the seeds until they interfere with one another. The result is a three-dimensional

---

EXERCISE 5.1   **A First Look at Igneous Rocks**

Name: _____   Section: _____

Course: _____   Date: _____

There are many ways to classify igneous rocks, but for now let's use three easily observable criteria: grain size, color, and specific gravity. Using the set of igneous rocks provided by your instructor, first group the specimens by grain size and record the specimen numbers in the appropriate column of the following table. Then group the specimens by color and specific gravity.

Grain size		Color		Specific gravity (heft)	
Coarse	**Fine**	**Light colored**	**Dark colored**	**Relatively high**	**Relatively low**

(a) Which two groupings are similar to one another? Which is different?

_____

_____

(b) Suggest an explanation for the similarities and the difference. *Think about the properties that might control the three variables in each rock and the processes by which igneous rock forms.*

_____

_____

_____

_____

Name: _____   Section: _____
Course: _____   Date: _____

We look first at rocks that have the same composition, because their different textures could only be caused by the different ways in which they cooled (**Fig. 5.2**). The samples in Figure 5.2 have the same minerals and thus similar compositions. Separate the light-colored igneous rocks in your set. Describe their textures, paying careful attention to the sizes and shapes of the grains and the relationships among adjacent grains. Use everyday language—the appropriate geologic terms will be introduced later.

Specimen	Textural description

interlocking texture found in *most* igneous rocks. But why do some igneous rocks have coarser (larger) grains than others?

Crystals grow as ions migrate through magma to the crystal seeds, so anything that assists ionic migration increases grain size in an igneous rock. **Cooling rate** is the most important factor controlling grain size. The slower a magma cools, the more time ions have to migrate to crystal seeds; the faster it cools, the less time there is and the smaller the grains will be. Another factor is a magma's **viscosity** (ability to flow). The *less* viscous a magma is (i.e., the more *fluid*), the easier it is for ions to migrate and the larger the crystals can become.

FIGURE 5.2 Grain sizes in light-colored igneous rocks.

(a) Very coarse grained (pegmatitic).

(b) Coarse grained (crystals approximately 1 cm across).

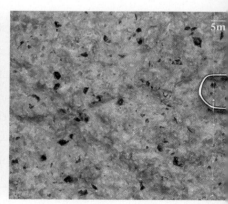

(c) Fine grained (most grains too small to see with the naked eye—less than 0.5 mm across).

Name: _____   **Section:** _____
Course: _____   **Date:** _____

It is a short step from understanding how cooling rate affects grain size to deriving the basic rules of how magma and lava cool in nature. Three simple thought experiments will help you understand the cooling process.

(a) Imagine two balls of pizza dough 20 cm in diameter. One is rolled out to form a crust 1 cm thick and 50 cm in diameter. Both pieces of dough still have the same mass and volume, but their shapes are very different. Both are baked in a 450° oven for 20 minutes and removed.

(i) Which has the greatest surface area for its volume, the crust or the ball? _____

(ii) Which will cool faster? _____ Why? _____

(iii) **Rule 1 of magma cooling:** A thin sheet of magma loses heat (faster / slower) than a blob containing the same amount of magma. This is because the amount of surface area available for cooling in the sheet is (larger / smaller) than the surface area available in a blob of the same volume.

(b) Equal amounts of hot coffee are poured into thin plastic and Styrofoam cups.

(i) Which cools faster, the coffee in the plastic cup or in the Styrofoam cup? _____ (*Hint:* How long could you hold the the Styrofoam cup before burning your fingers compared with the plastic cup?) Explain.

_____

_____

_____

(ii) **Rule 2 of magma cooling:** Magma loses heat much (faster / slower) when exposed to air or water than it does when surrounded by wall rock, because wall rock is a good insulator.

(c) Consider two cubes of steel measuring 1 m on a side. One is cut in half in each dimension to make eight smaller cubes.

(i) What is the surface area of the large cube? _____ cm^2, and of the eight small cubes? _____ cm^2

(ii) Imagine the large and small cubes are heated in a furnace to 500°C and removed. Which will cool faster, the large cube or the smaller ones? _____

(iii) **Rule 3 of magma cooling:** A small mass of magma loses heat (faster / slower) than a large one. Explain.

_____

_____

_____

(d) Now you can put the rules to use.

(i) In general, lava flows and shallow intrusive igneous rocks have (finer / coarser) grains than deep intrusive rocks.

(ii) Thick lava flows, sills, and dikes have (finer / coarser) grains than thin ones.

(iii) Which of the rocks in Figure 5.2 was probably extrusive?

(iv) Which were probably intrusive?

(v) How could it be possible for the rock in Figure 5.2a to have cooled more quickly than the rock in Figure 5.2b?

_____

_____

Name: _____   **Section:** _____
Course: _____   **Date:** _____

(vi) Many dikes, sills, and plutons have *chilled margins*—smaller grains at the contact with their wall rock than in their interiors. Explain how this happens.

_____

_____

_____

Cooling rate and viscosity cause some igneous rocks to have coarse grains (sometimes called a **phaneritic texture**) and others to have fine grains (**aphanitic texture**). But some igneous rocks have grains of two different sizes, one much larger than the other (**Fig. 5.3**). This is called a **porphyritic texture**. The larger grains are called

EXERCISE 5.4   **Interpreting Porphyritic Textures**

Name: _____   **Section:** _____
Course: _____   **Date:** _____

(a) Based on what you have deduced about magma cooling, how does a porphyritic texture form?

_____

_____

_____

(b) Describe the cooling histories of the rocks shown in Figure 5.3 in as much detail as you can.

5.3a

_____

_____

_____

5.3b

_____

_____

_____

5.3c

_____

_____

_____

5.3d

_____

_____

_____

FIGURE 5.3 Porphyritic texture (two different grain sizes).

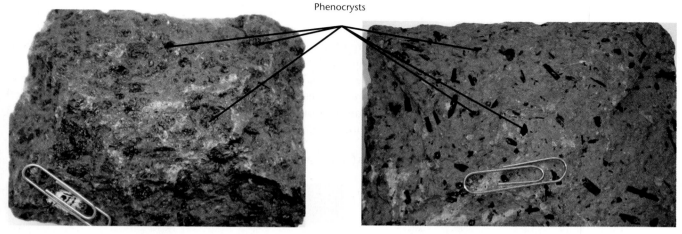

Phenocrysts

(a) Pyroxene crystals in fine groundmass.

(b) Hornblende crystals in very fine groundmass.

Phenocrysts   Small groundmass grains

(c) Dark-colored rock with fine-grained groundmass.

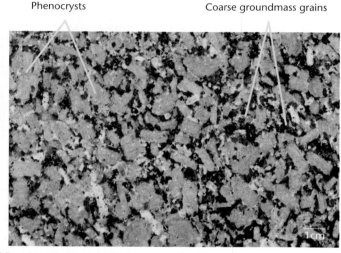

Phenocrysts   Coarse groundmass grains

(d) Light-colored rock with coarse-grained groundmass.

phenocrysts and the finer grains are called, collectively, the rock's *groundmass*.

5.2.2 Glassy Igneous Textures

Some magmas cool so fast or are so viscous that crystal seeds can't form. Instead, atoms in the melt are frozen in place, arranged haphazardly rather than in the orderly arrangement required for mineral grains. Because haphazard atomic arrangement is also found in window glass, we call a rock that has cooled this quickly **igneous glass**, and it is said to have a **glassy texture**. (**Fig. 5.4**). Like thick glass bottles, volcanic glass fractures conchoidally. Volcanic glass commonly looks black, but impurities may cause it to be red-brown or streaked.

Most igneous glass forms at the Earth's surface when lava exposed to air or water cools very quickly, but some may form just below the surface in the throat of the volcano. In addition, much of the ash blasted into the air during explosive eruptions is made of volcanic glass, formed when tiny particles of magma freeze instantly in the air.

FIGURE 5.4 Glassy texture (no mineral grains).

5.2.3 Porous (Vesicular) Textures

As magma rises toward the surface, the pressure on it decreases. This allows dissolved gases (H_2O, CO_2, SO_2) to come out of solution and form bubbles, like those that appear when you open a can of soda. If bubbles form just as the lava solidifies, their shapes are preserved in the lava (**Fig. 5.5**). The material between the vesicles may be fine-grained crystalline material (Fig. 5.5a) or volcanic glass (Fig. 5.5b). These rocks are said to have a **porous** (or **vesicular**) **texture**, and the individual bubbles are called **vesicles.** "Lava rock" used in outdoor grills and pumice used to smooth wood and remove calluses are common and useful examples of porous igneous rocks.

FIGURE 5.5 Porous (vesicular) igneous textures.

(a) Dark-colored porous igneous rock.

(b) Light-colored porous igneous rock.

5.2.4 Fragmental Textures

Violent volcanic eruptions can blast enormous amounts of material into the air. This material is collectively called **pyroclastic debris** (from the Greek *pyro*, meaning fire, and *clast*, meaning broken) or **tephra** (the Icelandic term). Pyroclastic debris includes large blocks or bombs erupted as liquid and cooled as fine-grained crystalline rock (**Fig. 5.6a**), crystals formed in the magma before eruption (**Fig. 5.6b**), tiny ash particles, and pieces of rock broken from the walls of the volcanic vent or ripped from the ground surface during an eruption (**Fig. 5.6c**).

In an **ash fall**, fine-grained tephra falls quietly like hot snow and blankets the ground. The ash may be compacted and turned to rock by the pressure of ash from later eruptions. In more violent eruptions, avalanches of ash called **pyroclastic flows**

EXERCISE 5.5   **Interpreting Fragmental (Pyroclastic) Textures**

Name: _____   Section: _____
Course: _____   Date: _____

Which rock in Figure 5.6 most likely formed in an ash fall? _____ In a pyroclastic flow? _____ Explain your reasoning.

_____

_____

_____

FIGURE 5.6 Fragmental (pyroclastic) igneous rocks in a range of grain sizes and textures.

(a) Coarse volcanic bombs in finer grained pyroclastic matrix at Kilauea in Hawaii.

Crystals (microphenocrysts) Glass shards (note the curved outlines that were once the walls of bubbles)

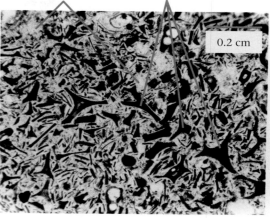

(b) Photomicrograph of glass shards welded together with a few tiny crystals.

Volcanic rock fragments

(c) Hand sample with volcanic rock fragments in a very-fine-grained brown ash.

rush down the side of a volcano while still so hot that the fragments weld together immediately to form rock.

5.2.5 Grain Shape

Some grains in igneous rocks are well-shaped crystals (phenocrysts in Fig. 5.3a-c) but others are irregular (Fig. 5.2a,b; groundmass grains in Fig. 5.3c,d). When magma begins to cool, its grains start as well-shaped crystals, but they interfere with one another as they grow, resulting in irregular shapes. This enables us to determine the sequence in which grains grew in an igneous rock: the well-shaped crystals formed early when there was no interference from others. Can you tell which grains grew first in Figure 5.3?

Name: _____   **Section:** _____

Course: _____   **Date:** _____

Examine the igneous rocks provided by your instructor. Apply what you have learned about the origins of igneous textures to fill in the "cooling history" column in the study sheets at the end of the chapter. Use the following questions as a guide to your interpretation.

- Which specimens cooled quickly? Which cooled slowly?
- Which specimens cooled very rapidly at the Earth's surface (i.e., are extrusive)?
- Which specimens cooled slowly beneath the surface (i.e., are intrusive)?
- Which specimens cooled from a magma rich in gases?
- Which specimens experienced more than one cooling rate?

You are now able to read the story recorded by igneous textures and can do so without identifying a single mineral or naming the rocks. Table 5.1 summarizes the origin of common igneous textures and the terms used by geologists to describe them.

TABLE 5.1 Interpreting Igneous rock textures.

Texture	Description	Intepretation	Example
Pegmatitic	Very large grains (>2.5 cm)	Very slow cooling or cooling from an extremely fluid magma (*usually the latter*)	Figure 5.2a
Coarse grained (phaneritic)	Individual grains are visible with the naked eye	Slow cooling, generally *intrusive*	Figure 5.2b
Fine grained (aphanitic)	Individual grains cannot be seen without magnification	Rapid cooling, generally *extrusive*	Figure 5.2c
Porphyritic	A few large grains (phenocrysts) set in a finer grained groundmass	Two cooling rates: slow at first, to form the phenocrysts, then more rapid to form the groundmass	Figure 5.3
Glassy	Smooth, shiny; looks like glass; no mineral grains present	Extremely rapid cooling, generally *extrusive*	Figure 5.4
Porous (vesicular)	Spongy; filled with large or small holes	Rapid cooling accompanied by the release of gases	Figure 5.5
Fragmental (pyroclastic)	Mineral grains, rock fragments, and glass shards welded together	Explosive eruption of ash and rock into the air	Figure 5.6

5.3 Igneous Rock Classification and Identification

You now know how to determine the conditions under which igneous rocks form by examining their textures. Igneous rock composition is the key to answering other questions, because it helps reveal how magma forms and why some igneous rocks occur in specific tectonic settings. We look first at how geologists use composition to classify igneous rocks and then at how igneous rocks are named by their composition and texture.

5.3.1 Igneous Rock Classification: The Four Major Compositional Groups

In Exercise 5.1 you saw that some of the igneous rocks in your set are relatively dark colored and others are light colored. Some have high specific gravities, others relatively low specific gravities. A rock's color and specific gravity are controlled mostly by its minerals, and they, in turn, are determined by its chemical composition. Oxygen and silicon are by far the two most abundant elements in the lithosphere, so it is not surprising that nearly all igneous rocks are composed primarily of *silicate minerals* like quartz, feldspars, pyroxenes, amphiboles, micas, and olivine.

There are many different kinds of igneous rock, but all fit into one of four major compositional groups—felsic, intermediate, mafic, and ultramafic—defined by how much silicon and oxygen (*silica*) they contain and by which other elements are most abundant (Table 5.2).

Felsic igneous rocks (from *fel*dspar and *si*lica) have the most silica and the least iron and magnesium. They contain abundant potassic feldspar and sodic plagioclase, commonly quartz, and only sparse ferromagnesian minerals—usually biotite or hornblende. Like their most abundant minerals, felsic rocks are light colored and have low specific gravities.

Intermediate igneous rocks have chemical compositions, colors, specific gravities, and mineral assemblages between those of felsic and mafic rocks: plagioclase feldspar with nearly equal amounts of calcium and sodium, both amphibole and pyroxene, and only rarely quartz.

Mafic igneous rocks (from *ma*gnesium and the Latin *f*errum, meaning iron) have much less silica, potassium, and sodium than felsic rocks but much more calcium, iron, and magnesium. Their dominant calcium-rich plagioclase and ferromagnesian minerals are dark green or black and have higher specific gravities than minerals in felsic rocks. Even fine-grained mafic rocks can therefore be recognized by their dark color and relatively high specific gravity.

Ultramafic igneous rocks have the least silica and the most iron and magnesium, with very little aluminum, potassium, sodium, or calcium. As a result, they contain mostly ferromagnesian minerals like olivine and pyroxene with very little, if any, plagioclase. Ultramafic rocks are very dark colored and have the highest specific gravities of the igneous rocks.

TABLE 5.2   **The four major groups of igneous rocks.**

Igneous rock group	Silica content (SiO_2 by weight)	Other major elements	Most abundant minerals
Felsic	64–75%	Aluminum (Al), potassium (K), sodium (Na)	K-feldspar, Na-plagioclase, quartz
Intermediate	52–63%	Al, Na, calcium (Ca), iron (Fe), magnesium (Mg)	Ca-Na plagioclase, amphibole, pyroxene
Mafic	45–51%	Al, Ca, Mg, Fe	Ca-plagioclase, pyroxene, olivine
Ultramafic	39–44%	Mg, Fe	Olivine, pyroxenes

5.3.2 Identifying Igneous Rocks

The name of an igneous rock is based on its mineral content *and* texture (**Fig. 5.7**). Each of the four igneous rock groups is shown by a column, each of which has coarse, fine, porphyritic, glassy, porous, and fragmental varieties. Rocks in a column may have exactly the same minerals but look so different that they are given different names. For example, granite and rhyolite are both felsic but look different because of their different textures. Although gabbro and granite have the same *texture*, they contain different minerals and therefore look different.

FIGURE 5.7 Classification of igneous rocks.

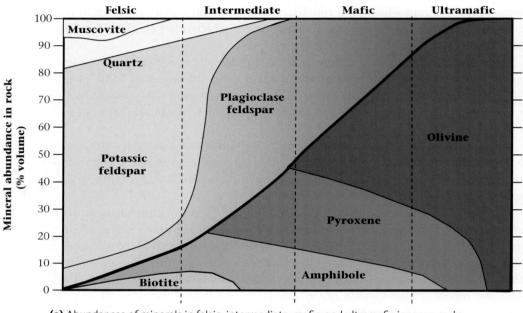

(a) Abundances of minerals in felsic, intermediate, mafic, and ultramafic igneous rocks.

		Light colored		Dark colored
Color		Low specific gravity		High specific gravity
Specific gravity		High silica content	Low silica content	
Silica (SiO₂) content				
Texture	Pegmatitic	Granitic pegmatite	Mafic pegmatite	
	Coarse grained	Granite	Diorite	Gabbro	**Dunite** (olivine only) **Pyroxenite** (pyroxene) **Peridotite** (olivine + pyroxene)
	Fine grained	Rhyolite	Andesite	Basalt	
	Porphyritic	Granite porphyry or* Rhyolite porphyry	Diorite porphyry or* Andesite porphyry	Gabbro porphyry or* Basalt porphyry	*Rocks with these textures and compositions are very rare*
	Glassy	Obsidian	Tachylite	
	Porous	Pumice	Scoria	
	Fragmental Fine	Rhyolite tuff	Andesite tuff	Basalt tuff	
	Fragmental Coarse	Volcanic breccia		

*Porphyritic rocks are named for the size of the groundmass grains. For example, a felsic porphyry in which the groundmass grains are coarse is called *granite porphyry*. If the groundmass grains are small, it is called a *rhyolite porphyry*.

(b) Choose a column based on the abundance of minerals and select the rock name based on the appropriate texture.

Identifying an igneous rock requires no new skills, just a few simple observations and your ability to identify the common rock-forming minerals.

If you are wondering why there aren't pictures of all the rock types to help you identify the specimens in your rock set, it's because granite may be gray, red, white, or even purple or black depending on the color of its feldspars. A picture can thus help only if it is exactly the same as the rock in your set. But if you understand the combination of minerals that defines granite, you will get it right every time.

5.4 Origin and Evolution of Magmas

To understand why there are four basic kinds of igneous rocks and why they can be used to interpret ancient plate tectonic events, we need to know (a) *where* and *why* rocks and minerals melt in the Earth, (b) *how* rocks and minerals melt, and (c) how plate tectonic settings produce specific rock types. The next three sections explore these issues.

5.4.1 *Where* and *Why* Do Rocks and Minerals Melt?

Some science fiction movies and novels suggest that the rigid outer shell of the Earth floats on a sea of magma. In fact, most of the crust and mantle is solid rock. Melting occurs only in special places, generally by one of the following three processes.

Decompression melting: Stretching of the lithosphere at a divergent plate boundary (ocean ridge) or continental rift lowers the confining pressure on the asthenosphere below. The amount of heat in the asthenosphere couldn't originally overcome the chemical bonds *and* the confining pressure, but once the pressure decreases, the heat already present in the rock is sufficient to overcome the bonds and cause melting. This process is called **decompression melting** *and requires no additional input of heat.*

Flux melting: Subducted oceanic crust contains hydrous minerals, some formed when ocean ridge basalt reacts with seawater shortly after erupting, others formed during weathering of island arc volcanoes. Water and other volatiles are released from these minerals when the subducted plate reaches critical depth (about 150 km) and rise into the asthenosphere where they help melt the asthenosphere in the process of **flux melting**.

Heat transfer melting: Iron in a blast furnace melts when enough heat is added to break the bonds between atoms, and some magmas form the same way in continental crust. Some very hot mafic magmas that rise into continental crust from the mantle bring along enough heat to start melting the crust, much as hot fudge does when poured onto ice cream. This process, called **heat transfer melting**, occurs mostly in three settings: (1) continental volcanic arcs where mafic magma transfers heat from the mantle to the continental crust of the arc; (2) during late stages of continent-continent collision when the base of the lithosphere peels off and the asthenosphere rises to fill that space, melts, and sends large amounts of basaltic magma upward; and (3) in continental rifts where mafic magma from the asthenosphere rises into the base of thinned continental crust.

5.4.2 *How* Do Rocks and Minerals Melt?

We can't directly observe melting in the mantle or a subduction zone, but we can study it in the laboratory. In the 1920s, N. L. Bowen and other pioneers melted minerals and rocks and chilled them at different stages of melting to learn how magma forms. They learned that magma melting (and crystallization) is very different from the way ice melts into water, and their results paved the way for understanding the origins of the different igneous rock groups.

- **Most rocks melt over a range of temperature**, because some of their minerals have lower melting points than others. The minerals with low melting points melt first while others remain solid. The entire rock melts only when the temperature rises enough to melt all the minerals—a range of several hundred degrees in some cases.

- **Some minerals, like quartz and potassic feldspar, melt simply, like ice.** When heated to their melting temperatures, these minerals melt completely to form a liquid with the same composition as the mineral. This is not surprising, but the next two findings are.

- **Plagioclase feldspars melt *continuously*, over a span of temperature.** When plagioclase containing equal amounts of calcium and sodium begins to melt, more sodium atoms are freed from the crystalline structure than calcium. The first liquid is thus more sodium-rich than the original plagioclase. More and more calcium is freed from the residual mineral as temperature rises, so the compositions of both the liquid and residual mineral change continuously. Only when the entire mineral has melted does the liquid have the same composition as the original mineral.

- **Ferromagnesian minerals melt *discontinuously*, and new minerals form in the process.** Melt ice and you get water; but melt pyroxene, for example, and when the first liquid forms, the remaining solid changes from pyroxene to olivine! Keep heating it and the olivine gradually melts. Similar things happen to the other common ferromagnesian minerals: biotite, amphibole, and olivine. As with plagioclase, the melt only has the same composition as the starting material when the process is complete.

Bowen summarized these findings graphically in what is now named Bowen's reaction series in his honor (**Fig. 5.8**). Note the relationship between the melting temperatures of the common rock-forming minerals and the assemblages that define the four major igneous rock groups.

5.4.3 Origin of the Igneous Rock Groups: Factors Controlling Magma Composition

Laboratory melting experiments last days, weeks, or months and are carried out in sealed containers to prevent contamination. Melting in the Earth is more complicated because it may take millions of years, and many things can happen to the magma. For example, material from the wall rocks can be absorbed by the melt, or magma can escape before melting is complete. Several factors control the

FIGURE 5.8 Bowen's reaction series.

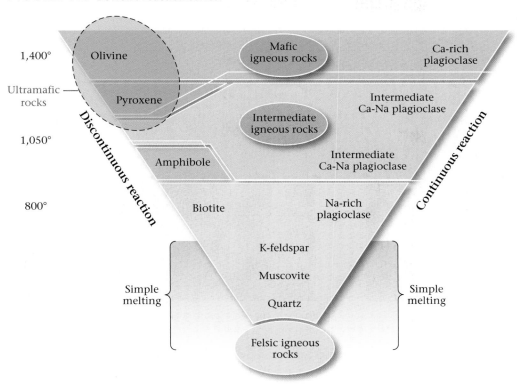

Name: _____   **Section:** _____

Course: _____   **Date:** _____

Results from melting experiments help explain several aspects of igneous rock texture and composition. Remember that *crystallization* is the exact opposite of melting. Answer the following questions.

(a) Figure 5.9 shows textural features displayed by plagioclase feldspar (a) and pyroxene and amphibole (b) found in many igneous rocks. Considering the continuous and discontinuous legs of Bowen's reaction series, suggest an explanation for the origin of these textures.

_____

_____

_____

(b) From Bowen's reaction series, is it likely that minerals with high melting points will occur in rocks with minerals with low melting points? Explain.

_____

_____

_____

(c) List the four major igneous rock groups in order from lowest magma temperature to highest.

_____

_____

_____

continued

Name: _____   Section: _____

Course: _____   Date: _____

FIGURE 5.9 Photomicrographs of common mineral textures in igneous rocks.

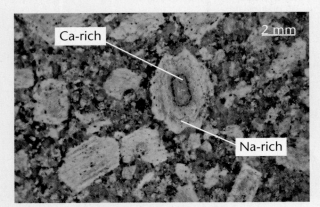

(a) Compositionally zoned plagioclase feldspar with Ca-rich core and progressively more Na-rich rims.

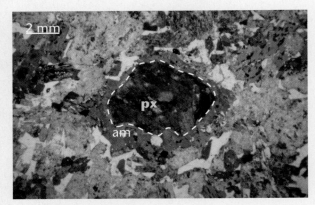

(b) Pyroxene grain (px; dark green) with amphibole rim (am; light green).

composition of a magma and the rocks that can form from it, but four processes play the most important roles.

Partial melting: Magma forms by **partial melting** of preexisting rock, not *complete* melting. When ferromagnesian minerals and plagioclase start to melt, the initial liquid has lower density than the source rock and therefore rises, escaping from the melting zone before the rock has melted entirely. In general, partial melting produces a magma that is more felsic (i.e., contains more silica and less iron and magnesium) than its source rock.

Thus, partial melting of ultramafic rock normally produces basalt, but different kinds of basalt could form depending on whether 10%, 20%, or 30% of the source rock melted before the magma escaped. (We said it was more complicated than in the lab!)

When the initial magma escapes, it rises to cooler levels and begins to crystallize. Even then it may not "follow the rules" of Bowen's reaction, because three other processes can change its composition and thus the minerals that can crystallize from it.

Magmatic differentiation (also called **fractional crystallization**): Early-formed minerals may separate from a magma, usually by sinking, because they are denser than the liquid. If early-crystallized olivine sinks to the bottom of the magma, it can no longer react with residual liquid to make pyroxene as it should according to Bowen's reaction series. Instead, the results are (a) dunite (an ultramafic rock composed entirely of olivine) at the bottom of the magma chamber and (b) a magma more silicic than the original magma. This residual magma might then crystallize minerals that would not have formed from the original. In extreme cases, differentiation of basalt magma can actually produce very small amounts of *granitic* magma.

Assimilation: As a magma rises, it may add ions by melting some of the surrounding rocks. As the new material is incorporated, the magma composition may change enough locally to enable minerals to crystallize that could not otherwise have been produced.

Magma mixing: Field and chemical evidence suggests that some intermediate rocks did not crystallize from an intermediate magma but rather formed when felsic and mafic magmas mixed.

EXERCISE 5.9 Explaining Igneous Features Found in the Field

Name: ______________________________ **Section:** ____________

Course: ______________________________ **Date:** ____________

Use your knowledge of melting and magmatic processes to explain the origin of the following features.

(a) Most continental rift zones contain basalt and rhyolite, but it is also possible to find small amounts of andesite in this setting. Suggest an origin for the andesite.

____

____

____

(b) The Palisades Sill on the west bank of the Hudson River in New Jersey is a shallow basaltic intrusive. The sill varies texturally as shown in **Figure 5.10**. Explain how the observed mineralogic and textural variations were produced by crystallization of the sill.

____

____

____

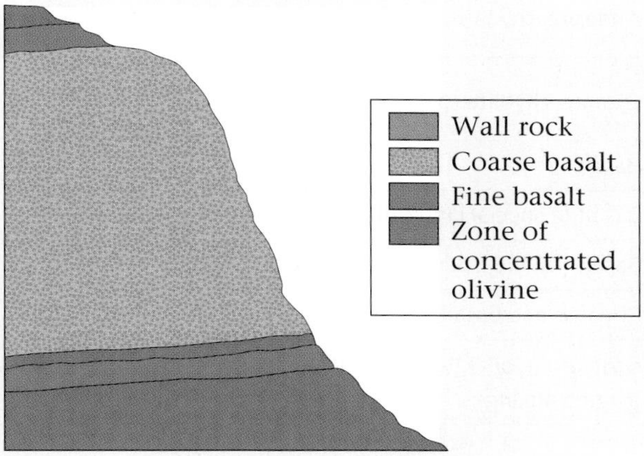

Wall rock
Coarse basalt
Fine basalt
Zone of concentrated olivine

FIGURE 5.10 Schematic cross section of the Palisades Sill.

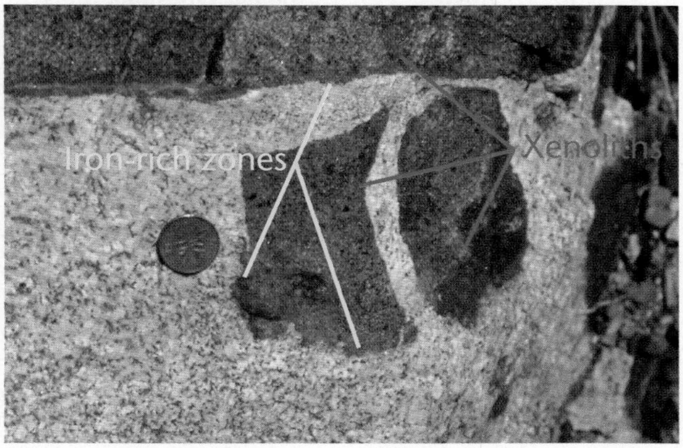

Iron-rich zones
Xenoliths

FIGURE 5.11 Mafic inclusions (xenoliths) in a granitic pluton.

(c) In **Figure 5.11**, a thin brown zone of weathering highlights iron-rich zones that separate the host granite from the mafic xenoliths. Based on the magmatic processes described above, suggest an origin for these zones.

____

____

____

5.5 Igneous Rocks and Plate Tectonics

Igneous rock types are not distributed equally or randomly on Earth, because melting occurs in different ways in different tectonic settings. As you read the next sections, remember that assimilation, magma mixing, and magmatic differentiation can produce exceptions to just about every generalization, so keep in mind the words "are generally found." With that warning, look at **Figure 5.12**, which illustrates the tectonic settings of igneous rocks, and **Table 5.3**, which lists the igneous rocks that are *generally* found in those settings.

FIGURE 5.12 Tectonic settings for major igneous rock types.

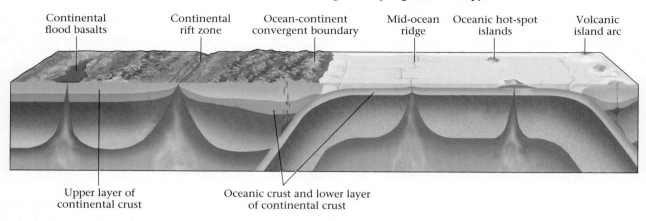

TABLE 5.3 Tectonic settings and associated igneous rock types.

convergent boundaries			
Subduction zones	**Volcanic island arcs (ocean-ocean convergence)**		Basalt, minor andesite, very minor rhyolite, basalt (and intrusive equivalents)
	Andean-type mountains (ocean-continent convergence)		Andesite, rhyolite, minor basalt (and intrusive equivalents)
	Continent-continent collision zones		Granite, rhyolite (peridotite in ophiolites)
Oceanic hot-spot islands			Basalt
Mid-ocean ridges			Basalt (a special kind called MORB—mid-ocean ridge basalt)
Ocean floors			Basalt (MORB); peridotite locally along faults
Continental rift zones			Rhyolite and basalt
Continents			Granite, rhyolite, basalt, and gabbro; andesite and diorite by magma mixing

5.5.1 Tectonic Settings of Ultramafic Rocks (Peridotites)

Several lines of evidence indicate that the mantle consists of the ultramafic rock peridotite, making it the most abundant rock on Earth. But peridotite is not a common rock because its magma is so much denser than the crust that it doesn't have the buoyancy to rise to the surface. It *is* found in divergent mid-ocean ridges and continental rifts where extreme crustal stretching has exposed the underlying mantle, and at convergent plate boundaries where mountain building has thrust slices of the upper mantle (called **ophiolites**) into the crust.

5.5.2 Tectonic Settings of Mafic Igneous Rocks (Basalt and Gabbro)

Basalt and gabbro are the most abundant igneous rocks in the lithosphere and occur in all tectonic settings—oceanic crust (formed at mid-ocean ridges), continental rifts,

Name: _____   **Section:** _____

Course: _____   **Date:** _____

Explain which type of melting (addition of volatiles, decompression, heat transfer) is responsible for mafic magmas in each of the following tectonic settings, and why the resulting magma is mafic.

(a) Mid-ocean ridges

_____

_____

(b) Continental rifts

_____

_____

(c) Oceanic and continental volcanic arcs

_____

_____

(d) Hot spots

_____

_____

oceanic and continental volcanic arcs, and hot spots. They are the result of partial melting in the asthenosphere, but the cause of melting is different in each setting.

5.5.3 Tectonic Settings of Intermediate Rocks (Andesite and Diorite)

In the mid-twentieth century, geologists were puzzled by the fact that nearly all andesites occur next to trenches in continental arcs and in some oceanic island arcs. Today we understand that most intermediate magmas are produced at subduction zones through a combination of the type of melting that occurs at a particular place and the composition of the lithosphere.

5.5.4 Tectonic Settings of Felsic Rocks (Granite and Rhyolite)

Granite and rhyolite are most abundant on the continents—in continental volcanic arcs, continent-continent collision zones, rifts, and where plumes rise beneath continents. They form largely by partial melting of the upper (granitic) layer of continental lithosphere and to a lesser extent by differentiation of mafic magmas. Some rhyolite and granite form in subduction zones by differentiation of mafic and intermediate magmas and/or by assimilation. Only very small amounts of rhyolite form in oceanic hot-spot islands by extreme fractional crystallization.

EXERCISE 5.11   **Origin of Intermediate Magmas in Subduction Zones**

Name: _____   **Section:** _____
Course: _____   **Date:** _____

Melting above the subducted slab produces mafic magma, as described above, yet intermediate rocks (andesite, diorite) are common in many subduction zones. Considering the magmatic processes discussed in section 5.4, and the difference between oceanic and continental lithosphere, explain how this intermediate magma forms.

(a) A relatively small amount of intermediate magma occurs in oceanic volcanic arcs. Remembering that melting above the subducted slab produces mafic magma, explain the origin of the intermediate magma.

_____

_____

_____

(b) Much more intermediate magma erupts in continental arcs. Why?

_____

_____

_____

EXERCISE 5.12   **Origin of Granite and Rhyolite in Continental Rifts**

Name: _____   **Section:** _____
Course: _____   **Date:** _____

Continental rifts typically contain large amounts of basalt and rhyolite. Your answer to Exercise 5.10b explained the origin of the basalt, but fractional crystallization of mafic magma can produce only a very small amount of felsic magma. Explain the origin of large volumes of felsic magma in continental rifts.

_____

_____

_____

_____

EXERCISE 5.13   **Interpreting Tectonic Setting of Igneous Rocks**

Name: _____   **Section:** _____
Course: _____   **Date:** _____

Based on the information in Table 5.3 and your answers to Exercises 5.10, 5.11, and 5.12, add possible tectonic settings for the igneous rocks in your study set to your rock identification charts.

5.5.5 Visiting Localities Where Igneous Rocks Are Forming or Have Formed in the Past

Software such as *Google Earth*™ and *NASA WorldWind* makes it possible to tour the world to see current igneous activity at these tectonic settings and view ancient igneous rocks revealed by erosion. For example, use either software to visit:

- **Yellowstone National Park in Wyoming, a continental hot spot.** About 2.2 million years ago, the northwestern corner of Wyoming exploded in a violent eruption that spread immense quantities of ash across the landscape. The sudden removal of magma caused the ground surface to collapse, forming the huge bowl-shaped depression in which Yellowstone National Park is located. The process continues today, with magma heating water for hot springs and geysers like Old Faithful.

- **the Palisades in New Jersey, a divergent plate margin.** About 180 million years ago, a supercontinent began to split apart to form the modern Atlantic Ocean. Rift valleys, like those in east Africa, formed along the east coast of North America from Nova Scotia to South Carolina. Part of the asthenosphere melted and basaltic magma moved upward toward those valleys. Some flowed as lava, but the Palisades along the Hudson River intruded nearly horizontal sedimentary rocks beneath the surface.

- **Hawaii, an oceanic hot spot.** The island of Hawaii is the youngest of a long chain of hot-spot volcanic islands that helps us track the motion of the Pacific Plate. The island consists of five huge volcanoes, one of which (Kilauea) has been erupting nearly continuously for almost 20 years.

- **the Central Rift Valley in Iceland.** Iceland is a unique place where a hot spot lies beneath a segment of an ocean ridge. The Central Rift Valley is the divergent boundary—Iceland is growing wider as the valley sides move apart, and the lava here and at nearby volcanoes comes from magma rising into the space caused by divergence.

- **the Cascade Mountains of Washington, Oregon, and northern California, a subduction-related continental volcanic arc.** The volcanoes of the Cascade Mountains are a continental arc that developed as a small plate subducted beneath North America. Prominent peaks such as Mt. Rainier, Mt. Hood, and Mt. Baker overlook densely populated areas, much as Mt. Vesuvius stood over Pompeii (and still towers over Naples) in Italy. Thankfully, the most recent eruption in the Cascades arc was of Mt. St. Helens, which is in a sparsely populated part of Washington.

- **the Sierra Nevada Mountains in California.** Large volumes of the continental crust under eastern and east-central California melted between 210 and 85 million years ago, possibly due to increased heat caused by crustal thickening. Most of the magma solidified below the surface, but tens of millions of years of erosion have exposed the resulting granitic rocks in the Sierra Nevada Mountains, as seen in the Yosemite, Kings Canyon, and Sequoia national parks.

IGNEOUS ROCKS STUDY SHEET

#	Texture	Minerals present (approximate %)	Name of rock	Cooling history; source of magma; tectonic setting

IGNEOUS ROCKS STUDY SHEET

#	Texture	Minerals present (approximate %)	Name of rock	Cooling history; source of magma; tectonic setting

CHAPTER 6

USING SEDIMENTARY ROCKS TO INTERPRET EARTH HISTORY

PURPOSE

- Become familiar with sedimentary rock textures and mineral assemblages.
- Learn how to use sedimentary rocks to interpret ancient geologic, geographic, and environmental settings.

MATERIALS NEEDED

- Set of sedimentary rocks
- Magnifying glass or hand lens, microscope, and thin sections of sedimentary rocks
- Mineral testing supplies (streak plate, glass plate, etc.)
- Glass beaker, concentrated Ca (OH)$_2$ solution, straws

6.1 Introduction

If you've ever seen shells on a beach, gravel along a river bank, mud in a swamp, or sand in a desert, you've seen sediment. By definition, **sediment** consists of loose grains derived from fragmented rock, of shells and shell fragments, of plant debris, or of mineral crystals precipitated from water bodies at the Earth's surface. **Sedimentary rock** is rock that forms at or near the Earth's surface by one of the following processes: compaction and cementing together of grains; accumulation of layers formed of minerals precipitated from water solutions; cementing together of shells and shell fragments; or accumulation and alteration of organic material. Sedimentary rocks preserve a record of past environments and ancient life, and thus tell the story of Earth history. For example, the fact that a type of sedimentary rock called limestone occurs throughout North America means that, in the past, a warm, shallow sea covered much of the continent. Geologists have learned to read the record in the rocks by comparing features in sedimentary rocks to those found today in environments where distinct types of sediment form, move, or accumulate. Remember, the present is the key to the past. In this chapter, you will learn to read this record, too.

6.2 Sediment Formation and Evolution

6.2.1 The Origin of Sediment

The material from which sedimentary rocks form ultimately comes from **weathering**—the chemical and/or physical breakdown of preexisting rock. Weathering produces both **clasts**, which are solid fragments of preexisting rock or minerals, and **dissolved ions**, which are charged atoms or molecules in a water solution. Once formed, clasts may be transported by water, wind, or ice to another location, where they are deposited and accumulate. Dissolved ions, meanwhile, enter streams and groundwater. Some of these ions precipitate from groundwater in the spaces between clasts and thereby form a **cement** that holds the clasts together. Others get carried to lakes or seas, where organisms extract them to form shells. Finally, some dissolved ions precipitate directly from water to form layers of new sedimentary minerals. In some environments, sediment may also include organic material, the carbon-containing compounds that remain when plants, animals, and microorganisms die.

EXERCISE 6.1   **What Kind of Sediment Can Be Produced from a Granite?**

Name: _____   Section: _____

Course: _____   Date: _____

Granite consists of about 35% quartz, 25% K-feldspar, and 30% plagioclase, with minor amounts (10%) of ferromagnesian minerals such as biotite and amphibole.

(a) If you could crush granite to form clasts that each contain only one mineral, what would be the proportions of minerals comprising the clasts?

_____

(b) Refer to Figure 5.7 or a comparable chart from your textbook. What clasts would you obtain if you crushed gabbro?

_____

(c) If you place granite in water so that a little of the granite dissolves, what ions would the resulting solution contain? (*Hint:* Look at the chemical formulas of the minerals in granite.)

_____

6.2.2 Weathering and Its Influence on Sediment Composition

The minerals that occur in sediment at a location depend both on the composition of the source rock(s) *and* on the nature of weathering to which the minerals were exposed. Here, we look at the weathering process and how it influences sediment composition.

Physical weathering, like hitting a rock with a hammer, breaks the rock into fragments (clasts) *but does not change the minerals* making up the rock. Initially, clasts are simply small rock pieces and may contain many grains or crystals that retain their original sizes and shapes. But when physical weathering progresses, the clasts break into smaller pieces, perhaps consisting only of a single mineral. While moving along in wind and water, clasts grind and crash against each other, and become progressively smaller. As a result, the size and shape of transported clasts do not tell us whether the source rock itself was coarse or fine, or whether grains in the source rock interlocked or were cemented together (**Fig. 6.1**). But because physical weathering does not change the composition of clasts, we can get a sense of the composition of the source rock by looking at these clasts. For example, sediment derived from physical weathering of granite would contain clasts of quartz, K-feldspar, plagioclase, and biotite, whereas sediment derived from physical weathering of basalt would contain clasts of plagioclase, pyroxene, amphibole, and olivine.

Chemical weathering is the process during which rock chemically reacts with air, water, and acidic solutions. In other words, chemical weathering involves **chemical reactions**, the breaking and forming of chemical bonds. These reactions can destroy some of the original minerals and can produce new minerals. Those minerals that are easily weathered are called *unstable* or *nonresistant*, whereas those that can survive weathering or are produced as a consequence of weathering are called *stable* or *resistant*. Chemical weathering has hardly any effect on quartz, because quartz is stable, but transforms feldspar, which is an unstable mineral, into clay and ions (K^+, Ca^{2+}, Na^+). Similarly, chemical weathering converts ferromagnesian minerals such as olivine, pyroxene, amphibole, and biotite into hematite and limonite (iron oxide minerals) along with ions of silicon and oxygen.

Relatively few minerals are stable at the Earth's surface, so *chemical weathering generally reduces the number of minerals in sediment* over time. As a result, weathering of very different source rocks can yield surprisingly similar sedimentary mineral assemblages (**Fig. 6.2**). Weathering of granite, for example, produces quartz, clay minerals, and iron oxides. Gabbro also weathers into clay minerals and iron oxides. Note that

FIGURE 6.1 Physical weathering of coarse-grained granite.

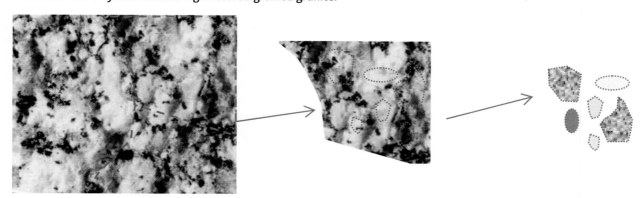

(a) A rock breaks along the dashed lines.

(b) The rock fragment preserves grain sizes and relationships.

(c) The fragment breaks into both smaller rock fragments and mineral grains.

FIGURE 6.2 Chemical weathering of felsic and mafic igneous rocks.

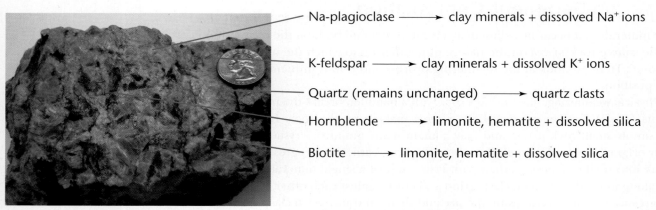

Na-plagioclase ⟶ clay minerals + dissolved Na⁺ ions

K-feldspar ⟶ clay minerals + dissolved K⁺ ions

Quartz (remains unchanged) ⟶ quartz clasts

Hornblende ⟶ limonite, hematite + dissolved silica

Biotite ⟶ limonite, hematite + dissolved silica

(a) Granite weathers to clay minerals, limonite, hematite, and quartz, plus dissolved potassium, sodium, and silica.

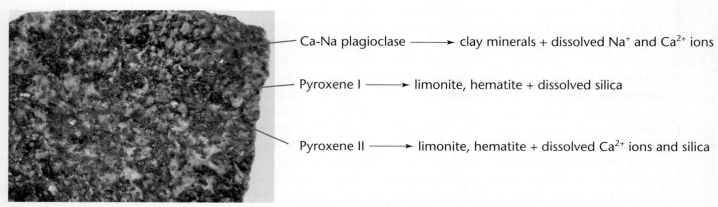

Ca-Na plagioclase ⟶ clay minerals + dissolved Na⁺ and Ca²⁺ ions

Pyroxene I ⟶ limonite, hematite + dissolved silica

Pyroxene II ⟶ limonite, hematite + dissolved Ca²⁺ ions and silica

(b) Gabbro weathers to clay minerals, limonite, and hematite, plus dissolved calcium, sodium, and silica.

EXERCISE 6.2   **Looking Again at Weathering Products**

Name: _____   Section: _____
Course: _____   Date: _____

In Exercise 6.1, you were asked to identify the minerals in crushed specimens of igneous rocks. Imagine that the pile of crushed sediment you've produced undergoes some chemical weathering.

(a) How would the mineralogy of the sediment change as a result of the chemical weathering?

_____

_____

_____

(b) Do you think the mineralogy of the resulting sediment would depend on the amount of chemical weathering? Explain your answer.

_____

_____

_____

only the proportions of minerals (more iron oxides in the weathered gabbro) and the occurrence of quartz (none in weathered gabbro) can distinguish between sediments derived from chemical weathering of granite and those derived from gabbro.

6.2.3 Mineralogical Maturity

Not all sediments undergo the same amount of weathering. We refer to sediments that still contain nonresistant minerals and rock fragments as **mineralogically immature**, whereas those in which all minerals have weathered to produce stable minerals (and ions) are **mineralogically mature**. Immature sediment exists where weathering has happened only for a relatively short time, or if there is not enough water for chemical reactions to occur.

EXERCISE 6.3   **Weathering History Recorded by Clasts in Sedimentary Rock**

Name: _____   Section: _____
Course: _____   Date: _____

(a) What weathering histories could account for each of the following mineral assemblages found in *clastic* sedimentary rocks? Think broadly. There may be more than one possible explanation.

   (i) All quartz grains

   _____

   _____

   (ii) Nearly equal amounts of quartz, K-feldspar, and Na-plagioclase with a small amount of hematite

   _____

   _____

   (iii) All fine-grained clay minerals with some hematite and limonite

   _____

   _____

   (iv) A mixture of quartz grains and clay minerals

   _____

   _____

   (v) Rock fragments composed of Ca-plagioclase and pyroxene

   _____

   _____

(b) Which of the above rocks are mineralogically mature? Which are immature?

   _____

   _____

   _____

   _____

(c) Rank the following environments in the order you would expect sediments to become mineralogically mature, from slowest to fastest. (*Hint:* Think about the amount of rainfall available.)

   desert _____   tropical rain forest _____   temperate climate _____

6.3 The Basic Classes of Sedimentary Rocks

Geologists distinguish among many kinds of sedimentary rocks, each with a name that is based on the nature of the material that the rock contains and the process by which the rock forms. To organize this information, we sort various rocks into classes. However, not all geologists use the same classification. We first introduce a simple scheme that groups rocks into three classes: clastic, chemical, and biogenic. Note that this classification scheme is not based on composition (the minerals present). We conclude this section by introducing an overlapping classification scheme that focuses on composition.

6.3.1 Clastic Sedimentary Rock

This class consists of rocks formed from clasts (mineral grains or rock fragments) derived from previously existing rocks. The majority of the most common clastic sedimentary rocks are derived from silicate rocks and thus contain clasts composed of silicate minerals. (To emphasize this, geologists sometimes refer to them as "siliciclastic rocks.") Formation of clastic sedimentary rock involves the following steps:

- **Weathering:** The process of weathering reduces solid bedrock into a pile of loose (unconsolidated) grains, which we refer to as clasts.
- **Erosion:** Moving water (streams and waves), moving air (wind), and moving ice (glaciers) pluck and/or pick up the clasts.
- **Transportation:** Moving water, wind, or ice carry clasts away from their source.
- **Deposition:** When moving water or wind slows, or when the ice melts, clasts settle out and accumulate. This happens in a great variety of **depositional environments** (such as the land surface, the sea floor, or a river bed).
- **Lithification:** Over time, accumulations of clasts are buried. When this happens, the weight of overlying sediment squeezes out air and/or water, thereby fitting the clasts more tightly together. This process is called **compaction**. As ion-rich groundwater passes through the compacted sediment, minerals precipitate and bind, or "glue," the clasts together. This process is called **cementation**, and the mineral glue between clasts is called **cement**. Cement in sedimentary rock typically consists of either calcite or quartz, and may contain minor amounts of hematite or pyrite.
- **Diagenesis:** Application of pressure, and circulation of fluids over time, may gradually change characteristics of sediments and sedimentary rock (e.g., grain size and composition of cement; the nature of grain boundaries), even at temperatures below those required for metamorphism. Any chemical or physical change that happens in a sedimentary environment subsequent to the original deposition of sediment is called **diagenesis**.

As we see later in this chapter, the names of clastic sedimentary rocks are based primarily on the size of the clasts they contain. Geologists refer to the Wentworth scale (**Fig. 6.3**) to distinguish among grain size. From coarsest to finest, sedimentary rock types include:

- **Conglomerate/breccia** consists of pebbles and/or cobbles. Clasts in conglomerate are rounded (have no sharp corners), whereas those in breccia are angular (have sharp corners).
- **Sandstone/arkose** consists of sand. Arkose differs from sandstone in that it is mineralogically immature (i.e., contains feldspar grains).
- **Siltstone** consists of silt.
- **Shale/mudstone** consists of clay. Shale and mudstone differ from each other in that the former tends to break into thin plates, whereas the latter does not.

FIGURE 6.3 Clast size terminology (Wentworth scale).

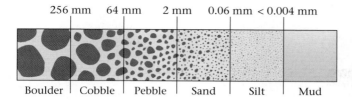

6.3.2 Chemical Sedimentary Rock

This class consists of rocks formed from mineral crystals precipitated directly from a water solution. Chemical sedimentary rock forms when water solutions of dissolved ions become oversaturated and excess ions bond together to form solid mineral grains. These crystals either settle out of the solution or grow outward from the walls of the container holding the solution. Exercise 6.4 allows you to simulate the chemical precipitation process in the lab in order to see how the textures of these rocks develop.

Groundwater, oceans, and saline lakes all contain significant quantities of dissolved ions and can serve as a source of chemical sedimentary rock. Precipitation to form chemical sedimentary rocks happens in many environments, including: (1) hot springs, where warm groundwater seeps out at the Earth's surface and cools;

EXERCISE 6.4   **Simulating Chemical Sedimentary Textures**

Name: _____   **Section:** _____
Course: _____   **Date:** _____

(a) Place a beaker with seawater (or homemade saltwater) on a hot plate and heat it gently until the water evaporates. Partially fill a second beaker with a clear, concentrated solution of calcium hydroxide ($CaOH_2$). Using a straw, blow *gently* into the solution until you notice a change.

    Describe what happened in each demonstration, and sketch the resulting textures.

(i)   (ii)

_____

_____

_____

_____

_____

_____

(b) Compare the texture in (i) with that of a granite, a rock formed by cooling of a melt. Describe and explain the similarities in texture.

_____

_____

_____

_____

FIGURE 6.4 Textures of chemical sedimentary rocks.

(a) Coarse, interlocking halite grains.

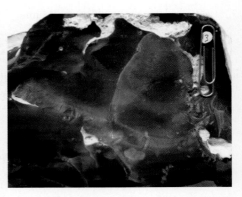

(b) Cryptocrystalline silica grains (chert). Note the almost glassy appearance and conchoidal fracturing.

(2) cave walls, where groundwater seeps out, evaporates, and releases CO_2; (3) the floors of saline lakes or restricted seas, where saltwater evaporates; (4) within sedimentary rocks, when reactions with groundwater result in the replacement of the original minerals with new minerals; and (5) on the deep sea floor, where the shells of plankton dissolve to form a gel-like layer that then crystallizes. The composition of a chemical sedimentary rock depends on the composition of the solution from which it was derived—some chemical sedimentary rocks consist of salts (e.g., halite, gypsum), whereas others consist of silica or carbonate. In some chemical sedimentary rocks, the grains are large enough to see (**Fig. 6.4a**). But in others, the grains are so small that the rock looks somewhat like porcelain (**Fig. 6.4b**). Such rocks are called **cryptocrystalline rocks**, from the Latin *crypta-*, meaning hidden.

Geologists distinguish among different types of chemical sedimentary rocks based primarily on composition:

- **Travertine** is composed of carbonate (generally formed at hot springs).
- **Chert** is composed of cryptocrystalline silica (formed from gels of silica on the sea floor, or by diagenesis in existing sedimentary beds).
- **Rock salt** is composed of halite (formed when saltwater evaporates).
- **Rock gypsum** is composed of gypsum (formed when saltwater evaporates).

6.3.3 Biogenic Sedimentary Rock

The life processes of organisms produce a great variety of materials by extracting elements from their surroundings. For example, clams, oysters, and some types of plankton and algae produce shells out of either carbonate or silica made from ions the organisms extract from water. Corals, similarly, produce mounds of solid carbonate. Sponges produce needle-like spines (called *spicules*). Trees and shrubs produce wood (from CO_2 extracted from the air). And all organisms produce proteins and fats. What happens to this material when the organisms die? Some simply dissolves or rots away, providing ions in water or CO_2 in air. But in some environments, such as those where the remains of organisms are buried quickly or accumulate in oxygen-free water, they may be preserved and become incorporated in sedimentary rock. Rock composed primarily of the remains of living organisms can be referred to, in a general sense, as **biogenic sedimentary rock**.

The texture (arrangement of grains) in biogenic sedimentary rock depends on the way the rock formed. For example, some biogenic rocks began as reefs that grew in place. Others are formed from wood fragments pressed together. Still others consist of shells that settled to the floor of a water body, or were transported and broken up by waves or currents before settling out, after the organisms died. Transported shells and shell fragments can be thought of as clasts—to emphasize this characteristic, geologists sometimes refer to rocks formed from transported shells and shell fragments as "bioclastic" rocks.

Diagenesis tends to modify biogenic rock significantly, so the rock we see today may look very different from the original material from which the rock formed. Over time, grains in biogenic rocks recrystallize, pressure causes some grains to dissolve, and ions from groundwater may substitute for ions in the rock's grains, causing the composition of the rock overall to change.

A great variety of names are used for various types of biogenic sedimentary rocks. Here are just a few of the major ones:

- **Limestone** is a general class of rocks formed from shells of calcite-secreting organisms; if fossil shells are visible, the rock may be called "fossiliferous limestone."
- **Coquina** is limestone composed of a mass of shells that have undergone minimal diagenesis.
- **Micrite** is limestone composed of very tiny calcite particles—in effect, it is solidified lime mud. The particles are probably biochemical in origin, and can include spiny parts (spicules) of sponges, as well as broken-down pieces of algae shells. Though now calcite, the particles probably started as aragonite, a polymorph.
- **Dolostone** is a rock containing significant amounts of dolomite, formed when diagenesis alters the calcite originally in a limestone.
- **Biogenic chert** is formed from silica gel that consists of the remains of silica-secreting plankton.
- **Coal** is composed primarily of carbon derived from plants.

6.3.4 Compositional Classes of Sedimentary Rocks

Note that the classes described above do not specify the chemical composition of rocks in their names. That's because each of the classes contains a variety of rocks with different compositions. To indicate chemical composition, geologists may use a different set of terms in relevant situations. Specifically, (1) **siliceous rocks** consist primarily of quartz; (2) **argillaceous rocks** consist mostly of clay; (3) **carbonate**

EXERCISE 6.5　　**Distinguishing among Compositional Classes**

Name: _____　　**Section:** _____
Course: _____　　**Date:** _____

Your instructor will provide a set of sedimentary rocks. Using the diagnostic tools that you have learned in the context of identifying minerals, determine the compositional class of each rock. Features that you should consider are hardness, reaction with acid, and taste.

Sample 1: _____

Sample 2: _____

Sample 3: _____

Sample 4: _____

rocks consist mostly of calcite or dolomite, both of which are carbonate minerals (i.e., contain CO_3^{2-} ions); (4) **evaporite rocks** consist of salts that precipitate when saltwater evaporates; and (5) **organic rocks** contain remnants of organic chemicals and may be more than 90% carbon.

6.3.5 Textural Classes of Sedimentary Rocks

Geologists distinguish between two general types of texture in sedimentary rocks: a sedimentary rock with a **clastic texture** is one in which discrete grains are held together by cement. Clastic textures form by compaction and lithification of once unconsolidated sediment. A sedimentary rock with a **crystalline texture** is one in which crystals are interlocking. Crystalline textures form either when the minerals in the rock grew by precipitation directly from a water solution, as occurs in chemical sedimentary rocks, or as a consequence of recrystallization during diagenesis. When describing textures, geologists commonly describe both the way in which grains are held together, and the grain size.

6.4 Identifying Sedimentary Rocks

At this point, you are familiar with the nature and origin of sediment, the various classes of sedimentary rocks, and the basic names of sedimentary rock types. Now, let's develop the skills needed to distinguish one type of sedimentary rock from another. To identify a sedimentary rock, we examine both its composition (e.g., Does the rock consist of quartz, calcite, dolomite, clay, or evaporite minerals?) and its texture (e.g., Is the rock clastic or crystalline? Is the rock coarse-grained or fine-grained?). We can organize our examination procedure into the following basic steps:

- *Hardness test:* If it's harder than glass, it is siliceous (e.g., it is a quartz sandstone, a quartz-rich conglomerate, or a chert). If not, it could be argillaceous, organic, carbonate, or a salt.

- *Reaction with acid:* Determine if the rock is a carbonate (limestone or dolostone) by watching how it reacts with hydrochloric acid. A limestone will bubble vigorously, a dolostone less so. You may have to scratch a dolostone first to get it to react.

- *Analysis of texture:* Determine whether the rock is clastic or crystalline by examining the texture—that is, does it consist of cemented-together grains or of intergrown crystals? Note that crystalline rocks are tricky to work with. Some limestones appear crystalline because of recrystallization during diagenesis, even though they may have originally started out as a bioclastic rock. It may be hard to see texture in very-fine-grained rocks (micrite, chert, mudstone), but other properties allow you to recognize these.

- *Analysis of grain size:* Compare the grain size to the sizes on a grain size chart (Fig. 6.3). From this, you can distinguish among conglomerate, sandstone, siltstone, and shale or mudstone. Note that you can't see the grains in a shale or mudstone, but you can confirm your identification of these rocks by checking their hardness—sandstone scratches glass, whereas you can scratch shale (composed of clay) with a nail. Also, you don't actually have to use a ruler to define the smaller clasts—if individual grains are easily visible with a hand lens, they are at least sand-sized; if you can't see the grains easily, rub the rock (gently!) across your teeth. If the rock feels gritty, the grains are silt-sized; if smooth, they are mud-sized.

- *Distinguishing between conglomerate and breccia:* These rocks are distinguished from each other by the shape of their clasts. Breccia clasts are angular, whereas conglomerate clasts are rounded.

- *Distinguishing among types of limestones:* If the rock is so fine-grained that you can't see the grains, it's micrite. If it is a mass of shells weakly cemented together, it's coquina. If it's a gray, somewhat recrystallized mass of calcite grains, it's just limestone, or fossiliferous limestone (if fossils are visible).

- *Recognizing evaporites:* Rock salt is grayish and the grains are transluscent. Also, it clearly has a crystalline texture and tastes salty. Gypsum is generally clear to white and is softer than a fingernail.

- *Recognizing coal:* If the rock is black, massive (meaning it has no visible layering) to subtlely layered, softer than glass, and has a relatively low specific gravity, it is coal. In detail, coal varies in hardness—soft coal looks like a compacted mass of soot, whereas hard coal breaks conchoidally and sometimes displays a rainbow-like reflection on its surface.

- *Recognizing chert:* If the rock is harder than glass, does not react with acid, and fractures conchoidally, it is chert. Chert comes in a variety of colors—black, white, or red.

Table 6.1 is a simplified classification scheme for sedimentary rocks based on texture and mineralogy, summarizing the above information. First use texture to determine whether the rock is dominantly of clastic, chemical, or biogenic origin, then look in the appropriate section of Table 6.1 to identify the rock. Figure 6.5 provides similar information in the form of a flow chart.

EXERCISE 6.6 **Identifying Sedimentary Rock Samples**

Name: _____ Section: _____
Course: _____ Date: _____

Examine the rock samples in your set that you classified as sedimentary rocks. Fill in the rock study sheets at the end of this chapter to identify each sample. Hold on to these samples and your study sheets until you go through the remainder of the chapter. At that point, you will be able to add an interpretation of the rocks' history.

6.5 Interpreting Clastic Sedimentary Textures

If you are given a sedimentary rock, identifying it is only part of the task. Examination of the texture of a clastic sedimentary rock can provide a rich source of information about the geologic history that led to the formation of the rock. Let's consider how to gain insight about depositional environments by looking at textures in clastic rocks.

6.5.1 Grain Size and Sorting

Clasts in sedimentary rocks range from the size of a house to specks so small they can't be seen without an electron microscope. As noted earlier, geologists use familiar words like *sand* and *pebble* to define clast size (Fig. 6.3). The size of grains in a rock reflects the kinetic energy of the agent that transports it—an agent with more kinetic energy can move bigger clasts. Kinetic energy depends both on the *viscosity* and the *velocity* of the agent. A denser material can move larger clasts, and a faster moving material can move larger clasts. Ice is denser than water, and water is denser than air. Thus, at a given velocity, ice can move larger clasts than running water or waves, and running water or waves can move larger grains than wind. For a given

TABLE 6.1 Classification of sedimentary rocks.

a. Clastic sedimentary rocks (individual grains set in cement or matrix)

Grain size name	Grain size (mm)	How to tell grain size	Grain shape		Compositional modifiers		
			Rounded	Angular	Quartz clasts	Quartz + feldspar clasts	Clasts mixed with mud matrix
Gravel	>2	Measure	**Conglomerate**	**Breccia**	Quartzose	Arkosic	
Sand	$\frac{1}{16}$–2	Grains visible	**Sandstone**		Quartzose (orthoquartzite)	Arkosic (or just *arkose*)	(Quartzwacke, arkosic wacke graywacke)
Silt	$\frac{1}{256}$–$\frac{1}{16}$	Feels gritty on the teeth	**Siltstone**		Too fine-grained to determine mineral content		
Mud	$<\frac{1}{256}$	Feels smooth on the teeth	**Mudstone** (**Shale** if rock breaks along parallel planes)		Too fine-grained to determine mineral content		

b. Chemical sedimentary rocks (chemical precipitates, either coarse or fine-grained)

Mineral	Texture	
	Visible interlocking crystals	**No visible crystals**
Halite (NaCl)	**Rock salt**	
Gypsum ($CaSO_4 \cdot 2H_2O$)	Satin spar or **rock gypsum**	**Alabaster**
Calcite ($CaCO_3$)	**Crystalline limestone**	**Micrite (micritic limestone)***
Dolomite [$CaMg(CO_3)_2$]	**Crystalline dolostone**	**Dolomicrite***
Quartz (SiO_2)	Microcrystalline, conchoidal fractures, scratches glass: **chert** or **flint**	
Calcite ($CaCO_3$)	Rock contains concentrically layered calcite spheres: **oolitic limestone**	

*Micrite and dolomicrite may result from biochemical processes or contain microscopic fossil remains.

c. Biogenic sedimentary rocks (composed predominantly of organic remains)

Texture and other properties	Rock name	
Very-fine-grained mass of *siliceous* shells of microscopic organisms; does not react with hydrochloric acid	**Diatomite**	
Very-fine-grained mass of *calcareous* shells of microscopic organisms; reacts strongly with hydrochloric acid	**Chalk**	Limestone
Open framework of broken shell fragments cemented together; reacts strongly with hydrochloric acid	**Coquina**	Limestone
Fossils and fossil fragments in a calcareous matrix; reacts strongly with hydrochloric acid	**Fossiliferous limestone**	Limestone
Fossils and fossil fragments in a noncalcareous matrix; does not react with hydrochloric acid	**Fossiliferous mudstone, siltstone,** or **sandstone** (depending on size of matrix grains)	
Soft, loose interlocking framework of brown, partially decayed woody material	**Peat**	
Moderately hard, dark brown to black, decomposed material in which some woody fragments are still visible	**Lignite**	Coal
Hard, black mass of decomposed plant material in which woody fragments are not generally preserved; relatively low specific gravity	**Bituminous coal**	Coal

FIGURE 6.5 Flowchart for identifying sedimentary rocks.

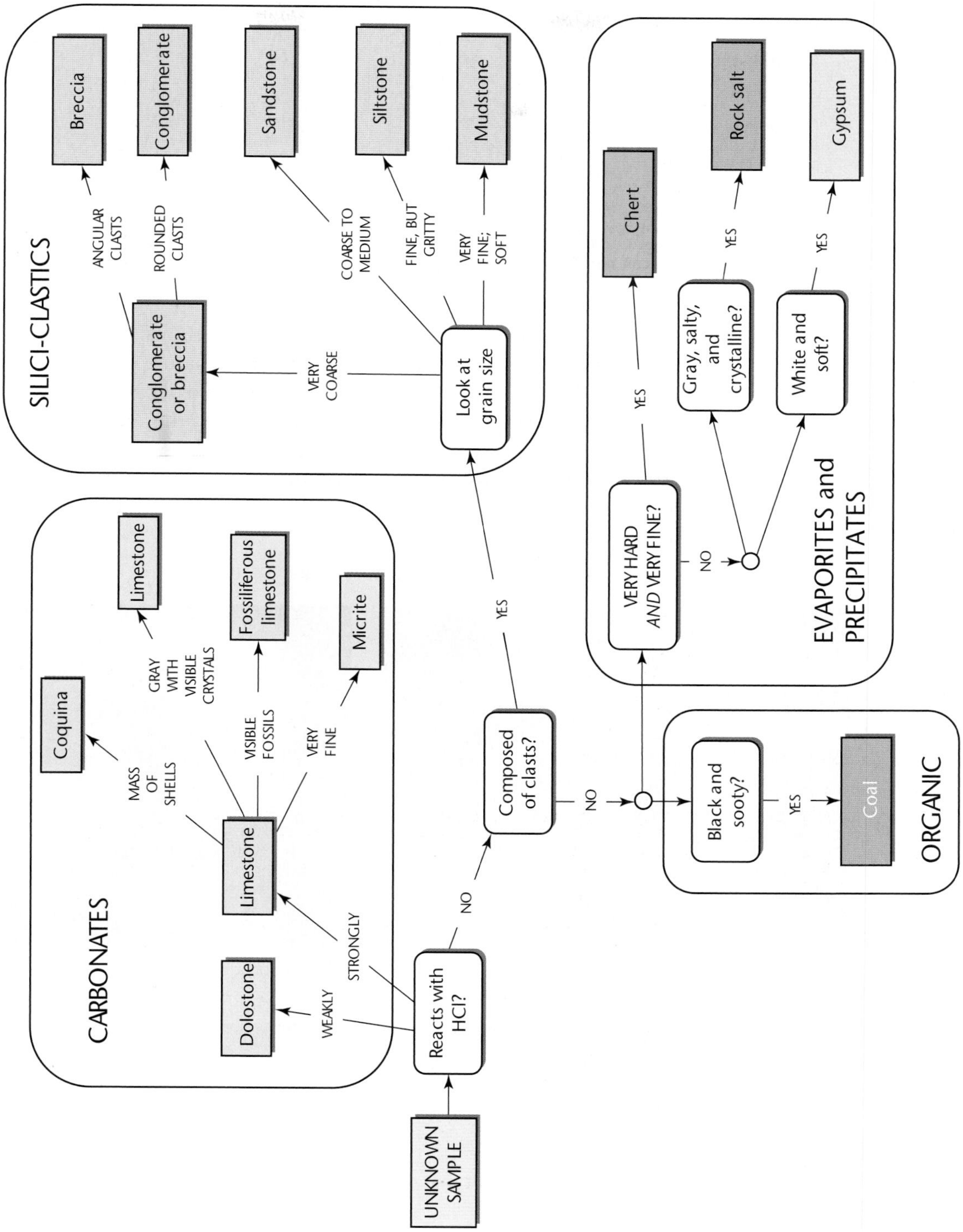

Match the agent of transport with the sediment size.

house-sized block	turbulent stream during a flood
boulder	very slow moving stream
cobble	ocean waves or desert winds
sand	glacial ice
silt	quiet water
mud	fast-moving stream

density, a faster moving material can move larger grains. Note that, in many cases, grain size alone may not completely constrain the nature of the transporting agent. For example, glaciers are not the only agent capable of carrying huge blocks—such blocks may also move in large debris flows in which they are buoyed by mud. Similarly, sand grains can be carried by ocean waves, streams, or strong winds.

The **sorting** of a clastic sedimentary rock is a measure of the uniformity of grain size (Fig. 6.6), and can also help to identify the transporting agent. For example,

FIGURE 6.6 Sorting in sedimentary rocks.

(a) Relatively good sorting.   **(b)** Poor sorting.

Look at the clasts in Figure 6.6. Suggest a transporting agent that could be responsible for each.

(a) Relatively good sorting:

_____

(b) Poor sorting:

_____

aeolian (wind-deposited) sediment is very well sorted because wind can only pick up clasts of a narrow size range. A fast-moving, turbulent stream can carry clasts ranging from mud- to boulder-sized. But as a stream slows, sediment of different sizes progressively settles out. The coarsest grains (boulders and cobbles) drop out first, then the mid-size grains (pebbles and sand), and only when the water becomes very slowly moving can silt and mud settle out. Thus, streams do sort clasts, but not as completely as the wind. Glacial ice is solid, so it can carry clasts of all sizes. As a result, glacially transported deposits are not sorted.

6.5.2 Grain Shape

The shape of clasts is a clue to the agent and distance of transportation. Grains that have not moved far from their source tend to have sharp edges and corners and are called **angular**. Angular clasts also occur in sediments deposited by glaciers, because these clasts are frozen into position and thus can't collide with one another. Clasts carried by water or wind collide frequently with one another as they move. As transport progresses, collisions knock off sharp corners and edges, eventually rounding the clasts. First the grains become **subrounded**, with smoothed edges and corners. Eventually, when they are almost spherical, they are called **rounded**. Thus, the farther streams and wind carry clasts, in general, the more spherical the grains become. The clasts also become smaller as a result of these collisions. Figure 6.7 shows different degrees of roundness in two sedimentary rocks.

FIGURE 6.7 Degrees of grain roundness in clastic sedimentary rocks.

(a) Rounded clasts.

(b) Angular clasts.

EXERCISE 6.9   **Recognizing the Difference between Breccia and Conglomerate**

Name: _____   Section: _____

Course: _____   Date: _____

Look back at the definition of breccia and conglomerate. Based on these definitions, which of these two rock types contains clasts that have been transported a longer distance?

_____

Name: _____   **Section:** _____
Course: _____   **Date:** _____

Fill in the following table to summarize the characteristics of sediment deposited by the different continental agents of erosion.

Textural feature	Agent of transportation		
	Streams	Wind	Glaciers
Grain size			
Sorting			
Grain shape			

6.5.3 Cements

In some cases, the cement that holds clasts together provides a clue to the depositional environment. For example, the formation of a hematite cement requires oxygen, so sedimentary rocks in which the cement contains hematite, and thus has a reddish color, come from environments where water contained dissolved oxygen. In contrast, sedimentary rocks containing pyrite in the cement formed in environments low in oxygen, because if oxygen had been present, the pyrite would have dissolved. Reddish sedimentary rocks, or *redbeds*, generally indicate deposition in terrestrial environments (e.g., rivers, alluvial fans).

Name: _____   **Section:** _____
Course: _____   **Date:** _____

How can you tell what minerals occur in a cement? Simply remember the basic physical properties of the minerals (e.g., hardness, the ability to react with acid, color). With this in mind, answer the following questions concerning sandstone, a type of sedimentary rock consisting of cemented-together grains of quartz sand.

(a) The cement (but not the grains) in a sedimentary rock reacts when in contact with dilute HCl. The cement consists of
_____.

(b) The rock, overall, has a reddish color. The cement contains _____.

(c) The cement is very strong, and when scratched with a steel needle it has the same hardness as the grains. The cement consists of _____.

6.6 Sedimentary Structures: Clues to Ancient Environments

6.6.1 Beds and Stratification

Gravity causes all sediment to settle to the floor of the basin in which it was deposited. Over time, layers of sediment (called **beds**) accumulate. Beds range from a millimeter to several meters thick, depending on the process involved. Each bed represents a single depositional event, and the different colors, grain sizes, and types of sediment from each event distinguish one bed from another (**Fig. 6.8**).

Many beds are fairly homogeneous, with uniform color, mineralogy, and texture, and have smooth surfaces. But some contain internal variations or have distinct features on their surfaces. These **sedimentary structures** provide important information about the rock's history. In the next section, we look at a few examples of sedimentary structures.

6.6.2 Sedimentary Structures

6.6.2a Graded Beds Graded beds are layers in which the clast size decreases progressively from the bottom to the top. These form when poorly sorted sediment avalanches down a slope in a body of water (**Fig. 6.9**). When the avalanche, called a *turbidity current*, slows, the coarsest grains settle first while the finer materials remain in suspension longer. With time, the finer sediments settle, with the finest ones settling last. To make a graded bed in the lab, put water, sand, silt, and gravel in a jar or graduated cylinder. Shake to mix thoroughly and watch as the grains settle.

6.6.2b Ripple Marks, Dunes, and Cross Beds Ripple marks are regularly spaced ridges on the surface of a bed. They form when sand grains are deposited by air or water currents. Small ripple marks are shown in **Figure 6.10**. Some ripple marks have systematically oriented steep and gentle sides (Fig. 6.10a). These **asymmetric ripple marks** are produced by a current that flowed from the gentle side toward the steep side. **Symmetric ripple marks** (Fig. 6.10b) have steep slopes on both sides and form from oscillating currents. Larger mounds of sediments that resemble asymmetric ripple marks in shape are called *dunes*. The largest dunes occur in deserts, where they build from wind-blown sand; however, dunes can also form under water.

FIGURE 6.8 Sedimentary rock beds.

(a) Horizontal beds of sandstone and siltstone in the Painted Desert of Arizona. White, dark red, and light red beds contain different amounts of hematite cement.

(b) Horizontal beds of sandstones, siltstones, and limestones in the Grand Canyon in Arizona. Beds are distinguished by differential resistance to weathering as well as color.

FIGURE 6.9 Graded bedding.

Top of bed
Silt and mud-sized clasts
Fine sand-sized clasts
Pebble- and granule-sized clasts
Bottom of bed

FIGURE 6.10 Ripple marks.

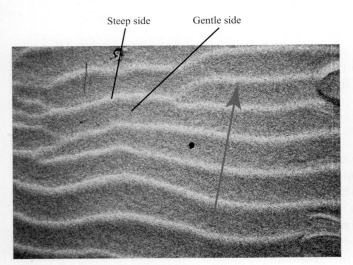

Steep side   Gentle side

(a) Asymmetric ripple marks in modern sand. The arrow indicates the direction of the current that formed the ripples.

(b) Symmetric ripple marks in ancient sandstone. Note the equally steep slopes on both sides of the ripples.

In some beds of sediment, subtle curving surfaces, delineated by coarser and/or denser grains, lie at an angle to the main bedding surfaces. These inclined surfaces are called **cross beds**. They form when sediment moves up the up-current side and then slips down the lee side as dunes and ripples build (**Fig. 6.11**).

6.6.2c Mud Cracks and Bedding-Plane Impressions Mud cracks are arrays of gashes in the surface of a bed formed when mud dries out and shrinks. Typically, mud cracks are arrayed in a honeycomb-like pattern. Sand deposited over the mud layer fills the cracks, so when the sediment lithifies, the shape of the cracks remains visible (**Fig. 6.12**).

Moving objects can leave impressions on unconsolidated sediment before the bed hardens into rock. The imprints of these objects can, in some instances, provide fascinating glimpses of animals living in the sediment or of processes acting on it. Imprints may be made by raindrops falling on a muddy floodplain (Fig. 6.12a), logs dragged by currents along the bottom of a stream, worms crawling along the sea floor in search of food (**Fig. 6.13a**), and the footprints of animals walking through a forest (**Fig. 6.13b**). Indentations also may be scoured into a bed surface by the turbulence of the fluid flowing over it.

FIGURE 6.11 Cross bedding.

Sand grains are deposited in inclined layers on the down-current side of a ripple. As beds accumulate, changes in wind or water direction are indicated by the slope of the cross-beds.

FIGURE 6.12 Formation of mudcracks.

(a) Mud cracks in modern sediment. The circular impressions were made by raindrops hitting the mud before it dried.

(b) Mud cracks preserved in a 180,000,000-year-old mudstone.

FIGURE 6.13 Impressions on bedding planes.

(a) Worm burrows on siltstone in Ireland (lens cap for scale).

(b) Footprints of different-sized dinosaurs.

6.6.3 Fossils: The Remnants of Past Life

Fossils are the remains or traces of animals and plants that are preserved in rocks—in effect, they are sedimentary structures. Generally, fossils reveal the shape of the hard parts (shells or skeleton) of an organism. Soft-bodied organisms are only preserved in special cases, where deposition happens before the organism decays (Fig. 6.14). Fossils are the subject of a whole sub-discipline of geology called paleontology, for

FIGURE 6.14 Fossils reveal remarkable details of life throughout geologic time.

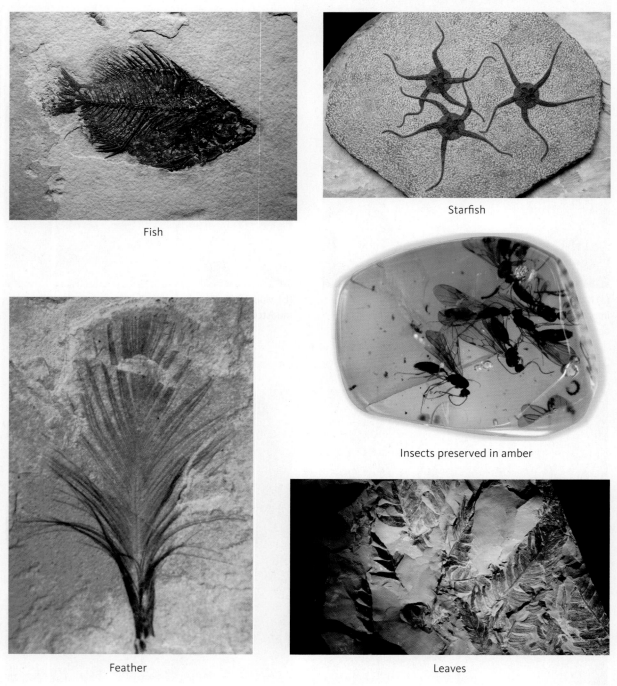

Fish

Starfish

Insects preserved in amber

Feather

Leaves

(a) Relatively fragile organisms.

FIGURE 6.14 Fossils reveal remarkable details of life throughout geologic time (con't).

Brachiopod

Brontosaurus

Tree trunks

(b) Relatively robust organisms.

the study of fossils reveals the history of life on Earth, and as discussed elsewhere in this book, provides clues to the age of strata. We mention them briefly here because most fossils are found in sedimentary rocks, as the environments in which sedimentary accumulation occurs are the same as those in which animals and plants lived.

Since different kinds of organisms live in different environments, the types of fossils in a sedimentary rock provide clues to the depositional environment in which the sediment comprising the rock accumulated. Applying the *principle of uniformitarianism,* we can deduce that sediments containing fossil corals probably formed in warm, clear seawater; sediments with fossils of tropical plants were probably deposited in equatorial latitudes; sediments with fossils of dinosaurs generally accumulated on land rather than in the deep ocean; and sediments containing fossil starfish accumulated in the oceans, not on land.

Name: _____   **Section:** _____

Course: _____   **Date:** _____

(a) Each sedimentary structure reveals something about the environment where it formed. What can you deduce about the geographic setting, climate, and depositional environment of the following?

   (i) Medium-grained sandstone containing fragments of tree trunks, branches, and leaves

_____

_____

_____

   (ii) Very-fine-grained limestone with fragments of coral

_____

_____

_____

   (iii) Hematite-cemented mudstone and sandstone containing dinosaur footprints, raindrop impressions, and mudcracks

_____

_____

_____

   (iv) Black shale containing small pyrite cubes

_____

_____

_____

   (v) Sandstone with oscillating ripple marks and fossil crabs

_____

_____

_____

(b) What can you deduce about the history of sedimentary rocks that contain the following fossils?

   (i) Fish

_____

_____

   (ii) Feather

_____

_____

   (iii) Starfish

_____

_____

continued

Name: _____   **Section:** _____

Course: _____   **Date:** _____

     (iv) Insect

_____

_____

     (v) Leaves

_____

_____

     (vi) Brachiopod

_____

_____

     (vii) Tree trunk

_____

_____

     (viii) Brontosaurus

_____

_____

6.7 Applying Your Knowledge to Stratigraphy

Ultimately, geologists use observational data about sedimentary rocks to interpret the depositional setting (conditions of deposition) in which the sediment forming the rock accumulated. Examples of depositional settings are listed in **Table 6.2**,

TABLE 6.2 **Examples of depositional environments.**

Glacial	Unsorted sediment containing angular clasts; commonly, such sediment may contain large clasts suspended in a much finer matrix
Fluvial (river)	Channels of sandstone or conglomerate, in some places surrounded by layers of mudstone; Hematite-containing cement
Beach	Well sorted, cross-bedded sandstone
Shallow marine	Fossiliferous limestone
Deep marine	Shale, chert, micrite, or chalk
Swamp	Coal
Alluvial fan	Conglomerate and arkose
Desert dune	Well-sorted sandstone, with large cross beds

Name: _____   Section: _____
Course: _____   Date: _____

(a) Look at the sedimentary rocks in **Figure 6.15**. Brief "field notes" are provided with each photograph. In each case, identify the rock and any sedimentary structures present, and interpret the depositional setting in which the rock formed.

FIGURE 6.15 Interpreting sedimentary outcrops.

(a) Entrance to Zion National Park in Utah. Large cliff face containing thick beds of medium-grained white, gray, and beige sandstones displaying excellent cross bedding.

(b) Excavation along a lumber road in eastern Maine. Very poorly sorted sediment with large clasts of granite, sandstone, and shale in a matrix of gravel, sand, and large amounts of clay.

(c) Sandstone with preserved dinosaur trackway and asymetric ripple marks. Note different orientation of ripple marks in the foreground.

continued

along with a simplified description of characteristics found there. Successions of beds, or strata, are like pages in a book that record the succession of depositional environments. Further study in geology will add additional detail to this image, but even with your basic knowledge, you can interpret outcrops in the field.

Name: _____   **Section:** _____

Course: _____   **Date:** _____

(b) Using *Google Earth*™, fly to the following coordinates. Describe what you see.

- Lat 24°44'40.42"S, Long 15°30'5.34"E

- Lat 23°27'03.08"N, Long 75°38'15.68"W

- Lat 67°3'38.31"N, Long 65°33'2.88"W

- Lat 25°7'48.94"N, Long 80°59'3.18"W

- Lat 30°10'2.73"N, Long 94°48'58.24"W

- Lat 5°24'54.48"N, Long 6°31'26.25"E

(c) What do regions underlain by sedimentary rocks look like? Fly to the following coordinates (Lat 47°48'02.31"N, Long 112°45'30.97"W) using *Google Earth*™ and see an examples of a region where sedimentary beds are well exposed. Note that the bedding appears as a series of parallel ridges if the layers are tilted. That's because softer beds (shale) weather faster than harder beds (sandstone).

Name _____

SEDIMENTARY ROCKS STUDY SHEET

Sample #	Texture (grain size, shape sorting)	Components			Minerals presents (approximate %)	Name of sedimentary rock	Rock history (transporting agent, environment, etc.)
		Clastic	Chemical	Biogenic			

SEDIMENTARY ROCKS STUDY SHEET

Sample #	Texture (grain size, shape sorting)	Components			Minerals presents (approximate %)	Name of sedimentary rock	Rock history (transporting agent, environment, etc.)
		Clastic	Chemical	Biogenic			

CHAPTER 7

INTERPRETING METAMORPHIC ROCKS

PURPOSE

- Become familiar with metamorphic rocks, textures, and mineral assemblages.
- Understand the processes that produce metamorphic rock.
- Determine the temperature and pressure conditions at which metamorphism occurs.

MATERIALS NEEDED

- A set of metamorphic rocks
- Magnifying glass or hand lens and, ideally, a microscope and thin sections of metamorphic rocks
- Standard supplies for identifying minerals (streak plate, glass plate, etc.)
- *PLAY-DOH*, small rods (plastic coffee stirrers cut into short pieces), pennies

7.1 Introduction

Biologists use the word *metamorphosis* (from the Greek *meta*, meaning change, and *morph*, meaning form) to describe what happens when a caterpillar turns into a butterfly. Geologists use the similar term, **metamorphism**, for the process of change that a rock undergoes when exposed to certain temperatures, pressures, and/or fluids that differ from those present when the rock first formed. The end product of metamorphism is a **metamorphic rock**. Note that we slipped the word *certain* into our definition of metamorphism. That's because not all changes that a rock can undergo are considered to be metamorphism. Specifically, metamorphism does *not* include weathering, diagenesis, or solidification from a melt; the changes that produce metamorphic rock take place at temperatures above those of diagenesis (i.e., at temperatures greater than about 250°C) and under conditions such that the rock remains solid (i.e., metamorphic rocks are not igneous). Because metamorphism takes place without either fragmentation or melting, we can say that metamorphism is a "solid-state" process.

Metamorphism can cause a great variety of changes in a rock. For example, during metamorphism, one or more of the following take place:

- The original minerals in a rock chemically react with one another to produce new minerals. In the terminology of a chemist, the products of the reaction (the **metamorphic minerals**) are "more stable" than the original ones under the conditions of metamorphism.

- The rock's texture may change as grains become larger or smaller and/or change shape.

- The rock may develop a **foliation**, or layering, due to the alignment of platy (pancake-like) or flattened minerals and/or because of the development of alternating bands of different composition. The rock may also develop a **lineation**, due to the alignment of elongate (needle-like or cigar-like) grains.

As a result of these changes, a metamorphic rock can be as different from its parent rock (its **protolith**) as a butterfly is from a caterpillar.

In this chapter, we begin by briefly reviewing the various causes of metamorphism. Then we examine the characteristic features of metamorphic rocks, the kinds of metamorphic rocks, and the information we can glean from a sample about the geologic conditions under which it was metamorphosed. Finally, we introduce the concept of metamorphic grade (the "intensity" of metamorphism), and describe how to interpret the grade of a specimen. Metamorphism is a very complex process, but we only have space to cover its basic aspects.

7.2 Agents of Metamorphism

Geologists refer to the heat, pressure, and fluids that cause metamorphism as **agents of metamorphism**. Let's consider how each of these may trigger change in a rock. Keep in mind, however, that even though we address each agent of metamorphism separately, all of them may act simultaneously in many geologic environments.

7.2.1 The Effect of Heat

A rock heats up when it either becomes buried to great depth, as can happen either during mountain building—when part of the crust is pushed up and over another part—or when magma intrudes nearby. As the temperature rises, atoms vibrate more rapidly and chemical bonds holding the atoms into the lattice of minerals begin to stretch and break. The freed atoms migrate slowly through the solid rock, a process called **diffusion**, much as atoms of ink spread when dropped into a glass of water. Eventually, the wandering atoms bond with other atoms to produce metamorphic minerals that are more stable under the higher temperatures. Put another way, heat drives chemical reactions that can produce a new assemblage of metamorphic minerals.

If the protolith has a very simple composition (e.g., quartz only, or calcite only), metamorphic reactions may produce new crystals of the same minerals. For example, metamorphism of a pure quartz sandstone (a sedimentary rock composed of SiO_2 grains) produces a metamorphic rock composed of new grains of quartz. But if the protolith contains a variety of elements, metamorphism can produce completely different minerals. For example, metamorphism of shale (composed of clay, which can contain Ca, Fe, Al, Mg, and K) produces a metamorphic rock composed of mica, quartz, and garnet. Diffusion is much slower in solid rock than in magmas or aqueous solutions, but over thousands to millions of years, it can completely change a rock. In some cases, new metamorphic minerals can grow to impressive sizes (**Fig. 7.1**).

FIGURE 7.1 Growth of metamorphic minerals.

(a) Garnet crystals.

(b) Andalusite crystals.

7.2.2 The Effect of Pressure

Two kinds of pressure are important in metamorphism (**Fig. 7.2**). **Lithostatic pressure** results from the burial of rocks within the Earth and, like hydrostatic pressure in the oceans or atmospheric pressure in air, it is equal in all directions. Lithostatic pressure can become so large at great depths beneath mountain belts, or in association with subduction, that it forces atoms closer together than the original mineral's structure can allow. As a consequence, a new mineral with a more compact atomic arrangement that is more stable under the high-pressure conditions can form. Such a change is called a **phase change**. Diamond, for example, is a metamorphic mineral formed at very high pressures from graphite.

When subjected to lithostatic pressure, the shape of mineral grains in a rock doesn't necessarily change, because the rock is being pushed *equally* from all directions (Fig. 7.2a). During mountain-building processes and along plate boundaries, bodies of rock undergo more intense squeezing in one direction than in others. To picture this vise-like squeezing, imagine what happens when you press a ball of clay or dough between a book and a table—it flattens into a pancake. In addition, the crust may move, producing shear, during which one part of a rock body moves sideways relative to another—to picture shear, imagine what happens when you smear a deck of cards across a table. Simplistically, we can refer to the type of pressure in such rock due to vise-like squeezing and/or shear as directed pressure, or **differential stress**. Under conditions of elevated temperature, differential stress can cause grains to change shape, flattening them in one direction and elongating or stretching in others (Fig. 7.2b).

FIGURE 7.2 Effects of lithostatic pressure and differential stress.

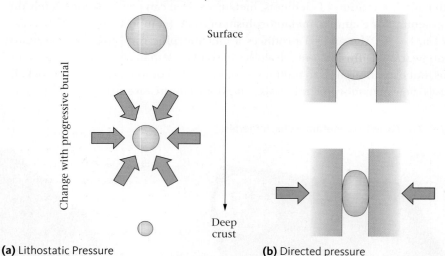

(a) Lithostatic Pressure
Pressure acts equally from all directions; Grain remains spherical but gets smaller.

(b) Directed pressure
Greater pressure in one direction than others; Spherical grain is flattened.

7.2.3 The Effect of Temperature and Pressure Combined

It's important to keep in mind that pressure changes in most metamorphic environments are accompanied by temperature changes, because as the depth of burial increases, *both* temperature and pressure rise. But the temperature change that occurs with increasing depth at one location is not necessarily the same as that which occurs at another. For example, at a given depth, the temperature near an intruding pluton of magma may be higher than in a region at a great distance from the magma. The particular minerals that form during metamorphism depend on both pressure and temperature. We can represent this fact on a graph, called a **phase diagram,** that has pressure on the vertical axis and temperature on the horizontal axis. **Figure 7.3** shows the phase diagram for the aluminum silicate minerals—three different minerals that exist in nature but have the same chemical formula. Note that, at Points Y and Z, the pressure is the same but the temperature is different—at a lower temperature, kyanite forms, whereas at a higher temperature, sillimanite forms. Point X is at the same

FIGURE 7.3 Phase diagram for aluminum silicate (Al_2SiO_5).

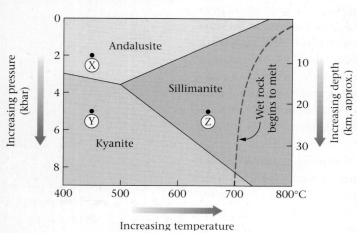

temperature as Y, but at a lower pressure; under these conditions, andalusite forms at low pressure, while kyanite forms at high pressure.

7.2.4 The Effect of Hydrothermal Fluids

In many environments, rocks contain high-temperature liquids, gases, and "super-critical fluids" (fluids that have characteristics of both a gas and liquid) composed of H_2O and/or CO_2 in which ions of other elements have been dissolved. Collectively, these fluids are called **hydrothermal fluids** (from *hydro,* meaning water, and *thermal,* meaning hot). They may be released from magma when the magma solidifies, may be produced by metamorphic reactions, or may be formed where groundwater has percolated deep into the crust. Hydrothermal fluids can be "chemically active" in that they can dissolve ions and transport them throughout the crust, and can provide ions that bond to other ions and produce new, metamorphic minerals. Thus, reactions with hydrothermal fluids tend to change the chemical composition of rocks—the fluids can be thought of as buses that pick up and drop off ions as they pass through a rock. In some cases, the fluids pick up certain chemicals and flush them entirely out of the rock. When the chemical composition of a metamorphic rock changes significantly due to reaction with hydrothermal fluids, we say that the rock has undergone **metasomatism.**

EXERCISE 7.1 **Effects of Metamorphic Agents**

Name: _____ Section: _____
Course: _____ Date: _____

(a) Andalusite, sillimanite, and kyanite are polymorphs of Al_2SiO_5 with identical chemical compositions but different crystal structures. Based on the phase diagram in Figure 7.3, and keeping in mind how pressure tends to affect the compactness of crystals, which of these minerals has the lowest specific gravity? Explain.

_____
_____
_____
_____

(b) During metamorphism at a given location, the overall amount of the element calcium in a metamorphic rock changes substantially. Which metamorphic agent is likely to have caused this metamorphism? Why?

_____
_____
_____
_____

(c) Would you expect to find aligned, flattened grains in a rock that has undergone heating but has not been subjected to differential stress during metamorphism? Why?

_____
_____
_____
_____

7.2.5 Environments of Metamorphism

Where does metamorphism occur? In a general sense, metamorphism takes place anywhere that a rock, formed under one set of environmental conditions, becomes subjected to agents of metamorphism. Geologists recognize the following four major types of metamorphism.

• **Contact metamorphism** (also called thermal metamorphism) occurs where rocks are subjected to elevated heat without a change in pressure and without the application of differential stress. This happens where an igneous intrusion comes in contact with a rock.

• **Regional metamorphism** (also called dynamothermal metamorphism) occurs where rocks in a large region of crust are subjected to increases in temperature and pressure, *and* are subjected to differential stress (squeezing and shearing). Such metamorphism typically happens during mountain-building processes.

• **Burial metamorphism** happens when rocks are simply buried very deeply by overlying sediment.

• **Dynamic metamorphism** occurs where rock undergoes differential stress in response to shear along a fault zone but does not undergo a change in temperature and/or lithostatic pressure. Hydrothermal fluids may be present in all metamorphic settings.

Table 7.1 summarizes the types of metamorphism and notes how the agents of metamorphism present in each tend to operate.

TABLE 7.1 Types of metamorphism.

Agent(s) of metamorphism	How the agents of metamorphism are applied	Type of metamorphism
Heat	Heat from cooling magma affects rocks adjacent to an intrusive body or lava flow.	Contact (thermal)
Lithostatic pressure and some heat	Gravity causes increased lithostatic pressure as rocks are buried deeper in the Earth; because temperature increases with depth the buried rock is also heated.	Burial
Differential stress and some heat	Two large blocks of rock grind past one another in a fault zone. The moving blocks generate the directed pressure; heat comes from depth in the Earth, and a very minor amount from friction.	Dynamic
Heat, lithostatic pressure, differential stress, and hydrothermal fluids	Lithosphere plates collide. Subduction or continent-continent collision creates intense differential stress; heat and lithostatic pressure come from depth within the Earth.	Regional (dynamothermal)
Chemically active hydrothermal fluids	Superheated water released from cooling magma or from metamorphic reactions carries dissolved ions into surrounding rock, and removes ions from rock, changing the rock's composition.	Metasomatism
Heat, differential stress	When a meteorite hits Earth, its enormous kinetic energy is converted instantaneously to heat and differential stress.	Impact*

*Because impact metamorphism happens so rarely on Earth, it is not discussed further in this chapter.

7.3 What Can We Learn from a Metamorphic Rock?

What do we want to discover when we study a metamorphic rock? First, we may want to determine the nature of the protolith. (For example, did the metamorphic rock we see today originate as a granite, sandstone, basalt, or shale?) Second, we may want to determine the agents of metamorphism responsible for causing the change. (For example, did metamorphism result from a change in temperature, pressure, or both? Was the rock subjected to differential stress during deformation, or not?) Third, we might want to determine the "intensity" of metamorphic conditions. In this context, the informal term *intensity* refers to the highest temperature and pressure at which metamorphism took place, and/or the overall amount of shearing or squeezing during deformation. Finally, we might want to define the tectonic setting in which metamorphism occurred. (For example, did metamorphism take place in a collisional mountain belt, along a fault zone, at the base of an accretionary prism, or next to a hot pluton?) The clues we need to answer all these questions come from identifying the metamorphic minerals in the rock, and from describing the rock's texture.

7.3.1 Identifying the Protolith: Compositional Classes

By looking at a metamorphic rock, can we say anything about the protolith from which it was derived? The answer is yes. We can often get a sense of the protolith by analyzing the *composition* of the metamorphic rock. Geologists distinguish among four main classes of metamorphic rock, based on the minerals the rock contains.

- **Aluminous** (or **pelitic**) **metamorphic rocks** contain a relatively large proportion of aluminum, silicon, and oxygen, with lesser amounts of potassium, iron, and magnesium. Such rocks were derived from shale or mudstone, which contain clay, an aluminum-rich mineral. Common metamorphic minerals of aluminous rocks include muscovite, biotite, chlorite, garnet, andalusite, kyanite, and sillimanite.

- **Quartzo-feldspathic rocks** consist mostly of silicon, oxygen, aluminum, potassium, and sodium, with lesser amounts of iron, magnesium, and calcium. Possible protoliths include felsic igneous rocks like granite or rhyolite, or sedimentary rocks like arkose. Typical metamorphic minerals in these rocks are quartz, potassic feldspar, and sodic plagioclase.

- **Calcareous metamorphic rocks** are composed mostly of calcium, carbon, and oxygen, and are derived from a protolith of limestone, which consists of calcite. During metamorphism, the calcite recrystallizes but remains calcite. If the protolith also contained clay and quartz, a variety of complex "calc-silicate" minerals also form.

- **Mafic metamorphic rocks** contain calcium, iron, magnesium, aluminum, silicon, and oxygen. As their name suggests, their protoliths were basalts and gabbros. Typical mafic metamorphic minerals include epidote, amphiboles, pyroxenes, and garnet.

The assemblage of metamorphic minerals found in a metamorphic rock depends on the minerals present in the protolith, because the metamorphic minerals are formed out of the chemical "ingredients" that occur in the protolith. In effect, during metamorphism, the atoms that comprise minerals in the protolith are rearranged into new crystal structures.

Name: _____   Section: _____
Course: _____   Date: _____

(a) Identify the minerals in the four metamorphic rocks provided by your instructor and record the data in the following table. Add the chemical compositions for these minerals from Table 3.1.

		Specimen #1	Specimen #2	Specimen #3	Specimen #4
Major minerals and their composition					

(b) Based on the minerals and their chemical compositions, what were the most abundant elements in the protoliths of these specimens?

Specimen #1	Specimen #2	Specimen #3	Specimen #4

(c) Is it likely that these metamorphic rocks came from the same parent rock? Explain.

_____

_____

_____

_____

Though we can learn a lot about a metamorphic rock's protolith from its mineral composition, analysis of chemistry alone cannot always constrain the identity of the protolith completely. Other clues come from examining the context in which the rock occurs. Geologists refer to the various features seen in the outcrop from which a rock sample was collected, and perhaps in the region around the outcrop, as the **field relations** of the sample. To get a sense of what you can learn from field relations, work through Exercise 7.4.

EXERCISE 7.3   **Identifying the Parent Material of a Metamorphic Rock**

Name: _____   Section: _____
Course: _____   Date: _____

(a) Compare the mineral content and most abundant ions from the four rocks in Exercise 7.2 with the information about the four chemical groups of metamorphic rock. What protoliths produced the four specimens you studied?

Specimen #1	Specimen #2	Specimen #3	Specimen #4

(b) How would your conclusion be affected if extensive metasomatism had occurred?

_____

_____

_____

_____

EXERCISE 7.4   **Using Field Relations to Constrain the Protolith**

Name: _____   Section: _____
Course: _____   Date: _____

A quartzo-feldspathic sample of metamorphic rock occurs as a distinct layer within a set of layers that include calcareous rocks, pure quartz rocks, and pelitic rocks. Is it more likely that the sample was derived from a granite protolith or an arkose protolith? Explain.

_____

_____

_____

_____

7.3.2 Interpreting Metamorphic Texture

During the process of metamorphism, the texture of a rock changes as the atoms are rearranged. In some cases, this process changes the size and/or shape of mineral grains without changing their composition. For example, during thermal or dynamo-thermal metamorphism of a quartz sandstone, the boundaries between grains and cement gradually disappears, and new, interlocking crystals of quartz grow. During dynamic metamorphism of quartz sandstone, spherical grains may flatten and stretch and then subdivide into very tiny grains. As we have mentioned, in rocks that start with minerals containing a great variety of different ions, totally different minerals such as garnet, amphibole, and mica may grow. If metamorphism occurs at higher temperatures, the crystals may grow larger, forming coarser-grained rock.

The orientation of the crystals in a metamorphic rock depends on the nature of the pressure acting on the rock. If the pressure is lithostatic, crystals grow in random orientations, so elongate (cigar-like) and/or platy (pancake-like) grains have no particular alignment. If, however, there is differential stress during metamorphism, two phenomena can happen. First, preexisting grains may be reoriented so that they align parallel with one another. (Exercise 7.5 demonstrates this process.) Second, new elongate or platy grains can grow in directions parallel to each other. Both phenomena result in **preferred orientation**, the alignment or parallism of grains.

EXERCISE 7.5   **Visualizing Preferred Orientation Due to Reorientation**

Name: _____   Section: _____

Course: _____   Date: _____

We can gain insight into how preferred orientation develops with a simple laboratory experiment.

(a) Randomly insert pennies (flat grains) and small sticks (elongate grains) into a ball of of PlayDoh. Flatten the dough into a pancake by pressing on it with a book (i.e., apply differential stress). What happens to the alignment of the objects? Are the sticks parallel to each other in the plane of the pancake?

_____

_____

(b) Clean off the pennies and sticks, and reinsert them into another ball of PlayDoh. This time, roll the dough into a cigar-shape. What happens to the alignment of the objects now? Are the sticks parallel to each other? How are they oriented relative to the length of the cigar?

_____

_____

Geologists distinguish between two kinds of preferred orientation based on analogies to familiar geometric shapes. If the fabric in a rock can be depicted as a stack of parallel planar horizons within the rock, we say that the rock has **metamorphic foliation**. Foliation can be due to the alignment (parallelism) of platy minerals (**Fig 7.4a**), or to the existence of alternating bands of different minerals (**Fig 7.4b**).

EXERCISE 7.6   **Thinking about the Geologic Significance of Foliation**

Name: _____   Section: _____

Course: _____   Date: _____

Refer back to Table 7.1.

(a) In what metamorphic agents would you expect foliation to develop?

_____

_____

(b) In what geologic setting would you expect foliation to be absent?

_____

_____

If, however, the fabric in a rock can be depicted by aligned rods or needles, we say that that the rock has a **mineral lineation**. Mineral lineations can reflect alignment of elongate crystals or grains (**Fig. 7.4c**). The occurrence of a lineation commonly means that the rock has been stretched or sheared in a direction parallel to the lineation.

FIGURE 7.4 Mineral alignment: foliation and lineation.

(a) Foliation of sheet-like grains (micas) in an aluminous schist. Muscovite flakes are aligned parallel to one another like pages in a book. As a result, they reflect light at the same time, producing a sheen.

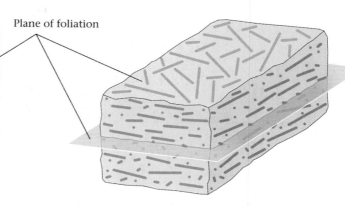

Plane of foliation

(b) Foliation can also exist if the long axes of rod-shaped minerals (e.g., amphibole) lie in the same plane, even if the rods are not parallel to one another.

(c) With lineation, rod-shaped minerals are aligned parallel to one another, producing a streaky appearance on some surfaces and a dotted pattern on others.

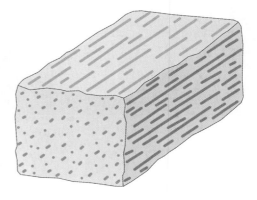

7.3.3 Granoblastic Texture

Metamorphic rocks with random grain orientations are said to have a **granoblastic texture** (**Fig. 7.5**). Granoblastic rocks lack foliation or lineation because *either* they were not affected by differential stress *or* they don't contain platy or rod-shaped minerals that could be aligned. A granoblastic texture in which the rock contains randomly oriented platy or rod-like minerals must have formed during contact or burial metamorphism, environments in which differential stress did not occur. If a rock with granoblastic texture contains only equant grains, it *might* or *might not* have experienced differential stress during metamorphism.

FIGURE 7.5 Granoblastic texture (random mineral alignment).

(a) Interlocking calcite, garnet (brown), and pyroxene (dark green) in coarse-grained marble.

(b) Photomicrograph of interlocking calcite grains in a marble.

(c) Interlocking quartz grains in quartzite.

(d) Photomicrograph of quartz grains fused together in quartzite.

7.3.4 Textures Produced by Dynamic Metamorphism

At temperatures high enough that rocks do not fracture and break up when sheared, dynamic metamorphism can transform a coarse-grained rock into a very-fine-grained rock with a strong foliation and lineation. The product is called **mylonite**. Such transformation occurs, simplistically, because the stress causes large crystals of quartz to subdivide into very small grains that fit together like a mass of soap bubbles. **Figure 7.6** shows how progressively more intense dynamic metamorphism affected a coarse-grained granite from the Norumbega fault system in east-central, Maine (Fig. 7.6a).

The process of transforming a rock into a mylonite takes place gradually, so we may see intermediate stages preserved in outcrops. [In fact, geologists sometimes distinguish between *protomylonite*, rock that has just started the transformation

Name: _____   **Section:** _____
Course: _____   **Date:** _____

(a) Summarize the relationship between mineral alignment and type of metamorphism by checking the appropriate box(es) in each row in the following table.

Texture	Type of metamorphism				
	Contact	Regional	Dynamic	Burial	Can't tell from texture alone
Foliation					
Lineation					
Granoblastic with micas					
Granoblastic without micas					
Layered but not foliated					
Layered and foliated					

(b) Examine the metamorphic rocks in your study set. Which are foliated? Lineated? Granoblastic? Write your answers in the appropriate column of the study sheets at the end of the chapter. Suggest, wherever possible, the type of metamorphism that produced each sample.

(Fig. 7.6b); *mylonite,* in which the transformation has affected most of the rock (Fig. 7.6c); and *ultramylonite,* in which the transformation is complete (Fig 7.6d).] Blocky feldspar grains tend to be aligned during the process, and grow tails of fine-grained mica, whereas some quartz grains smear into very thin ribbons (Fig. 7.6b). In some cases, relatively large feldspar grains remain after the surrounding rock has changed into a mass of very fine mica and quartz; these leftover grains of feldspar are called **porphyroclasts** (Fig. 7.6c).

7.3.5 Determining Metamorphic Grade

Metamorphic grade describes the approximate degree to which a rock has changed during dynamothermal or thermal metamorphism—the term doesn't really apply to dynamically metamorphosed rocks. In effect, the *grade* of a metamorphic rock reflects the *intensity* of metamorphism in cases where metamorphism involves heat. A *low-grade* rock has changed little, preserving much of the character of its parent rock; *moderate-grade* rock retains fewer aspects of its parent rock; and *high-grade* rock has changed so much that no traces of its parent rock's texture or minerals remain. As an example, consider a rock that starts out as a sequence of alternating layers of sandstone and shale. At low grade, the sandstone may have recrystallized and the shale may have developed a subtle foliation, but the bedding is still visible, and minerals found in the original rock still remain. At high grade, the rock may consist of a totally different assemblage of minerals with a very strong foliation and no hint of the original bedding. Grade depends on the magnitude of temperature and pressure. You can estimate the grade of a hand sample by examining the minerals it contains, and by looking at its texture.

FIGURE 7.6 Effects of dynamic metamorphism on coarse-grained granite.

(a) Undeformed granite showing original coarse-grained igneous texture.

Ribbon Tail

(b) Protomylonite: potassic feldspars (pink grains) are stretched and aligned parallel to the white line; quartz grains (gray) have coalesced to form continuous ribbons.

(c) Mylonite: grain size of the original granite has been reduced from a dark, very-fine-grained matrix with remnant feldspars smeared into ovoid grains.

(d) Ultramylonite: the intensely metamorphosed granite now consists entirely of fine grains and has strong foliation and lineation.

7.3.5a Textural Evidence for Metamorphic Grade. In a general sense, the longer a rock remains at its peak metamorphic temperature, the greater the amount of solid-state diffusion that occurs and the larger the metamorphic mineral grains that grow. Thus, grain size generally increases with increasing metamorphic grade. You can see this contrast in texture by comparing the fine-grained protoliths of **Figure 7.7** with their high-grade metamorphic equivalents (Fig. 7.5 a,b). In some cases, certain metamorphic minerals grow much larger than others during metamorphism. The particularly large grains are called **porphyroblasts** (**Fig. 7.8**); note that porphyro*blasts* grow during metamorphism, whereas porphyro*clasts* are large grains that resisted change during metamorphism. In Figure 7.8, large, nearly perfectly shaped red garnet crystals grew during metamorphism of a mafic igneous rock along with smaller grains of plagioclase feldspar (light-colored) and amphibole (dark green).

FIGURE 7.7 Fine-grained parent rock of a common metamorphic rock.

(a) Hand specimen of micritic limestone (compare with metamorphic equivalent in Fig. 7.5a).

(b) Photomicrograph of micritic limestone (compare with metamorphic equivalent in Fig. 7.5b).

(c) Hand specimen of a marble.

(d) Photomicrograph of a marble.

FIGURE 7.8 Porphyroblastic texture (red garnet porphyroblasts in a metamorphosed mafic igneous rock).

At high metamorphic grades, rock may develop compositional banding, defined by the alternation of light-colored and dark-colored layers. This compositional layering is a kind of foliation called **gneissic texture** (Fig. 7.9). If the rock contains mica, the mica flakes generally align with the foliation. But at very high grades, metamorphism may transform micas into other minerals, so that layers internally have a granoblastic texture.

FIGURE 7.9 Gneissic texture.

(a) Quartzo-feldspathic gneiss: light layers: quartz, K- and Na-feldspars; Dark layers: biotite.

(b) Calc-silicate gneiss: light layers: Wollastonite ($CaSiO_3$); dark layers: Ca-Al garnet, pyroxene.

(c) Mafic gneiss: light layers: Ca-Na plagioclase feldspar; dark layers: pyroxenes.

7.3.5b Mineral Evidence for Metamorphic Grade A progression from low-grade to high-grade metamorphic mineral assemblages corresponds primarily to an increase in temperature. Some minerals (e.g., chlorite) are stable only at relatively low grade (lower temperature), others (e.g., biotite) only at medium grade (moderate temperature), and still others (e.g., sillimanite) only at high grade (high temperature). Mineral indicators of metamorphic grade are called **metamorphic index minerals** (Table 7.2). Note that minerals such as quartz and feldspar, which are stable over a wide range of metamorphic conditions, are *not* helpful in determining metamorphic grade.

TABLE 7.2 Mineral assemblages as indicators of metamorphic grade.

Parent rock type	Increasing metamorphic grade ⟶				
	Low grade		**Medium grade**		**High grade**
Quartzo-feldspathic (quartz, potassic feldspar, plagioclase, ± micas)	Impossible to determine based on mineralogy alone				
Aluminous (muscovite, quartz, chlorite, garnet, staurolite, etc., ± quartz, feldspars)	Muscovite Chlorite	Muscovite Biotite	Muscovite Biotite Fe-Mg garnet Staurolite	Muscovite Biotite Fe-Mg garnet Kyanite or sillimanite	Sillimanite K-feldspar Fe-Mg garnet
Calcareous (calcite, dolomite, ± calc-silicate minerals such as those listed to the right)	Calcite Talc Olivine	Calcite Tremolite	Calcite Diopside Ca-Al garnet		Diopside Wollastonite Ca-Al garnet
Mafic (plagioclase + ferromagnesian minerals such as those listed to the right)	Chlorite Na-plagioclase Epidote	Actinolite Na-Ca plagioclase Biotite	Hornblende Ca-Na plagioclase Fe-Mg garnet		Pyroxene Ca-plagioclase Fe-Mg garnet

Each compositional class of metamorphic rock has its own set of index minerals due to its unique composition. Some minerals, such as chlorite, biotite, and garnet, can occur in more than one compositional class because the appropriate ions can be present in more than one kind of protolith. The complete *assemblage* of minerals in a rock, however, makes it clear what the parent rock type was. For example, biotite can be found in low- to moderate-grade aluminous and mafic metamorphic rocks. If biotite occurs in an assemblage along with muscovite and quartz, the protolith must have been aluminous, but if the assemblage includes plagioclase and actinolite, the protolith must have been mafic.

EXERCISE 7.8   **Estimating Metamorphic Grade**

Name: _____   Section: _____
Course: _____   Date: _____

Based on the textural and mineralogic criteria outlined above, determine whether each metamorphic rock in your study set represents low-, medium-, or high-grade metamorphism. Record your decision in the appropriate column of the study sheets at the end of the chapter.

7.3.5c Variations in Metamorphic Grade The pattern of variation in metamorphic grade depends on the type of metamorphism (**Fig. 7.10**) and can be predicted by using a little common sense. For example, metamorphic grade in wall rock due to thermal metamorphism is greatest at the contact between a pluton and its wall rock and decreases outward from the pluton (Fig. 7.10a).

FIGURE 7.10 Geographic distribution of metamorphic intensity.

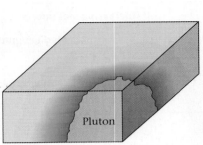

(a) Contact metamorphism: Intensity decreases outward from the intrusive pluton or lava flow.

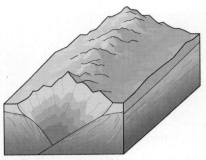

(b) Regional metamorphism: Intensity increases with depth and closeness to suture between colliding plates.

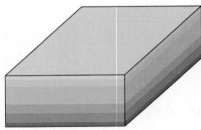

(c) Burial metamorphism: Intensity increases with depth.

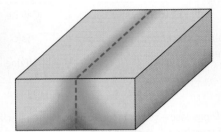

(d) Dynamic metamorphism: Intensity decreases outward from the fault and increases with depth.

Regional metamorphism occurs in a collisional mountain belt. In such settings, the crust squeezes together horizontally and thickens vertically. Rocks that were originally near the surface of one continental margin may end up at great depth when thrust beneath the edge of the other continent. The rocks that are carried to the greatest depth not only reach the highest temperature and the highest grade, but they also tend to be squeezed and sheared the most (Fig. 7.10b). Because burial metamorphism is due principally to lithostatic pressure associated with the thickness of overlying rocks, and to temperature due to the geothermal gradient, the burial metamorphic grade increases with depth (Fig. 7.10c).

EXERCISE 7.9   **Interpreting Metamorphic History**

Name: _____   Section: _____

Course: _____   Date: _____

The map below (**Fig. 7.11**) shows the distribution of rocks of different metamorphic grades. The local geologist has mapped a concentric pattern in which the highest metamorphic grade is in the center of the area, surrounded by rocks of progressively lower grade. All metamorphic rocks have a granoblastic texture, with the coarsest grains found in the central region. Although the granoblastic texture suggests contact metamorphism, the geologist has found no igneous rock anywhere in the region. Suggest an explanation for this metamorphic pattern.

_____

_____

_____

_____

continued

Name: _____   Section: _____
Course: _____   Date: _____

FIGURE 7.11 Mapped distribution of rocks with varying metamorphic grades.

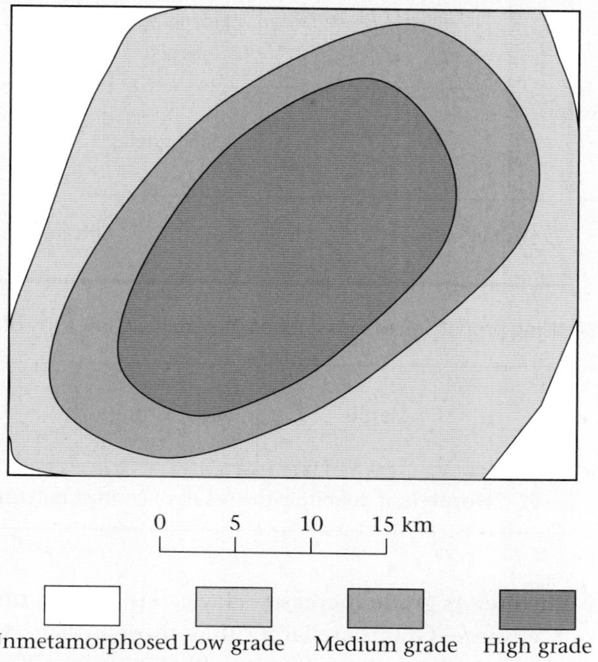

0   5   10   15 km

Unmetamorphosed   Low grade   Medium grade   High grade

7.4 Metamorphic Rock Classification and Identification

Metamorphic rocks are classified based on a their texture and composition. Table 7.3 presents a simplified classification scheme for common metamorphic rocks based on a combination of mineralogy and texture. The columns correspond to the five major compositional classes, and the rows to textural variations. Different names are given to some textural varieties depending on whether they result from low-, moderate-, or high-grade metamorphism. For example, slate, phyllite, and schist are all foliated aluminous rocks, but they represent different metamorphic grades.

To name a metamorphic rock, begin by identifying its minerals. This determines the compositional type (refer to Table 7.2) and places the rock in the correct column in Table 7.3. Move downward through the table until you come to the row with the appropriate texture. In some cases, there will be only one choice; in others, there may be a few possibilities. The following brief descriptions should help in either case. Note that the progression from slate, through phyllite, to schist is a progression from low-grade to high-grade metamorphism of shale.

Slate: A low-grade metamorphosed shale or mudstone composed of strongly aligned flakes of clay. The foliation is revealed by slate's tendency to split into thin plates. Typically, slate is gray, black, red, or green.

TABLE 7.3 Simplified classification scheme for metamorphic rocks.

Texture	Composition				
	Aluminous	**Calcareous**	**Mafic**	**Quartzo-feldspathic**	**Organic**
Foliated due to alignment of platy minerals	Slate Phyllite Schist	Calcareous schist	Greenschist Hornblende Schist		
Foliated due to compositional layering	Gneiss	Calc-silicate gneiss	Mafic gneiss	Gneiss	
Foliated, lineated, and extremely fine-grained due to dynamic metamorphism	Protomylonite Mylonite Ultramylonite → Increasing proportion of fine-grained material and intensity of foliation				
Nonfoliated	Granofels	Marble	Greenstone	Quartzite	Anthracite coal
Hornfels, if definitely the result of contact metamorphism					

(Between the first two rows, a downward arrow labeled "Increasing grade and grain size")

Phyllite: As grade increases, clay is replaced by tiny flakes of muscovite and/or chlorite. Grains are larger than those in slate but may still be too small to be seen without magnification. Phyllite has a silky sheen caused by reflection of light from the aligned platy minerals.

Schist: This rock is medium- to high-grade rock with a strong foliation defined by large mica flakes that can be identified with the naked eye. Porphyroblasts of other metamorphic minerals (e.g., garnet) may occur in schist. We distinguish among different types of schist based on composition. Aluminous schist contains mica and chlorite; mafic schist contains chlorite, talc, and hornblende; and calcareous schist contains Ca-mica and Ca-amphiboles.

Gneiss: A high-grade metamorphic rock of any composition in which minerals have segregated into light and dark layers. The foliation is defined by compositional banding.

Mylonite: Dynamically metamorphosed rock in which very tiny grains formed. The foliation is defined by flattened and stretched grains. (A *protomylonite* is only slightly sheared, so less then 50% of the rock is very fine-grained and remnant grains from the protolith may be present. An *ultramylonite* is intensely sheared, so more than 90% is extremely fine-grained.)

Granofels: Any metamorphic rock in which mineral grains are randomly oriented, i.e. have a granoblastic texture. The term **hornfels** may be applied if it can be proved that the rock formed by contact matamorphism.

Greenschist: Low-grade mafic metamorphic rock composed of chlorite and plagioclase feldspar with epidote and/or actinolite. Foliation is defined by alignment of chlorite and actinolite.

Greenstone: A nonfoliated (granoblastic) equivalent of greenschist.

Hornblende schist: A higher-grade version of greenschist in which foliated and often aligned hornblende crystals and plagioclase replace the lower-grade chlorite-epidote-actinolite assemblage. This rock is more commonly called *amphibolite.* Of note, some amphibolites have no lineation.

Quartzite: A monomineralic, granoblastic, metamorphic rock composed of quartz.

Marble: A calcareous metamorphic rock composed mostly of calcite or dolomite. Marble may also contain some of the calc-silicate index minerals. Marble is commonly granoblastic.

Anthracite coal: A black, shiny, hard rock made of carbon that has no visible mineral grains and has relatively low specific gravity. It may exhibit conchoidal fracture.

EXERCISE 7.10   **Interpreting Metamorphic Rock History**

Name: _____   **Section:** _____
Course: _____   **Date:** _____

Identify the metamorphic rocks in your study set and record their names on the study sheets at the end of the chapter. Complete your observations and conclusions about metamorphic history in the metamorphic grade column.

7.5 Applying Your Knowledge of Metamorphic Rocks to Geologic Problems

With your new skills, you can interpret the history of metamorphic rocks from textural and mineralogic data and information from the field. Now you're ready for a little detective work and geologic reasoning.

EXERCISE 7.11   **A Metamorphic Enigma**

Name: _____   **Section:** _____
Course: _____   **Date:** _____

The map below (**Fig. 7.12**) shows the distribution of aluminous metamorphic rocks with different textures and mineral assemblages. The field geologist who drew the map thinks all of the rocks were once the same shale that have experienced different metamorphic conditions and/or processes.

    Unfortunately, the geologist has worked with sedimentary rocks for the last ten years and doesn't remember much about metamorphism or metamorphic rocks. You've been brought in to help. A particularly puzzling feature of the map area is the fact that although there appears to be a concentric pattern based on metamorphic grade, only the outermost part of the area contains foliated rocks.

continued

EXERCISE 7.11 continued

Name: ______________________ **Section:** ____________
Course: ______________________ **Date:** ____________

FIGURE 7.12 Geologic map showing distribution of aluminous metamorphic rocks.

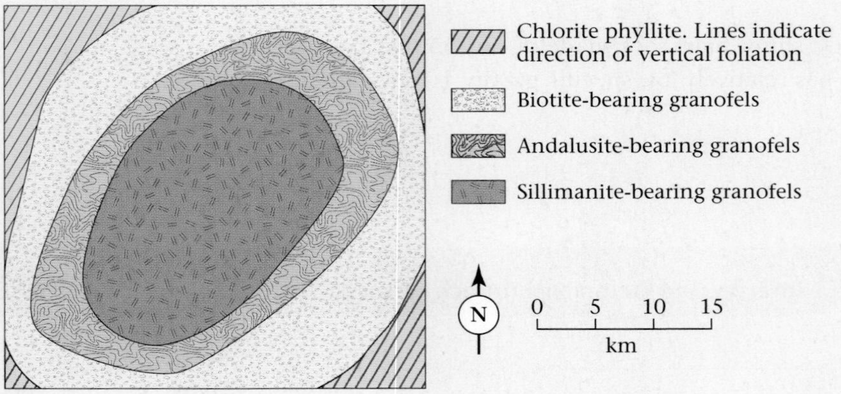

Chlorite phyllite. Lines indicate direction of vertical foliation
Biotite-bearing granofels
Andalusite-bearing granofels
Sillimanite-bearing granofels
N
0 5 10 15
km

(a) Describe the processes most likely responsible for producing the phyllite. Considering the orientation of the foliation, draw arrows on the map to indicate the directions in which pressure had been applied.

____

____

____

(b) Suggest an explanation for the distribution of metamorphic index minerals based on your knowledge of metamorphism and metamorphic rocks.

____

____

____

____

(c) What information is given about the grade of metamorphism?

____

____

____

____

continued

Name: _____ **Section:** _____
Course: _____ **Date:** _____

(d) If your colleague's hypothesis is correct that all the rocks had been the same at one time, explain why the inner parts of the concentric pattern are not foliated. (*Hint:* Consider the types of metamorphism and the processes that cause rocks to be foliated and those that cause rocks to be granoblastic.)

_____

_____

_____

_____

(e) Suggest a sequence of events that explains all of the information. (*Hint:* A geologic map is two-dimensional; remember the third dimension!)

_____

_____

_____

_____

Name _____

METAMORPHIC ROCKS STUDY SHEET

Sample #	Minerals present (compositional group)	Texture (grain orientation, size, shape)	Type of metamorphism	Metamorphic grade (low, medium, or high grade)	Rock name

Name _____

METAMORPHIC ROCKS STUDY SHEET

Sample #	Minerals present (compositional group)	Texture (grain orientation, size, shape)	Type of metamorphism	Metamorphic grade (low, medium, or high grade)	Rock name

CHAPTER 8

STUDYING EARTH'S LANDFORMS

PURPOSE

- Understand the different ways in which geologists study the surface of the Earth.
- Select the most appropriate image for studying an aspect of Earth's surface.
- Practice interpreting landform models.

MATERIALS NEEDED

- Clear plastic ruler with divisions in tenths of an inch or millimeters and
- Circular protractor, both in your tool kit at the end of the book
- A globe and maps provided by your instructor

8.1 Introduction

It is easier to study Earth's surface today than at any time in history. To examine surface features, we can now make detailed surface models from satellite elevation surveys and download images of any point on the planet with the click of a mouse using *Google Earth*™ and *NASA World Wind*. Geologists were quick to understand the scientific value of satellite imaging technology and have adopted new methods as quickly as they are developed. Some of the images in this manual were not even available to researchers a decade ago. The study of the Earth's surface is almost as dynamic as the surface itself!

And you don't have to be in a college laboratory to use this technology because much of it can be downloaded free of charge from the Internet. *Google Earth*™ and *NASA World Wind* provide satellite images of the entire globe and can be used to get three-dimensional views. MICRODEM can build realistic digital elevation models of the landscape, draw topographic profiles, measure straight-line distances and the length of meandering streams, and estimate slope steepness. Other resources are relatively inexpensive: National Geographic's TOPO! provides continuous topographic map coverage of each state for about $100 and also draws profiles and measures distances.

8.2 Images Used to Study Earth's Surface

The best way to study landforms is to fly over them to get a bird's-eye view and then walk or drive over them to understand them from a human perspective. That's not practical for a college course, so we have to bring the landforms to you instead. We use traditional tools like maps (**Fig. 8.1a**) and aerial photographs (**Fig. 8.1b**), a new generation of Landsat and other satellite images (**Fig. 8.1c**), digital elevation models (**Fig. 8.1d**), and *Google Earth*™ and *NASA World Wind* images that were still science fiction when we began planning this book.

Figure 8.1 shows a part of eastern Maine using four different imaging methods. A **topographic map** (Fig. 8.1a) uses contour lines to show landforms (see Chapter 9). Topographic maps used to be drawn by surveyors who measured distances, directions, and elevations in the field. They are now made by computers from aerial photographs and radar data. **Aerial photographs** (Fig. 8.1b), including U.S. Geological Survey (USGS) Orthophotoquads, are photographs taken from a plane and pieced together to form a mosaic of an area. **Landsat images** (Fig. 8.1c) are made by a satellite that takes digital images of Earth's surface using visible light and other wavelengths of the electromagnetic spectrum. Scientists adjust the wavelengths to color the image artificially and emphasize specific features. For example, some infrared wavelengths help reveal the amount and type of vegetation. **Digital elevation models** (DEMs; Fig. 8.1d) are computer-generated, three-dimensional views of landforms made from radar satellite elevation data spaced at 10- or 30-m intervals on the Earth's surface. A new generation based on 1-m data is now being released that provides a more accurate model of the surface than anything available to the public 5 years ago.

Each type of image is particularly helpful for different kinds of study, and each has its drawbacks, so no one imaging method is *the* right one. The latitude and longitude of the corners of a topographic map, DEM, or Landsat image are normally listed, along with a scale for measuring distance on the map and DEM. North is assumed to be at the top of all images unless otherwise indicated.

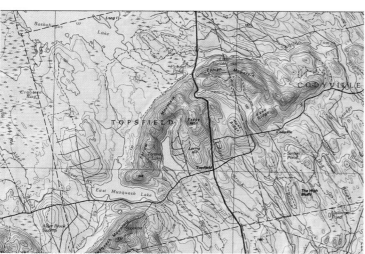

(a) Topographic map.

(c) Landsat image (artificial color).

(d) Digital elevation model.

The images in Figure 8.1 show the same area but look different because each emphasizes different aspects of the surface. Which is best? For learning the names of mountains or lakes, the topographic map is the obvious choice; for studying the network of roads in the area, the DEM would be useless. For analyzing landforms and their evolution, geologists can choose the image that works best for their particular project.

8.2.1 Map Projections

The images in Figure 8.1 are flat, two-dimensional pictures, but Earth is a nearly spherical three-dimensional body. The process by which a three-dimensional sphere is converted into a two-dimensional map is called making a **projection**. There are many different projections and each produces maps that look different from one

Name: _____   **Section:** _____
Course: _____   **Date:** _____

(a) Examine the images in Figure 8.1 and rank them (on a scale of 1 to 4) in **Table 8.1** by how well they show the map elements indicated (1 is most effective, 4 is least effective; ties are allowed).

(b) Which of the images enables you to recognize the topography most easily? Why?

_____

_____

TABLE 8.1 Evaluating landscape images.

	Topographic map	Aerial photograph	Landsat image	DEM
Location				
Direction				
Elevation				
Changes in slope				
Distance				
Names of features				

(c) Which is least helpful in trying to visualize the hills, valleys, and lakes? Why?

_____

_____

(d) Erosional agents often produce a "topographic grain," an alignment of elongate hills, ridges, and valleys. Which images show the topographic grain in this area most clearly? Once you've seen it on those images, can you recognize it on the others?

_____

(e) Which images show highways most clearly?

_____

(f) Which images show unpaved lumber roads most clearly?

_____

(g) Which image do you think is the oldest? The most recent? Explain your reasoning.

_____

_____

(h) Which image(s) would you want to have if you were planning a wilderness hike? Why?

_____

_____

_____

FIGURE 8.2 Four common map projections and their different views of the world.

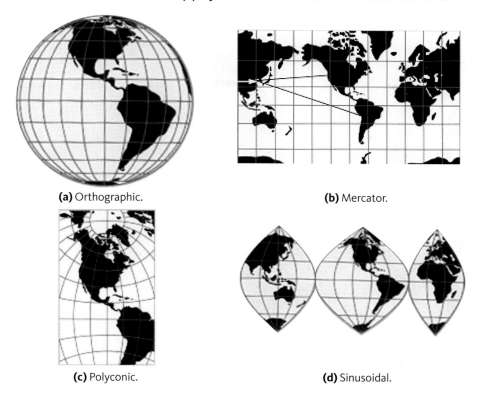

(a) Orthographic.

(b) Mercator.

(c) Polyconic.

(d) Sinusoidal.

another and are used for different purposes (**Fig. 8.2**). Some, like the Mercator projection, preserve accurate *directions* between points; others preserve the true *areas* of features or the true *distances* between points. All projections must to some extent distort three-dimensional reality to fit on a two-dimensional piece of paper.

8.3 Map Elements

All accurate depictions of Earth's surface must contain certain basic elements: *location*, a way to show precisely where the area is; a way to measure the *distance* between features; and an accurate portrayal of *directions* between features. It is also important to know *elevations* of hilltops and other features, and the *steepness of slopes*.

8.3.1 Map Element 1: Location

Road maps and atlases use a simple grid system to locate cities and towns (e.g., Chicago is in grid square A8). This is not very precise because many other places may be in the same square, but it is good enough for most drivers. More sophisticated grids are used to locate features on Earth precisely. Maps published by the USGS use three grid systems: latitude/longitude, the Universal Transverse Mercator (UTM) grid, and, for most states, the Public Land Survey System. The UTM grid is least familiar to Americans but is used extensively in the rest of the world.

8.3.1a Latitude and Longitude The latitude/longitude grid is based on location north or south of the **equator** and east or west of an arbitrarily chosen north-south line (**Fig. 8.3**). A *parallel of latitude* connects all points that are the same angular distance

FIGURE 8.3 The latitude/longitude grid.

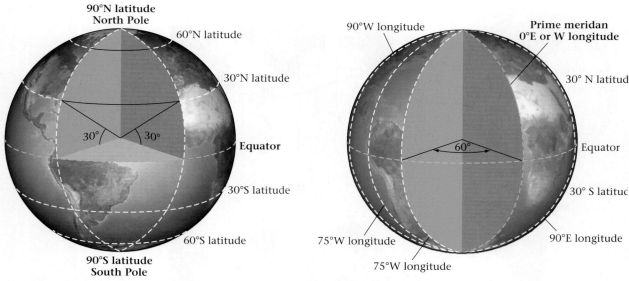

(a) Latitude is measured in degrees north or south of the equator.

(b) Longitude is measured in degrees east or west of the prime meridian (Greenwich, England).

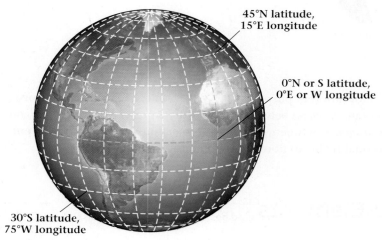

(c) Locating points using the completed grid.

north or south of the equator. The maximum value for latitude is 90°N or 90°S (the North and South Poles, respectively). A *meridian of longitude* connects all points that are the same angular distance east or west of the **prime meridian**, a line that passes through the Royal Observatory in Greenwich, England. The maximum value for longitude is 180°E or 180°W, the international date line. *Remember: You must indicate whether a point is north or south of the equator **and** east or west of the prime meridian.*

Latitude and longitude readings are typically reported in degrees (°), minutes ('), and seconds ("), and there are 60' in a degree and 60" in a minute (e.g., 40°37'44"N, 73°45'09"W). For reference, 1 degree of latitude is equivalent to approximately 69 miles (111 km), 1 minute of latitude is about 1.1 mile (185 km), and 1 second of latitude about 100 feet (31 m). The same kind of comparison can only be made for longitude *at the equator*, because the meridians merge at the poles and the distance between degrees of longitude decreases gradually toward the poles (Fig. 8.3b).

Name: _____   **Section:** _____
Course: _____   **Date:** _____

(a) With the aid of a globe or map, determine the latitude and longitude of your geology laboratory as accurately as you can. How could you locate the laboratory more accurately?

_____

_____

(b) If you have access to a GPS receiver, locate the corners of your laboratory building. Draw a map below (or on a separate piece of paper) showing the location, orientation, and distances between the corners.

(c) Locate the following U.S. and Canadian cities as accurately as possible.

Nome, Alaska _____   Seattle, Washington _____

Chicago, Illinois _____   Los Angeles, California _____

St. Louis, Missouri _____   Houston, Texas _____

New York, New York _____   Miami, Florida _____

St. Johns, Newfoundland _____   Ottawa, Ontario _____

Calgary, Alberta _____   Victoria, British Columbia _____

(d) Which of the cities in (c) do you think is closest in latitude to each of the following world cities? Predict first, without looking at a map or globe, then check. Were you surprised by any?

City	Predicted best match	Latitude and longitude	Actual best match
Oslo, Norway			
Baghdad, Iraq			
London, England			
Paris, France			
Rome, Italy			
Beijing, China			
Tokyo, Japan			
Quito, Ecuador			
Cairo, Egypt			
Cape Town, South Africa			

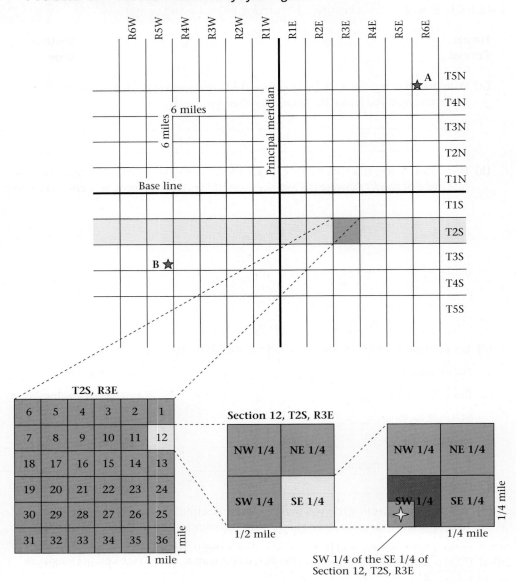

FIGURE 8.4 The Public Land Survey System grid.

Hand-held global positioning system (GPS) receivers and those used in cars and planes can locate points to within a second.

8.3.1b Public Land Survey System The Public Land Survey System was created in 1785 to provide accurate maps as America expanded from its thirteen original states. Most of the country is covered by this system, except for Alaska, Hawaii, Texas, and the southwestern states surveyed by Spanish colonists before the states joined the Union. Points can be located rapidly to within an eighth of a mile in this system (Fig. 8.4).

The grid is based on accurately surveyed north-south (**principal meridian**) and east-west (**base line**) lines for each survey region. Lines parallel to these at 6-mile intervals create grid squares 6 miles on a side forming east-west rows called **townships** and north-south columns called **ranges** (Fig. 8.4). Townships are numbered north or south of the base line, and ranges are numbered east and west of the principal meridian. Each 6-mile square is divided into 36 **sections**, each 1 mile on a side, numbered as shown in Figure 8.4. Each section is divided into **quarter sections**

Name: _____   **Section:** _____

Course: _____   **Date:** _____

(a) Determine the location of points A and B in Figure 8.4.

　　A _____   B _____

(b) Determine the location of points indicated by your instructor on topographic maps.

_____

_____

_____

½ mile on a side, and each of these is further quartered, resulting in squares ¼ mile on a side. The location of the yellow star in the orange box in Figure 8.4 is described in the series of blow-ups:

- **T2S, R3E** locates it somewhere within an area of 36 square miles (inside a 6 mi × 6 mi square).
- **Section 12, T2S, R3E** locates it somewhere within an area of 1 square mile.
- **SE ¼ of Section 12, T2S, R3E** locates it somewhere within an area of ¼ square mile.
- **SW ¼ of the SE ¼ of Section 12, T2S, R3E** locates it within an area of ¹⁄₁₆ square mile.

8.3.1c Universal Transverse Mercator Grid The UTM grid divides the Earth into 1,200 segments, each containing 6° of longitude and 8° of latitude (**Fig. 8.5**). North-south segments are assigned letters (C through X); east-west segments are called **UTM zones** and are numbered 1 to 60 eastward from the international date line (180° W). Thus, UTM zone 1 extends from 180° to 174° W longitude, zone 2 from 174° to 168° W longitude, and so on. The forty-eight conterminous United States lie within UTM zones 10 through 19, roughly 125° to 67° W longitude. Because of the polar distortion in the Mercator projection evident in the sizes and shapes of Greenland and Antarctica in Figure 8.5, the UTM grid covers only latitudes 80°N to 80°S.

To locate a point, begin with the grid box in which the feature is located. For example, the red box in Figure 8.5 is grid S22. UTM grid readings tell *in meters* how far north of the equator (*northings*) and east of the central meridian for each zone (*eastings*) a point lies. The central meridian for each UTM zone is a line of longitude that runs through the center of the zone (**Fig. 8.6**) and is arbitrarily assigned an easting of 500,000 m so that no point has a negative easting. Points east of a central meridian thus have eastings greater than 500,000 m, and those west of the central meridian have eastings of less than 500,000 m.

Figure 8.7 shows how the UTM grid appears on USGS topographic maps. Blue tick marks along the border of every map define the grid made up of boxes 1,000 m (1 km) on a side (Fig. 8.7a). In newer maps, the grid lines are drawn (Fig. 8.7b).

FIGURE 8.5 The UTM grid.

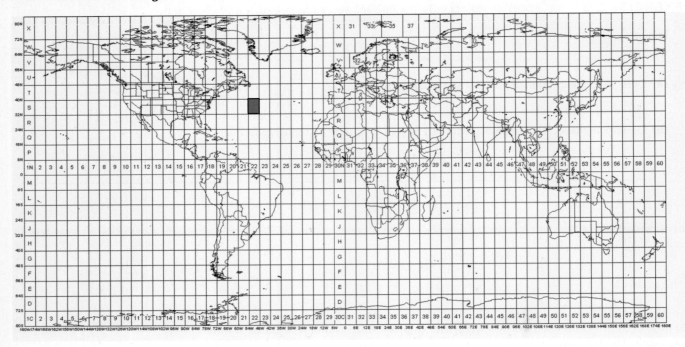

FIGURE 8.6 UTM zones for the forty-eight conterminous United States.

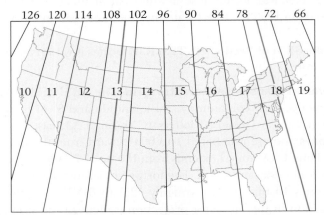

The red line is the central meridian (105° W) for UTM zone 13; the blue line is the central meridian for zone 18.

Grid labels across the top and bottom of a map are *eastings* (indicated by E), those along the side are *northings* (indicated by N). There are two kinds of labels—one is a shorthand version, the other is complete. In Figure 8.7b, $^{5}68^{000m}$E means 568,000 m east of the central meridian for the UTM zone. Remember that each grid box is 1,000 m square. The marker immediately west of $^{5}68^{000m}$E must be 567,000 m east of the central meridian, but only every tenth value is written fully. Abbreviations like $^{5}67$ are for the intervening grid markers. Northings are similar: the marker $^{50}25^{000m}$ N means 5,025,000 m north of the equator (5,025 km). And the marker $^{50}26$, 1,000 m further north, is 5,026,000 m north of the equator.

FIGURE 8.7 Using UTM grid marks to locate features.

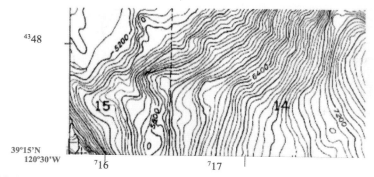

(a) The UTM tick marks (blue) along the sides and the latitude/longitude (red) at the corner of a topographic map.

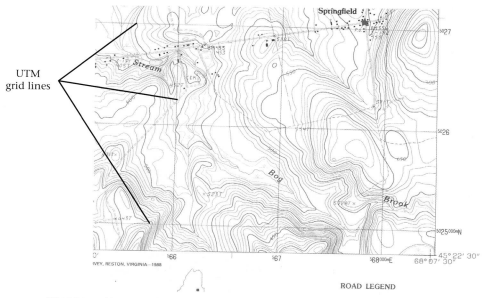

(b) UTM grid lines used to locate a road intersection.

To determine the location of the red star in Figure 8.7b, measure proportionally the distance east of ^{5}67 and north of ^{50}27 (567,670 m east, 5,027,165 m north). Rather than doing a lot of arithmetic, you can make a UTM tool for each major map scale (**Fig. 8.8**) and determine the location easily to within 10 m.

8.3.2 Map Element 2: Direction

Geologists use the **azimuth method** to indicate direction, based on the dial of a compass (**Fig. 8.9**). The red-tipped compass needle in Figure 8.9a is pointing northeast— that is, somewhere between north and east, but how much closer to north than to east? On an azimuth compass (Fig. 8.9b) 0° or 360° is due north, 090° due east, 180° due south, and 270° due west. The direction 045° is exactly halfway between north and east. The direction of the needle in Figure 8.9 can be read as 032°.

Use the circular protractor in your tool kit to determine the direction between any two points. Draw a line between the points, align the protractor's registration

FIGURE 8.8 Using a UTM grid tool.

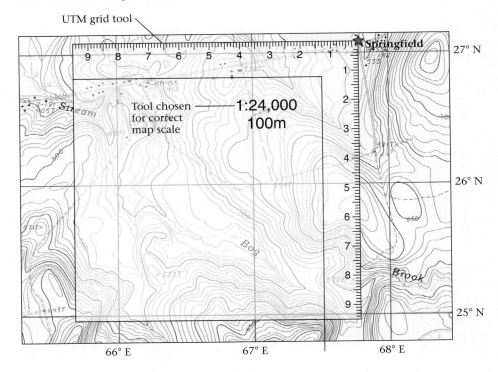

Locating Points with the UTM Grid

Name: _____   Section: _____

Course: _____   Date: _____

(a) Use the appropriate UTM grid tool from your tool kit to determine the location of the red star in Figure 8.8.

_____

FIGURE 8.9 Using the azimuth method to describe the direction between two points.

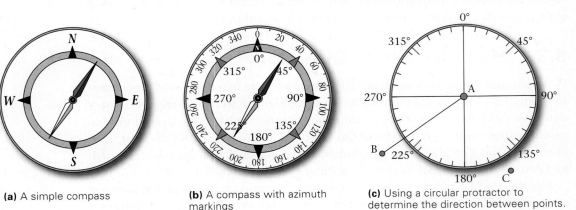

(a) A simple compass

(b) A compass with azimuth markings

(c) Using a circular protractor to determine the direction between points.

Use the circular (azimuth) protractor in your tool kit to determine the direction between the following two points.

(a) In Figure 8.9, from A to C _____°   C to A _____°   B to C _____°   C to B _____°

(b) In Figure 8.4, from A to B _____°

---

lines in a north-south or east-west position, and place its center point on one of the two points (Fig. 8.9c). The direction from the first point to the second is where the line intersects the azimuth scale; in this case, the direction from A to B is 235°.

8.3.3 Map Element 3: Distance and Scale

A map of the entire world or your school campus can fit onto one sheet of paper—if we scale the Earth down so it fits. A **map scale** indicates how much an area has been scaled down so we can relate inches or centimeters on the map to real distances on the ground. **Figure 8.10** shows three maps of the same general area made at different scales. The more we scale down an area, the more detail we lose; the closer the map is to the real size, the more detail we can see.

The three map segments in Figure 8.10 are approximately the same size on the page, but the area each covers is different because of their different scales. The map in Figure 8.10a has been scaled down more than 4 times as much as Figure 8.10c (1:100,000 versus 1:24,000) and therefore covers much more of the land surface in the same amount of space.

8.3.3a Different Ways to Describe Map Scale Map scale may be expressed verbally, proportionally, or graphically. A **verbal scale,** used on many roadmaps, uses words like "1 inch equals approximately 6.7 miles" to describe the scaling of map and real distances. A driver can estimate distances between cities, but not very accurately.

The most accurate way to describe scale is a **proportional scale**, one that tells exactly how much the ground has been scaled down to make the map. Proportional scales like 1:100,000 mean that distances on the map have been scaled down to a hundred thousandth of the ground distance. The larger the number, the more the ground has been scaled down. A proportional scale refers to all units. Thus, 1 cm on the map equals 100,000 cm (1 km) on the ground, or 1 inch on the map equals 100,000 inches (1.58 miles) on the ground, and so on.

The metric system is ideally suited for scales like 1:100,000,000 or 1:100,000 because it is based on multiples of 10. In the United States, we measure ground distance in miles but map distance in inches. Unfortunately, relationships among inches, feet, and miles are not as simple as in the metric system. There are 63,360 inches in a mile (12 in per foot × 5,280 ft per mile), so the proportional scale 1:63,360 means that 1 inch on a map represents exactly 1 mile on the ground. Old topographic maps use a scale of 1:62,500. For most purposes, we can interpret this scale to be approximately 1 inch = 1 mile, even though an inch on such a map

FIGURE 8.10 An area in central Maine mapped at three common map scales.

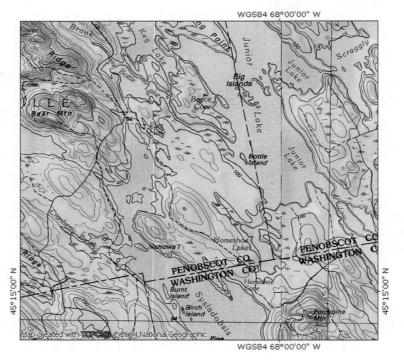

(a) Scale = 1:100,000.

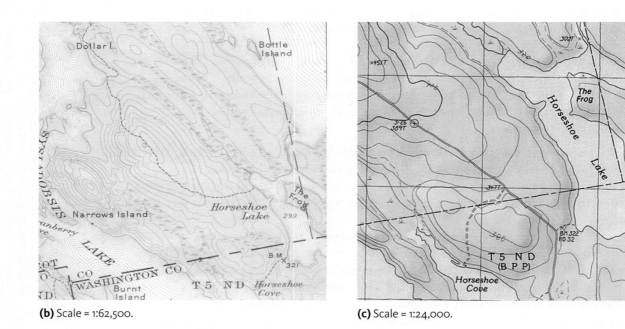

(b) Scale = 1:62,500.

(c) Scale = 1:24,000.

would be about 70 feet short of a mile. Other common map scales are 1:24,000 (1 in = _____ ft), 1:100,000 (see above), 1:250,000 (1 in = _____ mi), and 1:1,000,000 (1 in = _____ mi).

Map scale can also be shown **graphically**, using a **bar scale** (Fig. 8.11) to express the same relation as the verbal scale. Depending on how carefully you measure, a bar scale can be more accurate than a verbal scale, but not as accurate as a proportional scale.

FIGURE 8.11 Scale bars used with three common proportional scales.

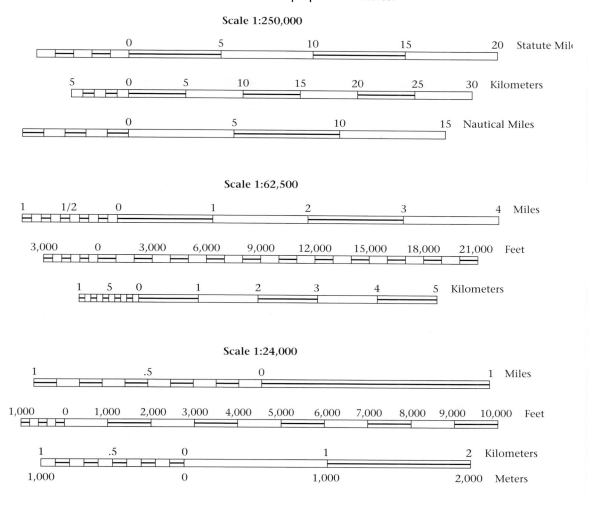

8.4 Vertical Exaggeration—A Matter of Perspective

DEMs show the land surface in three dimensions and must therefore use an appropriate *vertical* scale to indicate how much taller one feature is than another. It would seem logical to use the same scale for vertical and horizontal distances, but we don't usually do so because mountains wouldn't look much like mountains and hills would barely be visible. Landforms are typically much wider than they are high, standing only a few hundred or thousand feet above or below their surroundings. At a scale of 1:62,500, 1 inch represents about a mile. If we used the same scale to make a three-dimensional model, a hilltop 400 feet above its surroundings would be less than one-tenth of an inch high. A mountain rising a mile above its base would be only 1 inch high.

We therefore exaggerate the vertical scale compared to the horizontal to show features from a human perspective. For a three-dimensional model of a 1:62,500 map, a vertical scale of 1:10,000 would exaggerate apparent elevations by a little more than 6 times (62,500/10,000 = 6.25). A mountain rising 1 mile above its surroundings would stand 6.25 inches high in the model; a 400-foot hill would be about half an inch tall, which is more realistic than the 0.1 inch if the 1:62,500 horizontal scale had been used vertically.

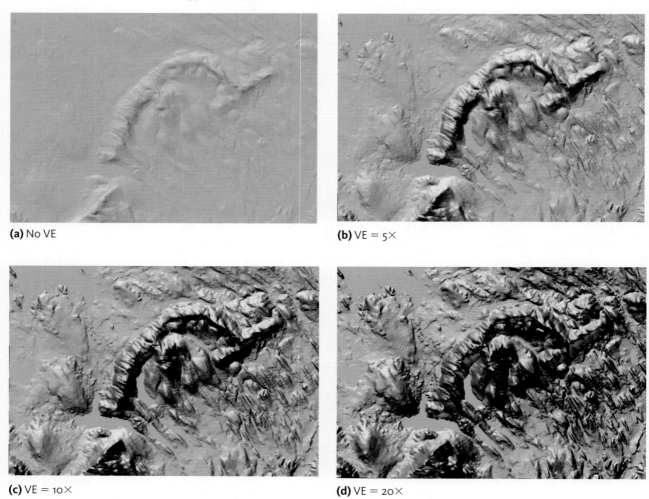

(a) No VE

(b) VE = 5×

(c) VE = 10×

(d) VE = 20×

The degree to which the vertical scale has been exaggerated is, logically enough, called the **vertical exaggeration**. **Figure 8.12** shows the effects of vertical exaggeration on a DEM. With no vertical exaggeration, the prominent hill in the center of Figure 8.12a is barely noticeable. One of the authors of this manual has climbed that hill several times and guarantees that climbing it is far more difficult than Figure 8.12a would suggest. On the other hand, Figure 8.12d exaggerates too much; the hill did not seem that steep, even with a pack loaded with rocks.

Is there such a thing as too much vertical exaggeration? The basic rule of thumb is not to make a mountain out of a molehill. Vertical exaggerations of 2 to 5 times generally preserve the basic proportions of landforms while presenting features clearly. We return to the concept of vertical exaggeration when we discuss drawing topographic profiles from topographic maps in Chapter 9.

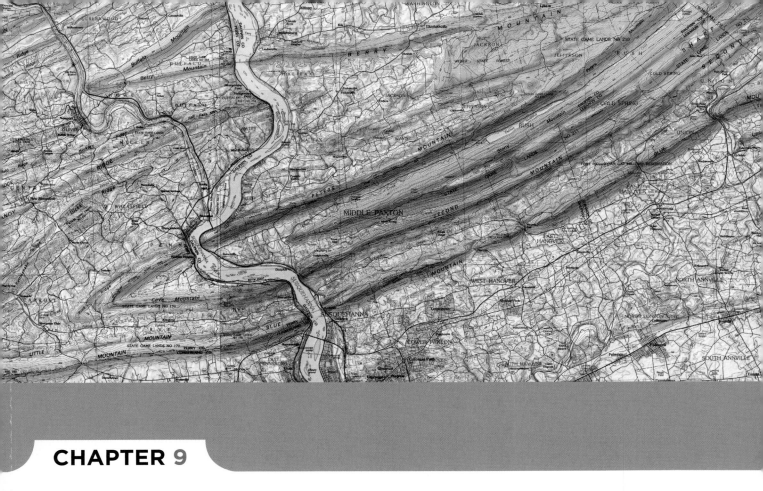

CHAPTER 9

WORKING WITH TOPOGRAPHIC MAPS

PURPOSE

- Understand how contour lines on topographic maps represent Earth's surface.
- Practice constructing topographic profiles to illustrate and interpret landforms.
- Use topographic maps to recognize potential natural disasters and solve economic problems.

MATERIALS NEEDED

- Pencil, ruler, protractor, and tracing paper
- Topographic maps provided by your instructor

9.1 Introduction

With the click of a mouse, *Google Earth*™ and *NASA World Wind* provide satellite images of any point on the planet, digital elevation models (DEMs) help visualize topography, and Geographic Information System (GIS) software can locate points, give elevations, measure lengths of meandering streams, and construct topographic profiles. But geologists still use topographic maps in the field instead of computers because they cost almost nothing, weigh much less than a digital device, and withstand rain, swarming insects, and being dropped better than computers. A recent topographic map gives the names and elevations of lakes, streams, mountains, and roads, and it outlines fields and distinguishes swamps and forests as well as a satellite image. With a little practice, you can learn more about landforms from topographic maps than from a satellite image or DEM. This chapter explains how topographic maps work and helps you develop map-reading skills for identifying landforms, planning hikes, solving environmental problems—and possibly saving your life.

9.2 Contour Lines

Like aerial photographs and satellite images, topographic maps show location, distance, and direction very accurately. Topographic maps show the shapes of landforms, elevations, and the steepness of slopes with a special kind of line called **contour lines**.

A contour line is a type of *isoline* (*iso*, meaning equal), a line that connects points of the same value for whatever is being measured. Contour lines can show many types of features on a map, like population density and average income, as well as elevation. You are already familiar with one common type of contour line on weather maps, which shows a change in temperature rather than a change in the elevation of Earth's surface (**Fig. 9.1**).

The contour lines in Figure 9.1 are *isotherms* (lines of equal temperature). Thus, the predicted temperature for every point on the 60° line is 60°F, every point on the 40° line 40°F, and so on. Each isotherm separates areas where the temperature is higher or lower than that along the line itself. Thus, all points in the area north of the 30° contour line have predicted high temperatures for the day lower than 30°F, those on the south side higher than 30°F. The map has a **contour interval** of 10°, meaning that contour lines represent temperatures at 10° increments.

FIGURE 9.1 Contour map showing predicted high temperatures for the United States.

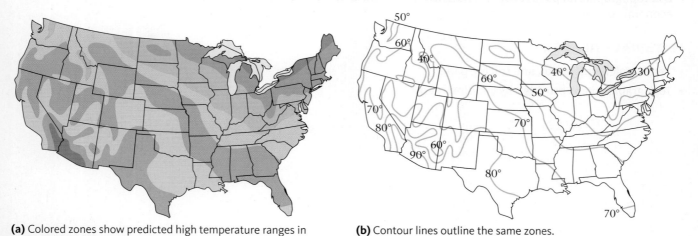

(a) Colored zones show predicted high temperature ranges in increments of 10°F.

(b) Contour lines outline the same zones.

Name: _____   **Section:** _____
Course: _____   **Date:** _____

Start with a weather map (**Fig. 9.2**) showing maximum temperatures for several cities, and contour the data using a contour interval of 10°F. Remember that temperatures on one side of a contour line are lower than the value for the contour, those on the other side are higher. You will need a ruler, pencil, and an eraser. Here are some hints on how to proceed:

- The contour lines will be multiples of 10° (80°, 90°, 100°, etc.).

- A contour line will pass through a dot whose predicted temperature is a multiple of 10°.

- You need to estimate where contour lines lie between temperatures that are not multiples of 10°. For example, the 80° contour must pass somewhere between San Francisco (76°) and Los Angeles (86°).

- Without information about how rapidly temperature changes between these cities, assume that the change is constant. Draw a light line between the two cities as shown in Figure 9.2. San Francisco is 4° cooler than the 80° contour, whereas Los Angeles is 6° warmer. The 80° contour will therefore be closer to San Francisco than Los Angeles. There is a total temperature difference of 10° along the line between the two cities. The distance from San Francisco to the contour line would be 4°/10°, or 4/10 of the distance along the line.

FIGURE 9.2   **Predicted high temperatures for a day in the United States.**

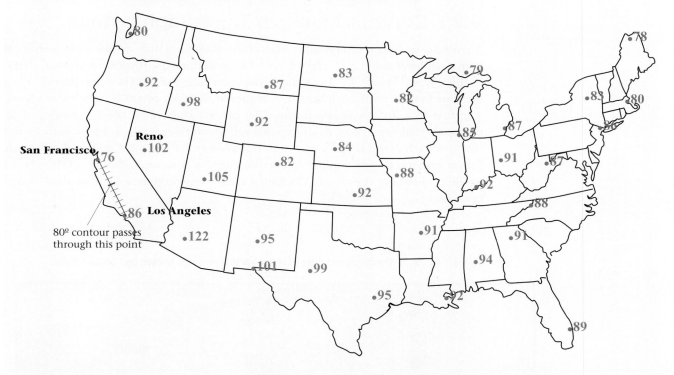

(a) Which contours must lie between San Francisco (76°) and Reno, Nevada (102°)? _____

Using a pencil, lightly sketch the segments of these contour lines between Reno and San Francisco. Then locate each more precisely with a ruler as you did with the San Francisco/Los Angeles data. With practice, your estimates will be accurate enough for our purposes on this map.

(b) Repeat for the contours between Los Angeles and Reno then connect the appropriate lines.

continued

Name: _____   Section: _____
Course: _____   Date: _____

(c) Do the same for all the remaining stations, lightly tracing the contour lines and checking their locations. When you are done, the area between any two contour lines should contain all temperatures that lie between those of the two contours.

The purpose of making a contour map is to be able to use it for some purpose: to analyze it to learn something or to clarify geographic relationships. You now have two contoured maps of predicted high temperatures—the one you just made and Figure 9.1—and can do a simple piece of analysis.

(d) Compare the two maps. What can you hypothesize about the time of year that each represents? Explain your reasoning.

_____

_____

_____

9.2.1 Contour Lines on Topographic Maps

A contour line on a **topographic** map connects points that are the same elevation above sea level, and the complete set of lines defines the three-dimensional shape of the surface. **Figure 9.3** shows the underwater shape of the Hawaiian Islands using contour lines to connect points that are some multiple of 1,000 m below sea level. The first contour line is easy to draw for any island, because the 0 m contour line connects points that are 0 m above sea level, which *is* sea level, and thus outlines the shape of each of the Hawaiian Islands. If sea level were to drop 1,000 m, the contour line now marked –1,000 m would become the 0 m contour and more of the islands would stand above the ocean. The initial shoreline would then become the 1,000 m contour line.

FIGURE 9.3 Underwater view of the Hawaiian Islands shown by contour lines.

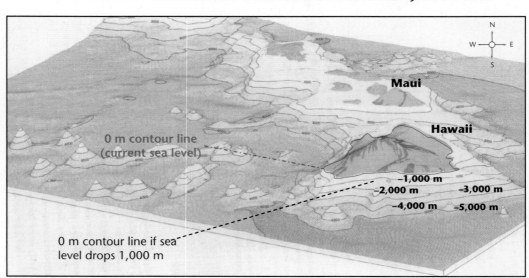

The elevation difference between contour lines is called the **contour interval.** Typical contour intervals on maps are 10, 20, 50, and 100 feet, depending on the topography. Small contour intervals are used where there isn't much change in elevation, large contour intervals are used in mountainous areas where the **relief** (difference in elevation) is high. In Figure 9.3, the sea floor is more than 5,000 m below the surface, so a large contour interval must be used. If the contour interval were 10 m, there would be 100 contour lines for every one shown in Figure 9.3 and they would be unreadable. The smaller the contour interval, the more detail can be shown; the larger the contour interval, the less detail is visible. **Figure 9.4** shows the above-water shape of the island of Hawaii, also using a 1,000-m contour interval.

FIGURE 9.4 Topographic map of the island of Hawaii.

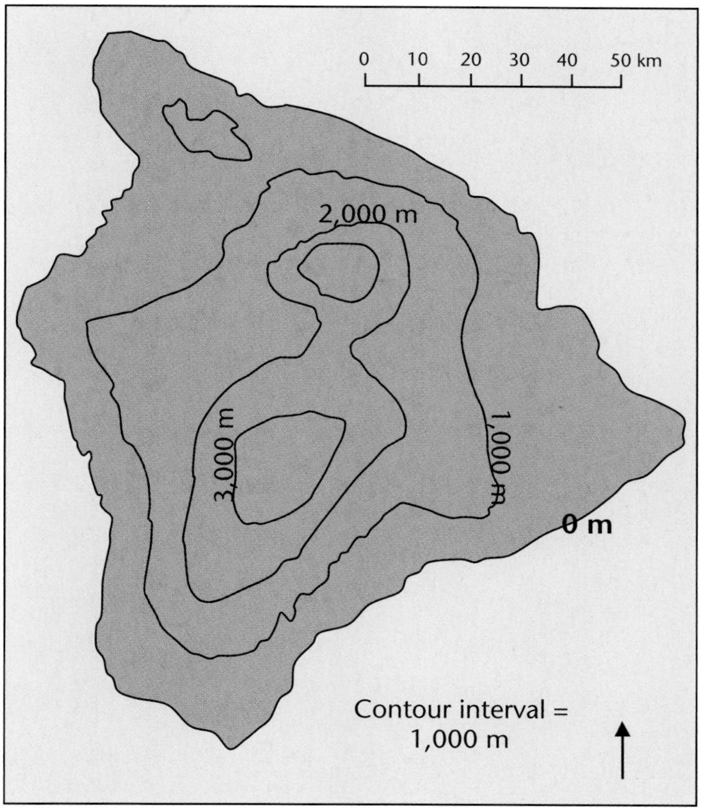

9.3 Reading Topographic Maps

Figures **9.5** and **9.6** are a topographic map and a DEM, respectively, of an area in eastern Maine. (Look at Topsfield, Maine, with *Google Earth*™ or *NASA World Wind.*) Both show the same hills, valleys, and ridges, as well as a few large lakes and several small ones. The topographic map shows hills, valleys, and ridges with brown contour lines, the DEM with shading and a perspective view.

Topographic maps use color to highlight forests (green) and bodies of water, streams, rivers, and lakes (blue)—an improvement over the DEM, in which everything is colored according to its elevation and slope, rather than the nature of the surface. On the topographic map, white areas are fields or marshes; urban areas, if there were any in this part of Maine, would be shown in pink. Human-built features like the roads and buildings in and around the town of Topsfield are shown in black.

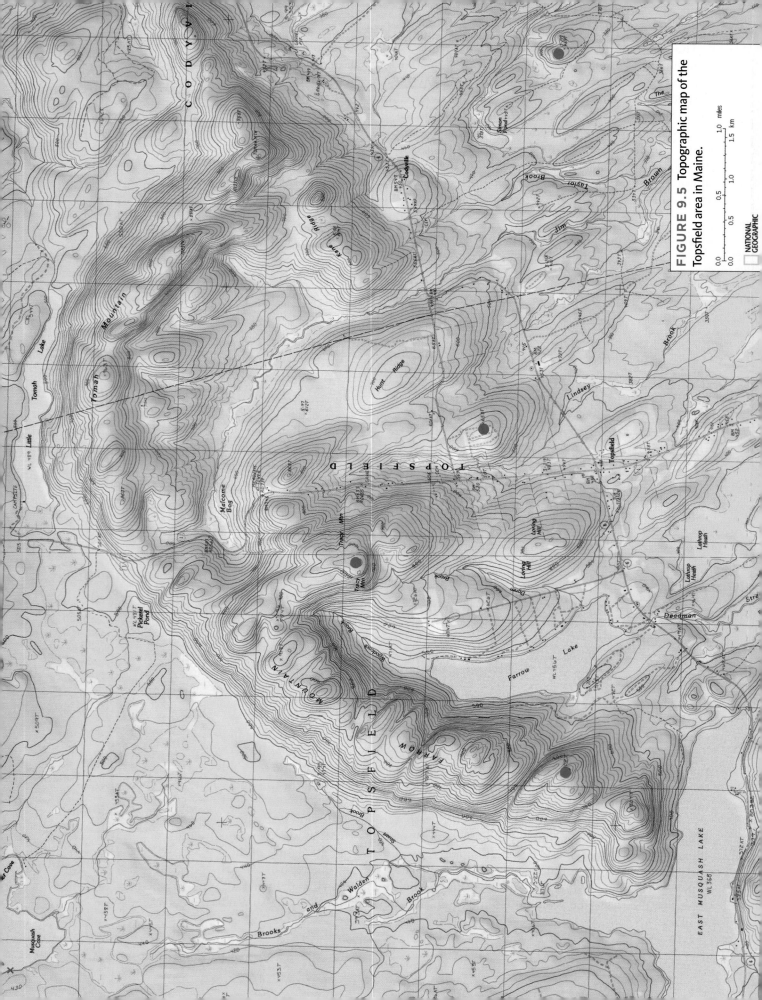

FIGURE 9.5 Topographic map of the Topsfield area in Maine.

FIGURE 9.6 DEM of the Farrow Mountain area in Maine .

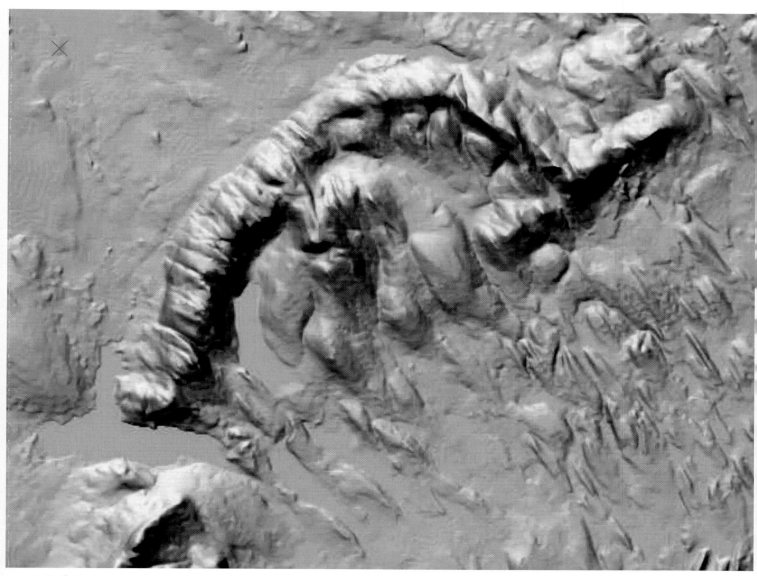

Scale = 1:24,000

The DEM uses shading to portray hills, valleys, lakes, and other flat areas. You can figure out how contour lines portray these features on a topographic map by comparing the two images (Exercise 9.2). We'll now focus on the basics of reading contour lines: interpreting slope steepness, estimating elevations, and determining which way streams are flowing.

EXERCISE 9.2   **How Topographic Maps Show Landform Shapes**

Name: _____   **Section:** _____
Course: _____   **Date:** _____

(a) Slope: Find a flat place on the DEM in Figure 9.6 and a place where the slope is steep. Now locate these places on the topographic map (Fig. 9.5). Compare the contour lines in the flat and steep places.

continued

Name: _____   Section: _____

Course: _____   Date: _____

(i) How does the spacing of the contour lines show the difference between gentle and steep slopes?

_____

_____

(ii) Describe the slopes on both sides of Farrow Lake. Are the slopes equally steep on both sides of the lake, or is one side steeper than the other? Explain your reasoning.

_____

_____

_____

(iii) Describe the slopes of Farrow Mountain in words and draw a sketch (on another piece of paper) showing what it would look like to climb over the mountain from northwest to southeast.

_____

_____

_____

_____

_____

_____

(b) Nested contour lines: Colored circles in Figure 9.5 identify places where there are a series of concentric (nested) contour lines. Look at these features on the DEM.

(i) What type of feature do nested contour lines indicate? _____

(ii) Mark similar features with colored dots.

9.3.1 Contour Lines and Elevation

Elevations of selected points, like hilltops, lake surfaces, and highway intersections, are given on a topographic map by a symbol and a number indicating the elevation, such as $\times_{1438'}$ or $\Delta_{561'}$. The latter symbol is for a *benchmark*, an accurately surveyed point marked on the ground by a brass plaque cemented in place and giving latitude, longitude, and elevation. There are only a few benchmarks on any given map, but you also know the elevation of any point on a contour line.

Check the contour interval at the bottom of the map for the difference in elevation between adjacent contours. Every fifth contour line is darker than those around it and has its elevation labeled. These are *index contours*, which represent elevations that are 5 times the contour interval (multiples of 5×20 in Fig. 9.5—300 feet, 400 feet, and so on). To determine the elevation of an unlabeled contour line, determine its position relative to an index contour. For example, a contour line immediately adjacent to the 500-foot index contour must be 480 or 520 feet—20 feet lower or higher than the index contour on a map where the contour interval is 20 feet.

What about the elevation of a point between two contour lines? All points on one side of a contour line are at higher elevations than the line itself, and those on the

FIGURE 9.7 Reading elevations of hills and valleys from topographic maps.

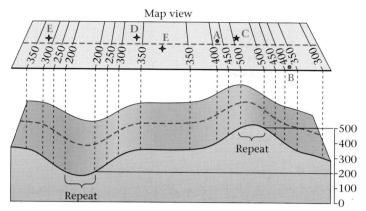

The elevation of Point A is between 400 and 450 feet. Because the slope of the hill is constant between the 400 and 450-foot contour lines and A is halfway between the contour lines, its elevation is 425 feet above sea level.

The elevation of Point B is between 350 and 400 feet. The slope of the hill is gentler close to the 350-foot line and steeper near the 400-foot line. We must therefore estimate elevation taking this into account. Approximate elevation of B is 355 feet above sea level.

The elevation of Point C is greater than 500 feet, but how much greater? If it was more than 550 feet above sea level, there would be another contour line (550 feet). Therefore, the most we can say about the elevation of C is that it is between 500 and 550 feet above sea level.

opposite side at lower elevations, but how much higher or lower? This is just as easy as finding the temperature between two isotherms on a weather map. The contour interval in **Figure 9.7** is 50 feet, so a point between two adjacent contour lines must be less than 50 feet higher or lower than those lines. Either use a ruler or estimate as you did in Exercise 9.1: a point midway between the 500- and 550-foot contour lines would be estimated as 525 feet above sea level.

9.3.1a Depressions Concentric contour lines may indicate a hill (as in Exercise 9.2b) or a depression. To avoid confusion, it must be clear that contour lines outlining a depression indicate progressively *lower* elevations toward the center of the concentric lines rather than the *higher* elevations that would indicate a hill. To show this, small marks called *hachures* are added to the line, pointing toward the lower elevation (see the innermost two lines in Fig. 9.8).

The feature in **Figure 9.8** could be a volcano with a crater at its summit. Note that the nested contour lines on the flanks of the feature indicate increasing elevation

FIGURE 9.8 Hachured contour lines indicate depressions.

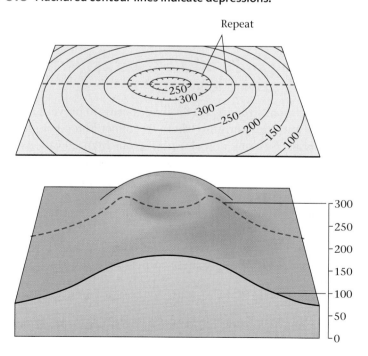

Name: _____   **Section:** _____
Course: _____   **Date:** _____

(a) What is the highest point on the rim of the crater in Figure 9.8? Explain.

_____

_____

_____

(b) What is the elevation of the lowest point of the crater? Explain.

_____

_____

_____

(c) The next three questions refer to Figure 9.6. A red "x" marks a spot on the shore of Baskahegan Lake but a printed DEM doesn't give information about its elevation. Use the topographic map (Fig. 9.5) to determine the elevation of that point as accurately as possible. _____ feet

(d) What is the elevation of:

- the highest point on Hunt Ridge? _____
- the highest point on Farrow Mountain? _____
- the intersection of US Route 1 and Maine Route 6 in Topsfield? _____
- Malcome Bog? _____

(e) What is the relief between Malcome Bog and the top of Tomah Mountain? _____

Between the crossroads in Topsfield and East Musquash Lake? _____

just like those of Figure 9.5, but that the 300-foot contour line is repeated, with the inner one hachured. This shows that the rim of the crater is higher than 300 feet but lower than 350 feet.

9.3.2 Contour Lines and Streams: Which Way Is the Water Flowing?

Geologists study the flow of streams and drainage basins to prevent the downstream spread of pollutants and to determine where to collect water samples to find traces of valuable minerals. Streams flow downhill from high elevations to low elevations. You could determine the flow direction by looking for benchmarks along the stream, or where different contour lines cross the stream, but there is an easier way as you will see in Exercise 9.4.

That's all you need to know to get started with topographic maps. The basic "rules" for reading contour lines are a matter of common sense, and you figured out the most important ones for yourself in Exercises 9.2, 9.3, and 9.4. Summarize these rules in Exercise 9.5 and apply them to making a topographic map of your own in Exercise 9.6.

Understanding Stream Behavior from Topographic Maps

Name: _____   **Section:** _____
Course: _____   **Date:** _____

(a) Refer to Figure 9.5 and look closely at Woodcock Brook at the north end of Farrow Lake. Based on the elevations of the contour lines that cross Woodcock Brook and the general nature of the topography, which way does this stream flow?

_____

(b) Apply the same reasoning to the stream at the east end of Malcome Bog.

_____

(c) Now look at the contour lines as they cross these two streams. Their distinctive V-shape tells which way the stream is flowing. Suggest a "rule of V" that describes how the direction of stream flow is revealed by the contour lines that cross it.

_____

_____

_____

(d) Based on your "rule of V," does the stream at the west side of Pickerill Pond flow into or away from the pond?

_____

(e) In what direction does Jim Brown Brook flow (in the southeast corner of the map)?

_____

EXERCISE **9.5** **Rules of Contour Lines on Topographic Maps**

Name: _____   **Section:** _____
Course: _____   **Date:** _____

In Exercise 9.2, you deduced for yourself the most important "rules" of contour lines, and you will use them in the next several chapters to study landforms produced by streams, glaciers, groundwater, wind, and shoreline currents. Complete the following sentences *using what you've just learned* to summarize the rules of contour lines.

(a) Two different contour lines cannot cross because _____.

(b) The spacing between contour lines on a map reveals the _____ of the ground surface. Closely spaced contour lines indicate _____ and widely spaced contour lines indicate _____ .

(c) Concentric contour lines indicate a _____ .

(d) Concentric *hachured* contour lines indicate a _____ .

(e) Contour lines form a V when they cross a stream. The open part of the V faces the (upstream/downstream) direction.

Name: _____   **Section:** _____
Course: _____   **Date:** _____

Now that you understand the basic rules of contour lines, you can make your own contoured map from elevation data. **Figure 9.9** shows elevation data for a coastal area. Construct a contour map showing the topography, remembering that nature has already drawn the first contour line: the coastline is, by definition, the 0-foot contour. You've already done something similar on the weather map, but topographic contours must follow the rules you have just learned. Use a contour interval of 20 feet. *Remember: Contour lines do not go straight across a stream—they form a "V".*

FIGURE 9.9 Topographic contouring exercise.

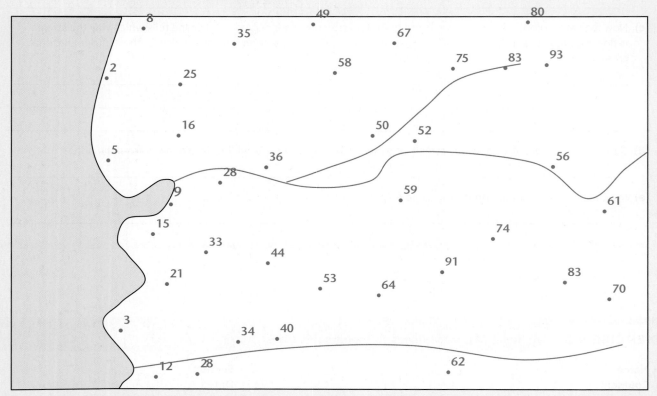

Contour interval = 20 feet

Knowing how to read a topographic map will do more than help you get a good grade in this course. The skills can save your life as shown in Exercise 9.7 or get you a good job (and a raise) as discussed below and in Exercise 9.8.

Name: _____   **Section:** _____
Course: _____   **Date:** _____

While on a short flight during a thunderstorm, a small plane carrying you and a friend crashes in the surf at the northeast corner of an island (**Fig. 9.10**). The nearest human beings are a lighthouse keeper and his family who live on the opposite side of the island as shown on the map. Unfortunately, the plane's radio was destroyed during the crash and your cell phones don't work. No one knows where you are, and the only way to save yourselves is to walk to the lighthouse.

continued

Name: _____   Section: _____
Course: _____   Date: _____

You and your friend were injured in the crash and can't climb hills taller than 50 feet or swim across rivers. You can only walk about 12 miles a day and you carry only enough water to last 3 days. It gets worse: the rivers have crocodiles, there's a large area filled with quicksand, and there's a jungle filled with, yes, lions and tigers and bears. The good news is that you have a compass and a topographic map showing the hazards (Fig. 9.10). You have figured out exactly where the plane crashed and know where you have to go. The map also shows a well where you can get drinking water—if you can get there in 3 days.

(a) With a pencil, protractor, and ruler, plan the shortest route from the crash site to the lighthouse, avoiding steep hills, rivers, quicksand, and hungry jungle carnivores. Record the direction (using the azimuth system) and distance of each leg of your trip in **Table 9.1**.

FIGURE 9.10 Survivor Island.

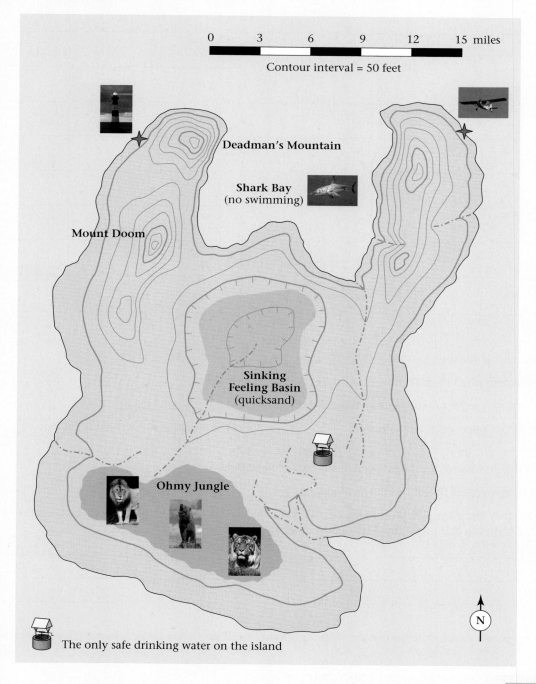

Name: _____   Section: _____
Course: _____   Date: _____

TABLE 9.1 Your route to safety on Survivor Island.

Leg #	Direction of leg (in azimuth degrees)	Length of leg (in miles)	Leg #	Direction of leg (in azimuth degrees)	Length of leg (in miles)
1			11		
2			12		
3			13		
4			14		
5			15		
6			16		
7			17		
8			18		
9			19		
10			20		
Total distance					

(b) How many days will the trip take? _____

(c) Do you have to stop for water? If so, on what day do you to get to the well? _____

(d) How many days will it take to get from the well to the lighthouse? _____

(e) Where is the highest point on the island? _____ Give its elevation as precisely as you can. _____

(f) What is the elevation of the lowest point in Sinking Feeling Basin? _____

(g) Which is the steepest side of Deadman's Mountain? _____

(h) Unfortunately, the plane didn't carry a life raft. If it had, what would have been the shortest route to sail or paddle from the crash site to the lighthouse? Give directions and distances for each leg.

_____

_____

_____

9.4 Topographic Profiles

A telecommunication company wants to place microwave relay towers in a new region to improve cell phone reception and plans to put a tower on the hilltop in the southeast corner of the map shown in **Figure 9.11**. Project managers are concerned that a prominent ridge might block the signal to areas to the northwest. It is not immediately obvious from the map if there will be a "dead zone." You have been hired as a consultant to answer this question. The best way to do so is to construct a topographic profile showing the hills and valleys.

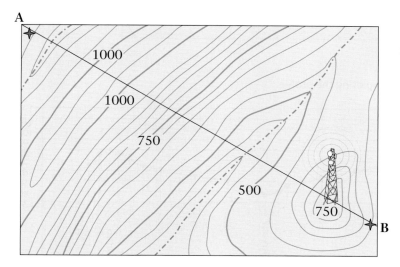

FIGURE 9.11 Map of the microwave tower project area.
Sacle = 1:62,500; contour interval = 50 feet

9.4.1 Constructing a Topographic Profile

One of the benefits of using a topographic map is that it is easy to construct an accurate **topographic profile**, a cross-section view of the topography. *National Geographic*'s TOPO!, MICRODEM, or other GIS software could draw the profile for you, but it is important to understand just what a profile can and cannot do. This is best learned by constructing profiles by hand—a simple process outlined below.

STEP 1: Place a strip of paper along the line of profile (A-B) shown in Figure 9.11 and label the starting and finishing points.

STEP 2: Draw a short line where the strip of paper crosses contour lines, streams, roads, and so on, and label each mark with its elevation (**Fig. 9.12**). For clarity, index contour labels are shown here.

STEP 3: Create a profile paper. Set up a vertical scale using graph paper or blank paper with a series of evenly spaced horizontal lines corresponding to the elevations represented by contour lines along the traverse. Label each line so its elevation is recognizable (right side of **Fig. 9.13**).

STEP 4: Place the strip of paper with the labeled lines at the bottom of the vertical scale and use a ruler to transfer the elevations to their correct positions on the profile paper. Place a dot where each contour line marker intersects the corresponding elevation line in the profile (Fig. 9.13).

STEP 5: Connect the dots, using what you know about contour lines to estimate the elevations at the tops of hills and bottoms of stream valleys along the profile (Fig. 9.13).

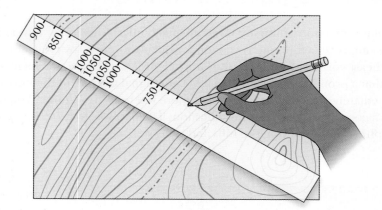

FIGURE 9.12 Constructing the profile.

FIGURE 9.13 Completing the profile.

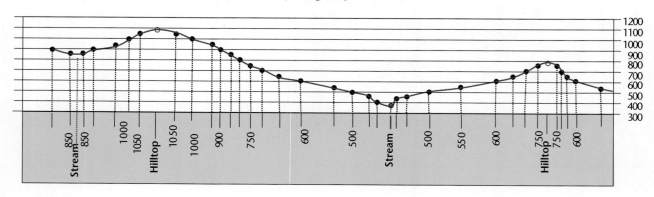

9.4.2 Vertical Exaggeration

Constructing a topographic profile is quick, even without a computer, and can yield important information that can't be obtained just by looking at a map. But care must be taken to make sure that the profile is a realistic view of the topography. It is possible to minimize a mountain so that it looks like an anthill or exaggerate an anthill to make it look like Mt. Everest, depending on the vertical scale you choose. A profile must have not only a horizontal scale to agree with the map but also a vertical scale to show topography. The elevation lines in Figure 9.13 actually define a vertical scale. When the horizontal (map) and vertical (profile) scales are the same, the profile is a perfect representation of topography. If the scales are different, the result is **vertical exaggeration**. Vertical exaggeration overemphasizes the vertical dimension with respect to the horizontal. This is illustrated by three profiles drawn at different vertical exaggerations across the same part of the Hanging Rock Canyon quadrangle in California (**Fig. 9.14**).

To find out how much the profile in Figure 9.13 exaggerates the topography, we need to know the vertical scale used in profiling. First, measure the vertical dimension with a ruler; this shows that 1 *vertical* inch on the profile represents nearly 900 feet of elevation, a proportional scale of 1:10,800 (900 ft × 12 in per foot = 10,800 in). To calculate vertical exaggeration, divide the map proportional scale by the profile proportional scale.

$$\textbf{Vertical exaggeration} = \frac{\text{Map scale}}{\text{Profile scale}} = \frac{62,500}{10,800} = \textbf{5.8×}$$

FIGURE 9.14 The effect of vertical exaggeration on topographic profiles.

(a) DEM of the Hanging Rock Canyon quadrangle showing the line of profile.

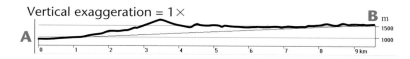

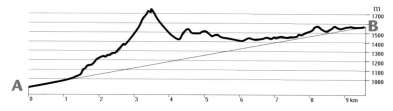

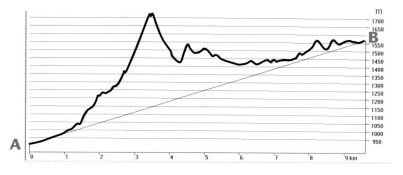

(b) Topographic profiles along the line A-B drawn with different vertical exaggerations. Note that the horizontal scale is the same in all three profiles; only the vertical scale changes.

Name: _____   Section: _____
Course: _____   Date: _____

(a) Construct a profile along the line indicated in Figure 9.11.

(b) What is the vertical exaggeration in your profile?_____ ×

(c) Is there a direct line of sight from the proposed location of the relay tower to the northwest corner of the map? If not, what obstruction(s) are there?

_____

_____

_____

(d) How tall would the tower have to be at the proposed location to eliminate the dead zone? Explain how you made this estimate.

_____

_____

_____

9.5 Additional Information Shown on Topographic Maps

Topographic maps come in many sizes and scales, cover areas of different sizes, and use several different contour intervals. Before looking at the topography on a map, look at the map borders for several useful kinds of information. **Figure 9.15** is a guide for finding that information on the most recent generation of USGS topographic maps, at a scale of 1:24,000. Appendix 9.1 shows standard map symbols for natural features, such as swamps, forests, reefs, permanent and intermittent streams, and for man-made features, such as political boundaries, railroad lines, highways, paved roads, dirt roads, trails, houses, and airports.

9.6 Using Topographic Maps to Predict and Prevent Natural Disasters

The next five chapters show how topographic maps can illustrate the processes that created landforms and landscapes and can identify sites of potential hazards like floods, hurricanes, and earthquakes. The usefulness of topographic maps in identifying potential natural disasters and lessening or even preventing damage is illustrated by a classic landslide.

FIGURE 9.15 Useful information outside the borders of a topographic map.

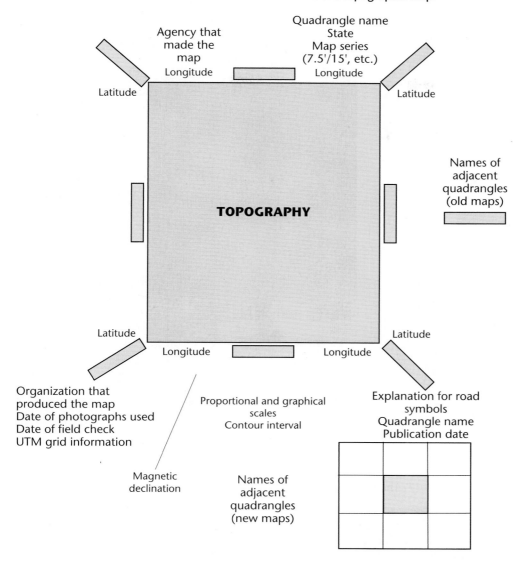

Landslides are rapid forms of mass wasting, the result of gravity-driven downhill movement of rock and regolith. The Slumgullion landslide first moved about 700 years ago when millions of tons of volcanic rock and regolith rushed downhill into the Gunnison River in the San Juan Mountains of Colorado (**Fig. 9.16a**). This slide was in a sparsely inhabited area. The La Conchita landslide in California shows what can happen in a more densely populated region (**Fig. 9.16b**). Thankfully, this was a much smaller slide than Slumgullion, otherwise the neighborhood in its path could have been wiped out.

Avoiding potential landslide areas when planning public buildings, businesses, and homes can save lives, property, and lots of money. Modern GIS technology enables us to estimate slope steepness and therefore recognize areas where it would be dangerous to build. **Figure 9.17** shows a Micro DEM slope analysis map of a mountainous area, highlighting slide-prone areas in shades of red (slopes greater than 50°). The critical slope value would be different in more humid areas, but the software can be adjusted to key on those values just as easily.

FIGURE 9.16 Comparison of two landslides.

(a) The Slumgullion slide in Colorado seen from across the Gunnison River and Lake San Cristobal.

(b) The La Conchita slide in California.

FIGURE 9.17 The Hanging Rock Canyon quadrangle in California.

(a) Digital elevation model.

(b) Slope analysis map.

Name: _____   **Section:** _____
Course: _____   **Date:** _____

Examine the topographic map of the Slumgullion landslide (**Fig. 9.18**) and answer the following questions.

(a) What evidence is there that other landslides had occurred in the area prior to the Slumgullion slide?

_____

_____

_____

(b) Outline the source of the slide with a colored pencil.

(c) Outline the slide mass itself with a colored pencil.

(d) How far did the toe of the slide travel? _____ miles

(e) How far did it travel vertically? _____ feet

(f) There is evidence on the map that significant time elapsed between the date of the slide and the date the map was created, even if you didn't know the dates. Explain how you could have figured this out. *Hint:* Lake San Cristobal didn't exist before the slide.

_____

_____

_____

(g) How has the Gunnison River adjusted its course after the slide? What evidence is there that it is still in the process of making those adjustments?

_____

_____

_____

(h) Based on the topographic setting that led to the Slumgullion slide, indicate other potential landslide areas.

_____

_____

_____

(i) Sketch a profile across the Gunnison River valley along the line A–B–C. Which side is steeper and would therefore be most likely to generate a landslide?

_____

_____

_____

_____

_____

_____

continued

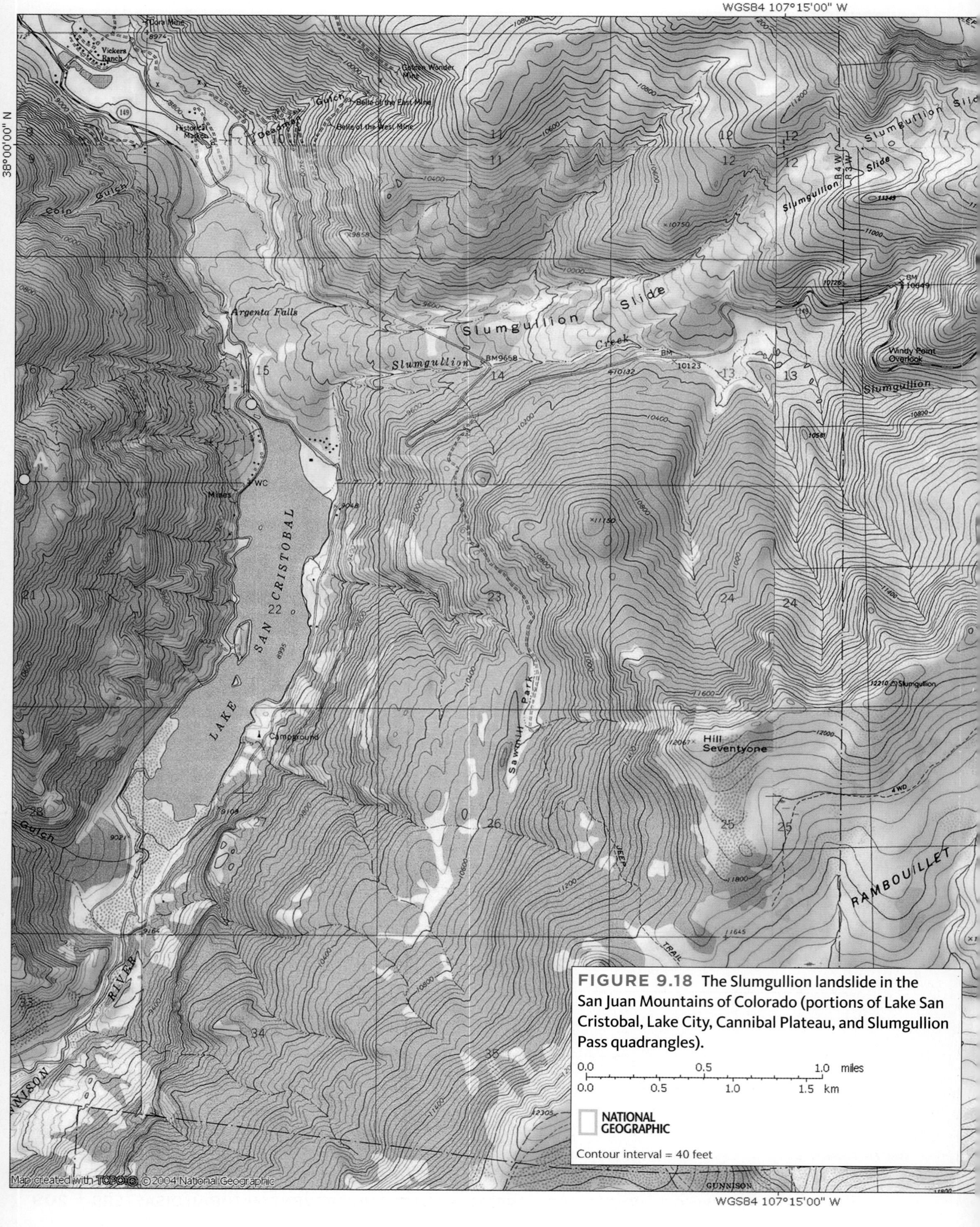

FIGURE 9.18 The Slumgullion landslide in the San Juan Mountains of Colorado (portions of Lake San Cristobal, Lake City, Cannibal Plateau, and Slumgullion Pass quadrangles).

Contour interval = 40 feet

NATIONAL GEOGRAPHIC

Map created with TOPO!® © 2004 National Geographic

Name: _____   **Section:** _____

Course: _____   **Date:** _____

(j) Satellite and ground-based measurements show that much of the upper portion of the Slumgullion slide is moving slowly downhill. The difference between the active and inactive parts of the slide mass can be seen in Figure 9.16a. Sketch the boundary between the active and inactive parts of the Slumgullion slide mass. What clues did you use to do so?

_____

_____

_____

_____

_____

_____

(k) What potential hazards would further motion of the slide pose for the area?

_____

_____

_____

(l) **Figure 9.19** is a detailed view of the Gunnison River just south of Lake San Cristobal (the southern end of the lake is at the top of the map). Had people been around, should they have been surprised that the Slumgullion slide happened? Explain your reasoning.

_____

_____

_____

_____

continued

FIGURE 9.19 Close-up view of the Gunnison River south of the Slumgullion slide in Colorado.

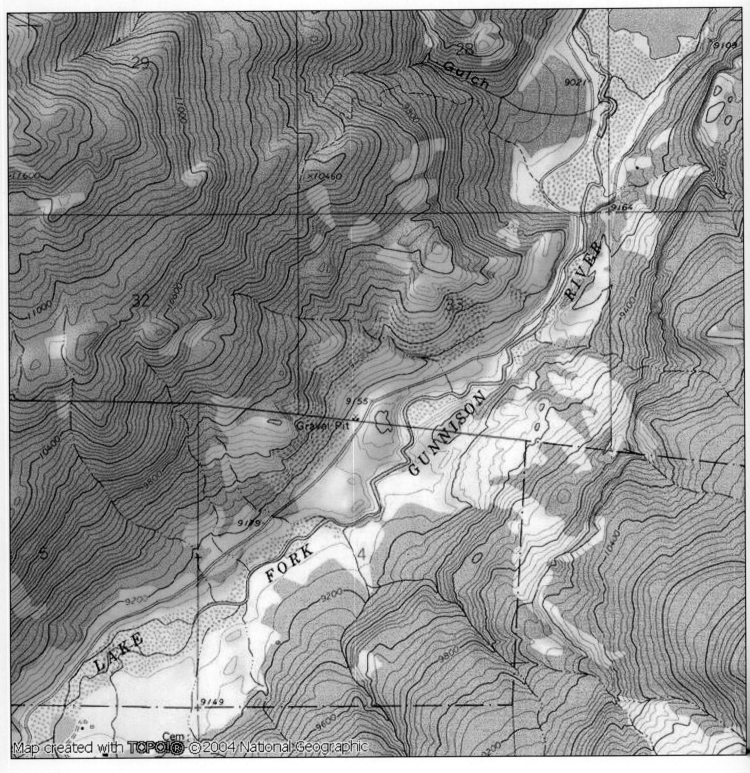

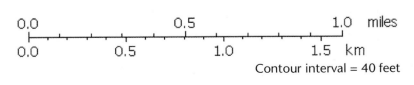

Contour interval = 40 feet

04/23/09

TOPOGRAPHIC MAP SYMBOLS

BATHYMETRIC FEATURES

Area exposed at mean low tide; sounding datum line***	
Channel***	
Sunken rock***	

BOUNDARIES

National	
State or territorial	
County or equivalent	
Civil township or equivalent	
Incorporated city or equivalent	
Federally administered park, reservation, or monument (external)	
Federally administered park, reservation, or monument (internal)	
State forest, park, reservation, or monument and large county park	
Forest Service administrative area*	
Forest Service ranger district*	
National Forest System land status, Forest Service lands*	
National Forest System land status, non-Forest Service lands*	
Small park (county or city)	

BUILDINGS AND RELATED FEATURES

Building	
School; house of worship	
Athletic field	
Built-up area	
Forest headquarters*	
Ranger district office*	
Guard station or work center*	
Racetrack or raceway	
Airport, paved landing strip, runway, taxiway, or apron	
Unpaved landing strip	
Well (other than water), windmill or wind generator	
Tanks	
Covered reservoir	
Gaging station	
Located or landmark object (feature as labeled)	
Boat ramp or boat access*	
Roadside park or rest area	
Picnic area	
Campground	
Winter recreation area*	
Cemetery	

COASTAL FEATURES

Foreshore flat	
Coral or rock reef	
Rock, bare or awash; dangerous to navigation	
Group of rocks, bare or awash	
Exposed wreck	
Depth curve; sounding	
Breakwater, pier, jetty, or wharf	
Seawall	
Oil or gas well; platform	

CONTOURS

Topographic

Index	
Approximate or indefinite	
Intermediate	
Approximate or indefinite	
Supplementary	
Depression	
Cut	
Fill	
Continental divide	

Bathymetric

Index***	
Intermediate***	
Index primary***	
Primary***	
Supplementary***	

CONTROL DATA AND MONUMENTS

Principal point**	
U.S. mineral or location monument	
River mileage marker	

Boundary monument

Third-order or better elevation, with tablet	
Third-order or better elevation, recoverable mark, no tablet	
With number and elevation	

Horizontal control

Third-order or better, permanent mark	
With third-order or better elevation	
With checked spot elevation	
Coincident with found section corner	
Unmonumented**	

CONTROL DATA AND MONUMENTS – *continued*

Vertical control

Third-order or better elevation, with tablet	BM ✕ 5280
Third-order or better elevation, recoverable mark, no tablet	✕ 528
Bench mark coincident with found section corner	BM + 5280
Spot elevation	✕ 7523

GLACIERS AND PERMANENT SNOWFIELDS

Contours and limits	
Formlines	
Glacial advance	
Glacial retreat	

LAND SURVEYS

Public land survey system

Range or Township line	
Location approximate	
Location doubtful	
Protracted	
Protracted (AK 1:63,360-scale)	
Range or Township labels	R1E T2N R3W T4S
Section line	
Location approximate	
Location doubtful	
Protracted	
Protracted (AK 1:63,360-scale)	
Section numbers	1 - 36 1 - 36
Found section corner	+
Found closing corner	
Witness corner	WC +
Meander corner	MC
Weak corner*	+

Other land surveys

Range or Township line	
Section line	
Land grant, mining claim, donation land claim, or tract	
Land grant, homestead, mineral, or other special survey monument	
Fence or field lines	

MARINE SHORELINES

Shoreline	
Apparent (edge of vegetation)***	
Indefinite or unsurveyed	

MINES AND CAVES

Quarry or open pit mine	✕
Gravel, sand, clay, or borrow pit	✕
Mine tunnel or cave entrance	
Mine shaft	
Prospect	X
Tailings	*Tailings*
Mine dump	
Former disposal site or mine	

PROJECTION AND GRIDS

Neatline	39°15′ 90°37′30″
Graticule tick	55′
Graticule intersection	
Datum shift tick	

State plane coordinate systems

Primary zone tick	640 000 FEET
Secondary zone tick	247 500 METERS
Tertiary zone tick	260 000 FEET
Quaternary zone tick	98 500 METERS
Quintary zone tick	320 000 FEET

Universal transverse metcator grid

UTM grid (full grid)	273
UTM grid ticks*	269

RAILROADS AND RELATED FEATURES

Standard guage railroad, single track	
Standard guage railroad, multiple track	
Narrow guage railroad, single track	
Narrow guage railroad, multiple track	
Railroad siding	
Railroad in highway Railroad in road Railroad in light duty road*	
Railroad underpass; overpass	
Railroad bridge; drawbridge	
Railroad tunnel	
Railroad yard	
Railroad turntable; roundhouse	

RIVERS, LAKES, AND CANALS

Perennial stream	
Perennial river	
Intermittent stream	
Intermittent river	
Disappearing stream	
Falls, small	
Falls, large	
Rapids, small	
Rapids, large	
Masonry dam	
Dam with lock	
Dam carrying road	

RIVERS, LAKES, AND CANALS – *continued*

Perennial lake/pond	
Intermittent lake/pond	
Dry lake/pond	
Narrow wash	
Wide wash	
Canal, flume, or aqueduct with lock	
Elevated aqueduct, flume, or conduit	
Aqueduct tunnel	
Water well, geyser, fumarole, or mud pot	
Spring or seep	

ROADS AND RELATED FEATURES

Please note: Roads on Provisional-edition maps are not classified as primary, secondary, or light duty. These roads are all classified as improved roads and are symbolized the same as light duty roads.

Primary highway	
Secondary highway	
Light duty road Light duty road, paved* Light duty road, gravel* Light duty road, dirt* Light duty road, unspecified*	
Unimproved road Unimproved road*	
4WD road 4WD road*	
Trail	
Highway or road with median strip	
Highway or road under construction	*Under Const*
Highway or road underpass; overpass	
Highway or road bridge; drawbridge	
Highway or road tunnel	
Road block, berm, or barrier*	
Gate on road*	
Trailhead*	

SUBMERGED AREAS AND BOGS

Marsh or swamp	
Submerged marsh or swamp	
Wooded marsh or swamp	
Submerged wooded marsh or swamp	
Land subject to inundation	*Max Pool 431*

SURFACE FEATURES

Levee	*Levee*
Sand or mud	*Sand*
Disturbed surface	
Gravel beach or glacial moraine	*Gravel*
Tailings pond	*Tailings Pond*

TRANSMISSION LINES AND PIPELINES

Power transmission line; pole; tower	
Telephone line	*Telephone*
Aboveground pipeline	
Underground pipeline	*Pipeline*

VEGETATION

Woodland	
Shrubland	
Orchard	
Vineyard	
Mangrove	*Mangrove*

* USGS-USDA Forest Service Single-Edition Quadrangle maps only.

In August 1993, the U.S. Geological Survey and the U.S. Department of Agriculture's Forest Service signed an Interagency Agreement to begin a single-edition joint mapping program. This agreement established the coordination for producing and maintaining single-edition primary series topographic maps for quadrangles containing National Forest System lands. The joint mapping program eliminates duplication of effort by the agencies and results in a more frequent revision cycle for quadrangles containing National Forests. Maps are revised on the basis of jointly developed standards and contain normal features mapped by the USGS, as well as additional features required for efficient management of National Forest System lands. Single-edition maps look slightly different but meet the content, accuracy, and quality criteria of other USGS products.

** Provisional-Edition maps only.

Provisional-edition maps were established to expedite completion of the remaining large-scale topographic quadrangles of the conterminous United States. They contain essentially the same level of information as the standard series maps. This series can be easily recognized by the title "Provisional Edition" in the lower right-hand corner.

*** Topographic Bathymetric maps only.

Topographic Map Information

For more information about topographic maps produced by the USGS, please call:
1-888-ASK-USGS or visit us at http://ask.usgs.gov/

CHAPTER 10

LANDSCAPES FORMED BY STREAMS

PURPOSE

- Learn how streams erode and deposit material.
- Become familiar with landforms formed by stream erosion and deposition.
- Interpret active and ancient stream processes from landscape features.

MATERIALS NEEDED

- Thin string
- Ruler with divisions in tenths of an inch or millimeters
- Graph paper for constructing topographic profiles
- Colored pencils
- Magnifying glass or hand lens to read closely spaced contour lines

10.1 Introduction

Water flowing in a channel is called a **stream** whether it is as large as the Amazon River or as small as the smallest creek, run, rill, or brook. Streams are highly effective agents of erosion and may move more material after one storm in an arid region than wind does all year. This chapter explores why not all streams behave the same way and how streams can produce very different landscapes.

10.2 How Do Streams Work?

All streams operate according to a few simple principles regardless of their size.

- Water in streams flows downhill because of gravity.

- Streams normally flow in a well-defined **channel**, except during floods when the water overflows the channel and spills out across the surrounding land.

- The motion of water gives a stream kinetic energy, enabling it to do the geologic work of erosion and deposition. The amount of energy depends on the amount of water and its velocity (remember: kinetic energy = $\frac{1}{2}mv^2$), so big, fast-flowing streams erode more than small, slow-flowing streams.

- Its kinetic energy allows a stream to transport sediment, from the finest mud-sized grain to small boulders. These particles slide or roll on the bed of the stream, bounce along, or are carried in suspension within the water.

- The flow of water erodes unconsolidated sediment from the walls and bed of the channel, which can abrade solid rock.

EXERCISE 10.1 **Differences between Streams**

Name: _____ Section: _____

Course: _____ Date: _____

The basic rules listed above are the same for all streams, but the rules can be applied differently, resulting in streams that look very different from one another (**Fig. 10.1**). In your own words, describe how these streams differ.

_____

_____

_____

FIGURE 10.1 A tale of two streams.

(a) Yellowstone River in Wyoming.

(b) Cascapedia River in Quebec.

- Streams deposit sediment when they lose kinetic energy by slowing down (or evaporating). The heaviest particles are deposited first, then the smaller grains, as the energy wanes.

Now for a few geologic terms: A stream **channel** is the area within which the water is actually flowing. A stream **valley** is the region within which the stream has eroded the land. In some cases, the channel completely fills the bottom of the valley; in others, it is much narrower than the broad valley floor. Some valley walls are steep, others gentle.

EXERCISE 10.2   **Getting Familiar with Properties of Streams**

Name: _____   Section: _____
Course: _____   Date: _____

(a) Which stream in Figure 10.1 has the wider channel? _____

(b) Which stream has the broader valley? _____

(c) Which stream has the most clearly developed valley walls? _____

(d) Describe the relationship between valley width and channel width for both streams.

_____

_____

_____

(e) Which stream has the straighter channel? _____ Which has a more *sinuous* (meandering) channel?

_____

(f) Which stream appears to be flowing faster? _____ What evidence did you use to determine this?

_____

_____

_____

(g) Which stream appears to be flowing more steeply downhill? _____ *Note:* The steepness of a stream channel is called its *gradient* and is a major factor in stream behavior.

(h) Look carefully at the valley of the Cascapedia River. What evidence is there that the river had more energy at one time than it does now? Where could that energy have come from?

_____

_____

_____

_____

The questions in Exercise 10.2 begin to look at factors that control stream activity. Streams are complex dynamic systems in which changes in one factor bring about changes in others, affecting the way the stream looks and behaves. For example, changes in a stream's gradient can completely change the nature of erosion and deposition, the width of the valley, and the degree to which the channel meanders.

10.2.1 Stream Erosion: Downward or Sideways

A brief lesson in stream anatomy helps to explain stream erosion and deposition. A stream begins at its **headwaters** (or head), and the point at which it ends—by flowing into another stream, the ocean, or a topographic low—is called its **mouth**. The headwaters of the Mississippi River are in Lake Itasca in Minnesota, and its mouth is the Gulf of Mexico in Louisiana. The *longitudinal profile* of a stream from headwaters to mouth is generally a smooth, concave-up curve (**Fig. 10.2**). The gradient (steepness) may vary from a few inches to hundreds of feet of **vertical drop** per mile and is typically steeper at the head than at the mouth.

A stream can erode its channel only as low as the elevation at its mouth, because if it cut deeper it would have to flow uphill to get to the mouth. The elevation at the mouth thus controls erosion along the entire stream and is called the **base level**. Sea level is the ultimate base level for streams that flow into the ocean; base level for a *tributary* that flows into another stream is the elevation where the tributary joins the larger stream.

One difference between the Yellowstone and Cascapedia Rivers is the straightness of their channels: the Yellowstone has a relatively straight channel, whereas the Cascapedia channel meanders across a wide valley floor. The **sinuosity** of a stream measures how much it meanders, as shown in the following formula. Because sinuosity is a ratio of the two lengths, it has no units. An absolutely straight stream would have a sinuosity of 1.00 (if such a stream existed), whereas streams with many meanders have high values for sinuosity (**Fig. 10.3**).

FIGURE 10.2 Longitudinal stream profile.

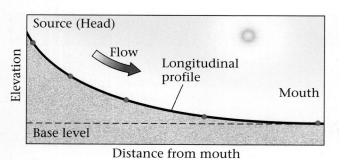

$$\text{Sinuosity} = \frac{\text{Length of stream channel (meanders and all)}}{\text{Straight-line distance between the same points}}$$

FIGURE 10.3 Stream sinuosity.

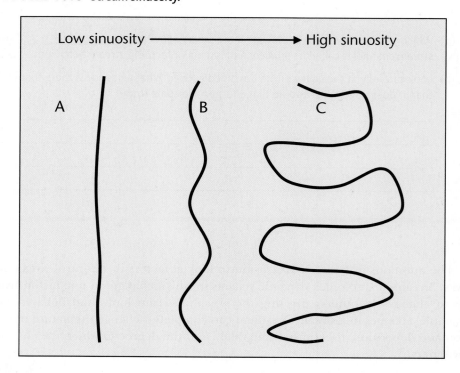

Name: _____   Section: _____
Course: _____   Date: _____

(a) Compare the course of the Big Horn River between points A and B with that of its tributary between points C and D (**Fig. 10.4**). Fill in the table below.

	Big Horn River	**Unnamed tributary**
Channel length		
Straight line length		
Sinuosity		
Highest elevation		
Lowest elevation		
Vertical drop		
Gradient		

(b) What is the apparent relationship between a stream's gradient and whether it has a straight or meandering channel?

_____

_____

_____

(c) Test this hypothesis on the Genesee River of New York (**Fig. 10.5**) and the Casino Lakes area of Idaho (**Fig. 10.6**). Complete the following table and describe how the Genesee River differs from the Idaho streams.

	Genesee River	**Casino Lakes area**	
		Stream A–B	**Stream C–D**
Valley shape (V-shaped or broad with flat bottom)			
Gradient			
Valley width			
Channel width			
Valley width/channel width			
Sinuosity			

continued

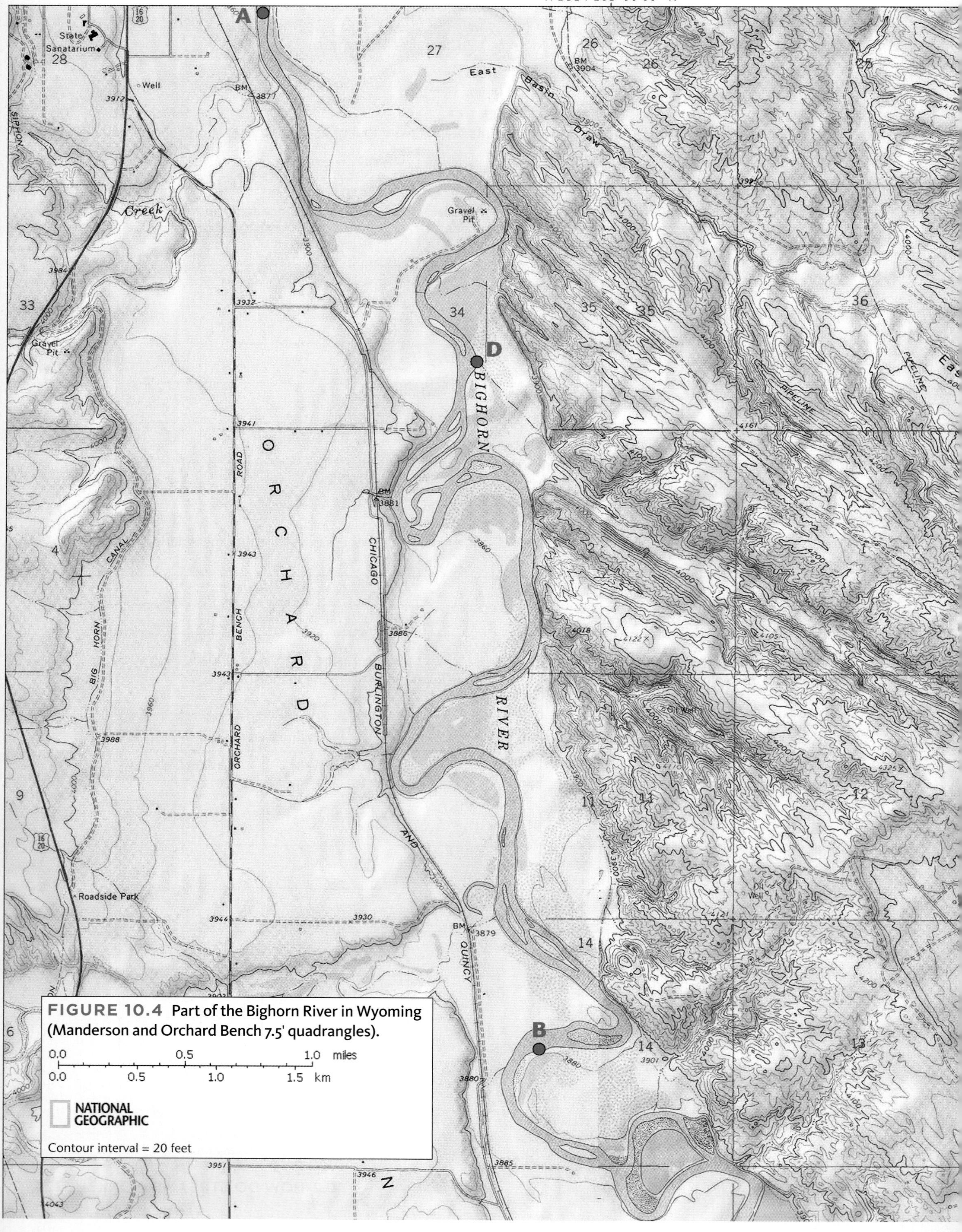

FIGURE 10.4 Part of the Bighorn River in Wyoming (Manderson and Orchard Bench 7.5' quadrangles).

0.0   0.5   1.0   miles
0.0   0.5   1.0   1.5   km

NATIONAL
GEOGRAPHIC

Contour interval = 20 feet

FIGURE 10.5 The Genesee River south of Rochester, New York.

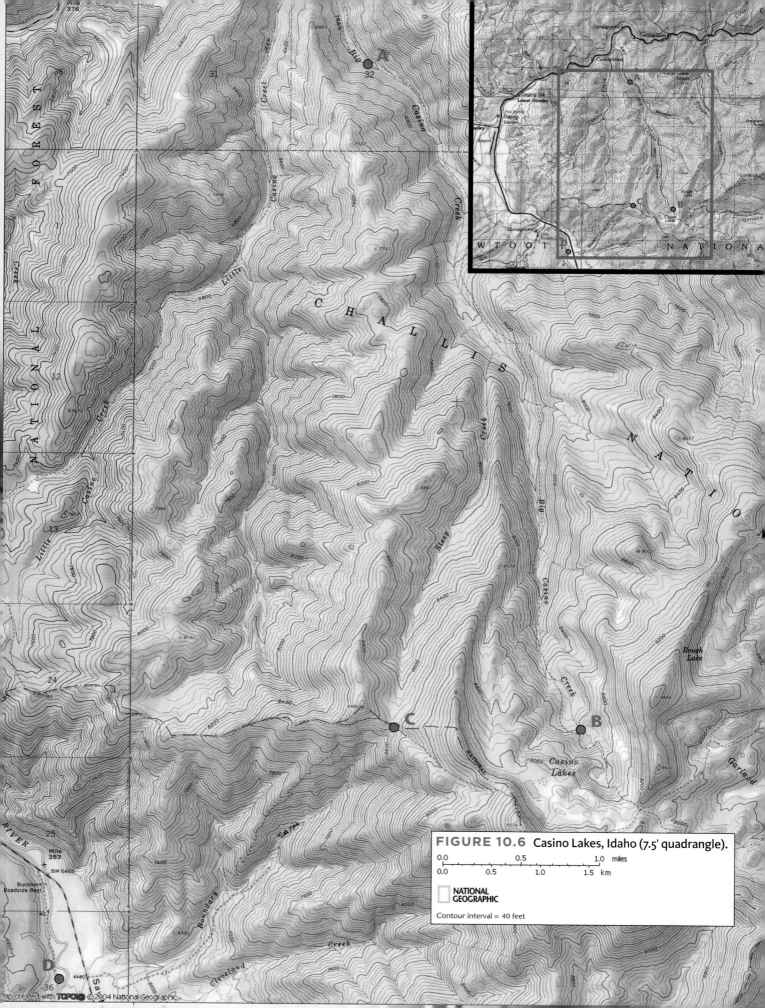

FIGURE 10.6 Casino Lakes, Idaho (7.5' quadrangle).

NATIONAL
GEOGRAPHIC

Contour interval = 40 feet

Name: _____ Section: _____

Course: _____ Date: _____

(d) Did these maps support your hypothesis about the relationship between meandering (sinuosity) and gradient? Explain.

_____

_____

_____

(e) What is the apparent relationship between sinuosity and the valley width/channel width ratio?

_____

_____

_____

(f) What is the apparent relationship between stream gradient and the shape of a stream valley?

_____

_____

_____

Now apply what you've learned to the streams in Figures 10.1 and 10.3.

(g) Which probably has the steeper gradient—the Cascapedia River or the Yellowstone River? Explain your reasoning.

_____

_____

_____

(h) Which of the streams sketched in Figure 10.3 probably has the steepest gradient? The gentlest gradient? Explain your reasoning.

_____

_____

_____

_____

10.3 Stream Valley Types and Features

The Yellowstone and Cascapedia Rivers in Figure 10.1 illustrate the two most common types of stream valleys: steep-walled, V-shaped valleys whose bottoms are occupied fully by the channel, and broad, flat-bottomed valleys much wider than the channel and within which the stream meanders widely between the valley walls. **Figure 10.7** shows how these valleys form.

When water is added to a stream in a V-shaped valley, the channel expands and fills more of the valley. When more water enters a stream with a broad, flat valley and a relatively small channel, it spills out of the channel onto the broad valley floor in a

FIGURE 10.7 Evolution of stream valleys.

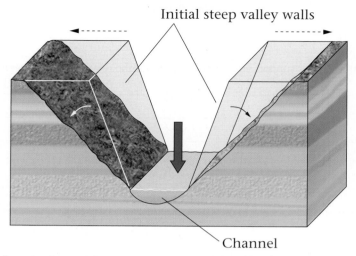

Initial steep valley walls

Channel

(a) Steep, V-shaped valley: Vertical erosion carves the channel and valley downward vertically (large blue arrow), producing steep valley walls. Mass wasting (slump, creep, landslides, and rock falls) reduces slope steepness to the angle of repose (curved arrows) and widens the top of the valley (dashed arrows).

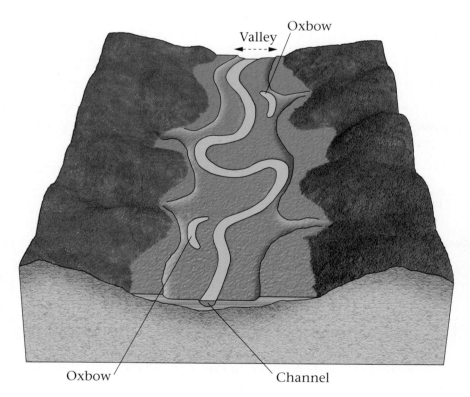

Valley
Oxbow

Oxbow
Channel

(b) Broad, flat-bottomed valley: As the stream meanders, it widens the valley (arrows). Mass wasting gentles the slopes of the valley walls as in (a). Oxbows mark the position of former meanders.

FIGURE 10.8 Floodplain and associated features.

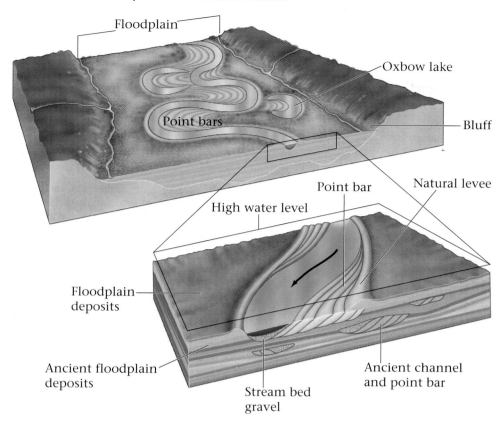

flood. Sediment carried by the floodwater is deposited on the **floodplain**, and other depositional and erosional features can be recognized easily on topographic maps or photographs (**Fig. 10.8**).

10.3.1 Features of Floodplains

Natural levees are ridges of sediment that outline the channel, and they form when a stream overflows its banks and deposits its coarsest sediment next to the channel. Several generations of natural levees are visible in Figure 10.8, showing how the meanders changed position with time. **Point bars** form when water on the inside of a meander loop slows down, causing sediment to be deposited. At the same time, erosion occurs on the outside of the meander loop because water there moves faster. The result is that meanders migrate with time, moving outward (toward their convex side) and downstream. Sometimes a stream cuts off a meander and straightens itself. The levees that formerly flanked the meanders help outline the former position of the river, leaving **meander scars**. Meander scars that have filled with water are called **oxbow lakes**.

10.4 Changes in Streams over Time

Streams erode *vertically* by leveling the longitudinal profile to the elevation of the mouth. Streams also erode *laterally*, broadening their valleys by meandering. Most streams erode both laterally and vertically at the same time, but the balance between vertical and lateral erosion commonly changes as the stream evolves.

Headwaters of a high-gradient stream are much higher than its mouth, and stream energy is used largely in vertical erosion, lowering channel elevation all along its profile. Over time, the gradient lessens as erosion lowers the headwater area. The stream still has energy with which to alter the landscape, some of which it uses to erode laterally. The valley then widens by a combination of mass wasting and meandering. Even when headwaters are lowered to nearly the same elevation as the mouth, a stream still has energy for geologic work, but because it cannot cut vertically below its base level, most energy at this stage must be used for lateral erosion.

Initially, a stream meanders within narrow valley walls, but with time it erodes those walls farther and farther, eventually carving a very wide valley. As its gradient decreases, a stream redistributes sediment that it had deposited, moving it back and forth across the floodplain.

It is important to understand that some streams have very gentle gradients and meander widely from the moment they begin. Streams on the Atlantic and Gulf coastal plains are good examples of this kind of behavior.

EXERCISE 10.4   **Interpreting Stream Behavior**

Name: _____   Section: _____

Course: _____   Date: _____

Figure 10.9 shows three meandering streams, each of which balances energy use differently between vertical and lateral erosion.

(a) From their valley width/channel width ratios alone, which stream would you expect to have the steepest gradient? The gentlest gradient? Explain your reasoning.

_____

_____

_____

(b) Which stream do you think is doing the most vertical erosion? The least? Explain.

_____

_____

_____

(c) What indicates former positions of the meandering rivers? For all three? Explain.

_____

_____

_____

continued

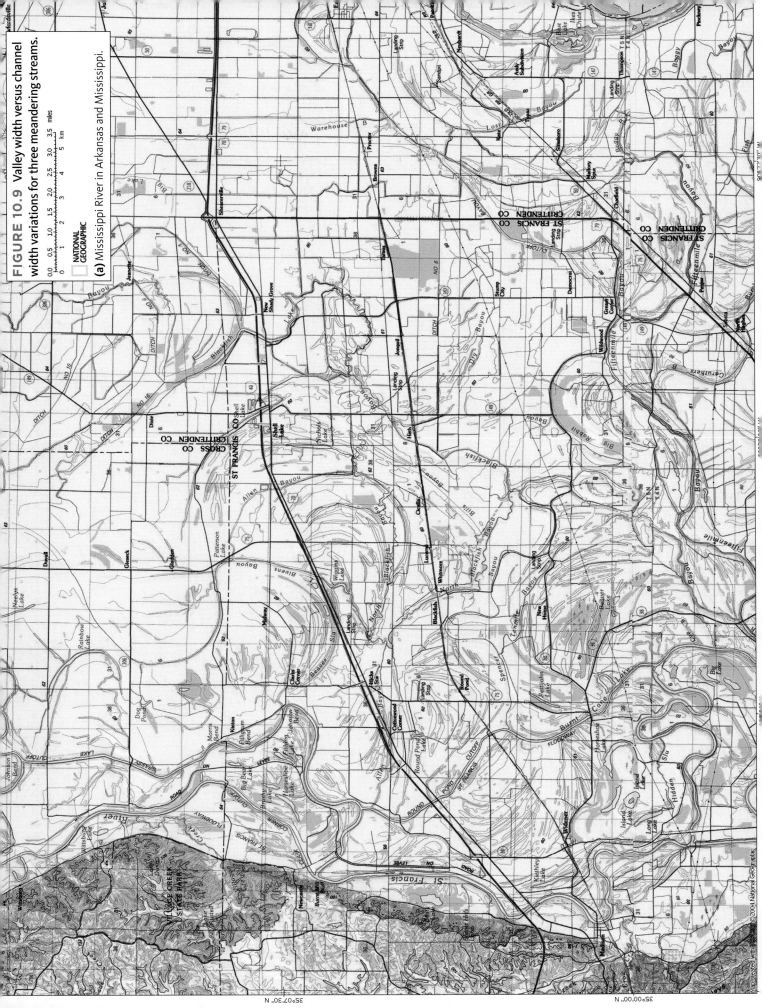

FIGURE 10.9 Valley width versus channel width variations for three meandering streams.

(a) Mississippi River in Arkansas and Mississippi.

FIGURE 10.9 continued.

(b) Meadow River in West Virginia.

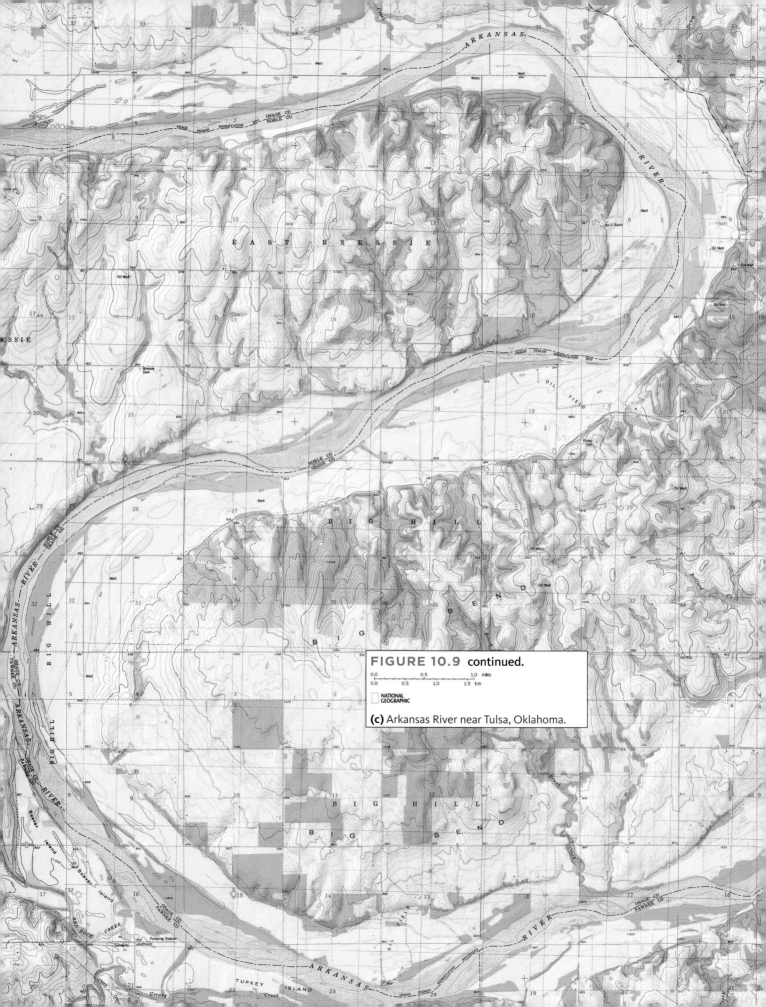

FIGURE 10.9 continued.

0.0   0.5   1.0   miles
0.0   0.5   1.0   1.5   km

NATIONAL
GEOGRAPHIC

(c) Arkansas River near Tulsa, Oklahoma.

Name: _____   **Section:** _____

Course: _____   **Date:** _____

Label as many erosional and depositional fluvial landforms as you can in Figures 10.4, 10.5, 10.6, and 10.9. What do these features reveal about the active and ancient stream processes in their areas?

_____

_____

_____

_____

10.5 Stream Networks

Streams are particularly effective agents of erosion because they form networks that cover much of Earth's surface. Rain falling on an area runs off into tiny channels that carry water into bigger streams and eventually into large rivers. Each stream—from tiniest to largest—expands headward over time as water washes into its channel, increasing the amount of land affected by stream erosion. Understanding the geometric patterns of stream networks and the way they affect the areas they drain is the key to understanding how to prevent or remedy stream pollution, soil erosion, and flood damage.

10.5.1 Drainage Basins

The area drained by a stream is its **drainage basin**, which is separated from adjacent drainage basins by highlands called **drainage divides**. The drainage basin of a small tributary may cover a few square miles, but that of the master stream may be hundreds of thousands of square miles. **Figure 10.10** shows large-scale drainage basins whose waters flow into the Pacific, Arctic, and Atlantic Oceans, the Hudson Bay, and the Gulf of Mexico. The Mississippi drainage basin is the largest and drains much of the interior of the United States. The *continental divide* separates streams that flow into the Atlantic Ocean from those that flow to the Pacific. The Appalachian Mountains are the divide separating the Gulf of Mexico and direct Atlantic Ocean drainage; the Rocky Mountains separate Gulf of Mexico and Pacific drainage. A favorite tourist stop in Alberta, Canada, is a *triple* divide that separates waters flowing north to the Arctic Ocean, west to the Pacific, and south to the Gulf of Mexico.

10.5.2 Drainage Patterns

Master and tributary streams in a network typically form one of six geometric patterns (**Fig. 10.12**). **Dendritic** patterns (from the Greek *dendros*, for veins in a leaf), develop where surface materials are equally resistant to erosion. This may mean horizontal sedimentary or volcanic rocks; loose, unconsolidated sediment; or igneous and metamorphic areas where most rocks erode at the same rate. **Trellis** patterns form where ridges of resistant rock alternate with valleys underlain by weaker material. **Rectangular** patterns indicate zones of weakness (faults, fractures) perpendicular

FIGURE 10.10 Major drainage basins of North America.

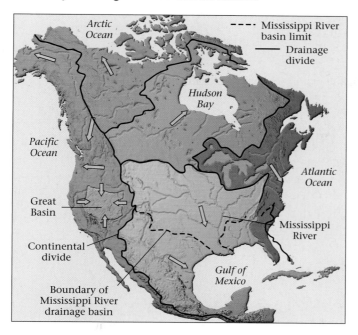

EXERCISE 10.6   **Drainage Basins and Stream Divides**

Name: _____   Section: _____
Course: _____   Date: _____

Name: _____   Section: _____
Course: _____   Date: _____

Figure 10.11 is a map showing several tributaries on the north and south sides of the Missouri River near Jefferson City, Missouri. One large tributary, the Osage River, joins the Missouri from the south, near the eastern margin of the map, but most of the tributaries on the north are much smaller.

(a) With a colored pencil, trace one of the tributary creeks feeding *directly* into the Missouri from the north. With the same pencil, trace tributaries that flow directly into this creek, and then the tributaries of these smaller streams.

(b) With a different color, trace an adjacent tributary of the Missouri and its tributaries. Repeat for more streams and their tributaries on the north side of the Missouri River, using a different color for each stream.

(c) You have just outlined most of the drainage on the north side of the Missouri. Now, with a red pencil, trace the divides that separate the individual drainage basins for each master stream. This should be easy since you've already identified streams in each drainage basin with a different color.

Note that some divides are defined sharply by narrow ridges, but others are more difficult to locate within broad upland areas where most of the headwaters are located.

(d) With dotted lines, suggest how headward erosion might extend each main stream in the future.

(e) What do you think will happen when the headwaters of two streams meet during headward erosion?

_____

_____

_____

(f) Local residents are worried that a recent toxic spill at an electrical substation (asterisk in Fig. 10.11) will work its way into the drainage system. Based on your drainage basin analysis, shade in areas that might be affected. Be conservative: if there is any doubt, err on the side of including areas rather than excluding them.

FIGURE 10.11 Drainage divides of the Missouri River near Jefferson City, Missouri.

0.0 0.5 1.0 1.5 2.0 2.5 3.0 3.5 miles

0 1 2 3 4 5 km

NATIONAL GEOGRAPHIC

Contour interval = 20 feet

to one another. In **radial** patterns, streams flow either outward (**centrifugal**) from a high point (e.g., a volcano) or inward (**centripetal**) toward the center of a large basin. **Annular** drainage patterns occur where there are concentric rings of alternating resistant and weak rocks—typically found in structures called domes and basins.

FIGURE 10.12 Common drainage patterns.

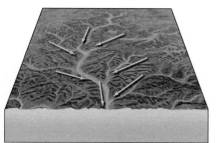

Dendritic

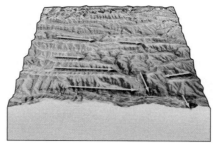

Trellis

Rectangular

Radial: centrifugal

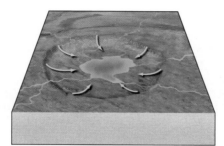

Radial: centripetal

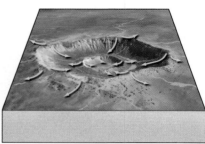

Annular

EXERCISE 10.7   **Recognizing Drainage Patterns**

Name: _____   Section: _____

Course: _____   Date: _____

(a) What drainage pattern is associated with the Mississippi River drainage basin? What does that tell you about the materials that underlie the central part of the United States?

_____

_____

_____

(b) What drainage patterns are associated with the areas shown in Figures 10.4, 10.5, 10.6, 10.11, 10.15, 10.16, and 10.17? What do these indicate about the rocks underlying those areas?

_____

_____

_____

_____

_____

_____

10.6 Changes in Stream-Carved Landscapes with Time

Just as a single stream or entire drainage network changes over time, so too do **fluvial** (stream-created) *landscapes*. Consider a large block of land uplifted to form a plateau. Several things will change with time: the highest elevation, the number of streams, the stream gradients, and the amount of flat land relative to the amount of land that is part of valley walls. These changes reflect the different ways in which stream energy is used as the landscape is eroded closer to base level. **Figure 10.13** summarizes these idealized changes. In the real world, things rarely remain constant long enough for this cycle to reach its end: sea level may rise or fall due to glaciation or tectonic activity, the land may be uplifted tectonically, and so on. Nevertheless, the stages in Figure 10.13 are typical of landscapes produced by stream erosion and can be recognized on maps and other images.

FIGURE 10.13 Idealized stages in the evolution of a stream-carved landscape.

Stage 1: A fluvial landscape is uplifted, raising stream channels above base level.
- Few streams, but with relatively steep gradients.
- Broad, generally flat divides between streams.
- Channels are relatively straight.
- Relief is low to modest (not much elevation difference between divides and channels).

Stage 2: Main streams cut channels downward and tributaries form drainage networks.
- Numerous tributaries develop with moderate gradients, dissecting most of the area.
- Stream divides are narrow and sharp; most of area is in slope, with little flat ground.
- Main streams and some tributaries meander moderately.
- Relief is high, and little, if any, land is at the original uplifted elevation.

Stage 3: Stream erosion has lowered land surface close to base level.
- Few streams as in Stage 1, but with gentle gradients.
- Stream divides are broad and flat; relief is low.
- Streams meander broadly.

Name: _____   **Section:** _____
Course: _____   **Date:** _____

(a) Examine the topographic maps in Figures **10.14**, **10.15**, and **10.16**. Fill in the following table and explain your reasoning for each decision.

	Southeast Texas (Fig. 10.14)	Colorado Plateau (Fig. 10.15)	Appalachian Plateau (Fig. 10.16)
Greatest relief (distance between highest and lowest points)			
Greatest number of streams			
Greatest land area involved in valley slopes			
Steepest stream gradients			
Stream divides: flat versus broadly rounded versus angular			
Stage of stream dissection (Stage 1, 2, or 3)			

(b) With this practice, look at the other fluvial landscapes in this chapter and suggest which stages of erosion each map represents.

_____

_____

_____

_____

_____

_____

(c) Arrange the areas shown in Figures 10.14, 10.15, and 10.16 in order from earliest stage of erosion to most advanced.

Earliest stage   Latest stage

_____

_____

_____

continued

FIGURE 10.14 Stream dissection in southeast Texas (Fred, Spurger, Magnolia Springs, and Potato Patch Lake quadrangles).

NATIONAL GEOGRAPHIC

Contour interval = 5 feet

FIGURE 10.15 Stream dissection in the Colorado Plateau (Del Muerto quadrangle in Arizona).

NATIONAL GEOGRAPHIC

Contour interval = 20 feet

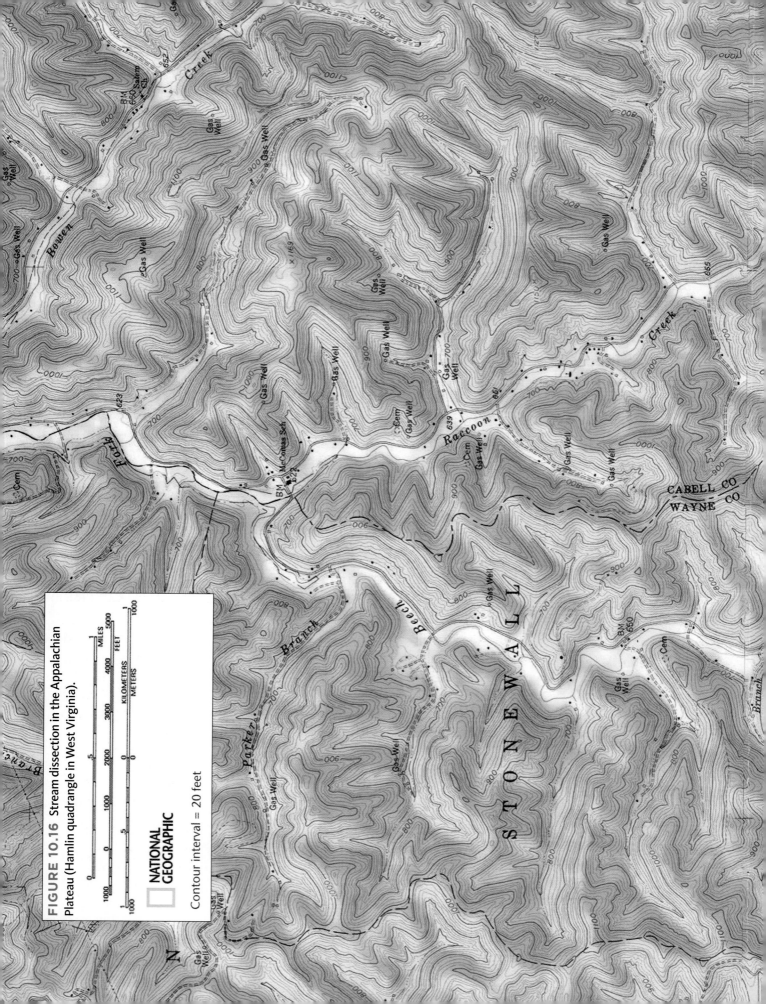

FIGURE 10.16 Stream dissection in the Appalachian Plateau (Hamlin quadrangle in West Virginia).

Contour interval = 20 feet

10.7 When Streams Don't Seem to Follow the Rules

Most stream erosion and deposition follow the principles you just deduced, but there are some notable exceptions. Actually, these streams aren't violating any rules; they are following them to the letter, but their situations are more complex than the basic ones we have examined.

Consider, for example, the Susquehanna River as it flows across the Pennsylvania landscape shown in **Figure 10.17**. This area is part of the Valley and Ridge Province of the Appalachian Mountains and is characterized by elongate ridges and valleys made of resistant and nonresistant rocks, respectively. Streams are the dominant agent of erosion in the area.

EXERCISE 10.9   **Deducing the History of the Susquehanna River**

Name: _____   Section: _____
Course: _____   Date: _____

(a) Figure 10.17 shows a part of the Valley and Ridge Province in Pennsylvania. What is unusual about the relationship between the Susquehanna and Juniata Rivers and the valley-and-ridge topography?

_____

_____

Most of the small streams flow in the elongate valleys, but the Susquehanna and its tributary, the Juniata River, cut across the ridges at nearly right angles. It is tempting to think that the big streams had enough energy to cut through the ridges while the small ones couldn't, but this is not the case. The answer lies in a multistage history of which only the last phase is visible today.

(b) Why do most of the smaller streams flow in the elongate valleys?

_____

_____

_____

(c) Suggest as many hypotheses as you can to explain why the two larger rivers cut across the valley-and-ridge topography. *Hint:* How might the landscape have been different at an earlier time?

_____

_____

_____

_____

_____

Rivers with enough energy to cut through the ridges should certainly have been able to simply meander around them, but the Susquehanna and Juniata Rivers didn't take the easy way out. It's almost as if they didn't even know the ridges and valleys were there. As a matter of fact . . . but no more help.

continued

Name: _____   Section: _____

Course: _____   Date: _____

(d) With that clue, suggest a series of events that explains the behavior of the Susquehanna and Juniata Rivers. *Hint:* This type of stream is called a *superposed* stream.

_____

_____

_____

_____

FIGURE 10.17 The Susquehanna River cutting across the Appalachian Valley and Ridge Province in Pennsylvania.

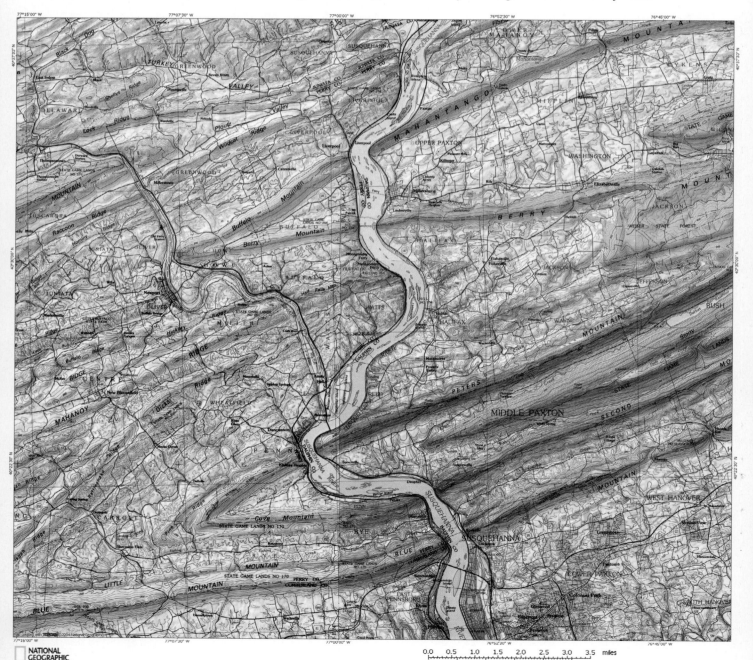

NATIONAL GEOGRAPHIC

Name: _____   **Section:** _____

Course: _____   **Date:** _____

The Green River is in the Colorado Plateau, an area where meanders of many rivers cut deeply (are *incised*) into the bedrock (**Fig. 10.18**). The most famous is the Colorado River itself, particularly where it flows through the Grand Canyon. This behavior is totally unlike that of the meandering streams encountered earlier in this chapter. The Green River seems to violate rules of stream behavior but, as with the Susquehanna, it is following them perfectly. Some geologic detective work will let you figure out the difference.

(a) Describe the path of the Green River as shown in Figure 10.18.

_____

_____

(b) What is the sinuosity of the Green River? _____

(c) When a river meanders with such sinuosity, how is it using most of its energy—in lateral or vertical erosion?

_____

_____

(d) What evidence is there that the Green River is eroding laterally?

_____

_____

_____

(e) What evidence is there that the Green River is eroding vertically?

_____

_____

_____

(f) What is the probability that the Green River will straighten its path by cutting through the walls of the Bowknot Bend? Explain.

_____

_____

_____

(g) The Green River flows into the Colorado, which flows into the Gulf of Southern California. How far above base level is the river in this area? Is this what you expect for a meandering stream? Explain.

_____

_____

_____

(h) Suggest an origin for the incised meanders. (Don't forget possible effects of tectonic activity.)

_____

_____

_____

continued

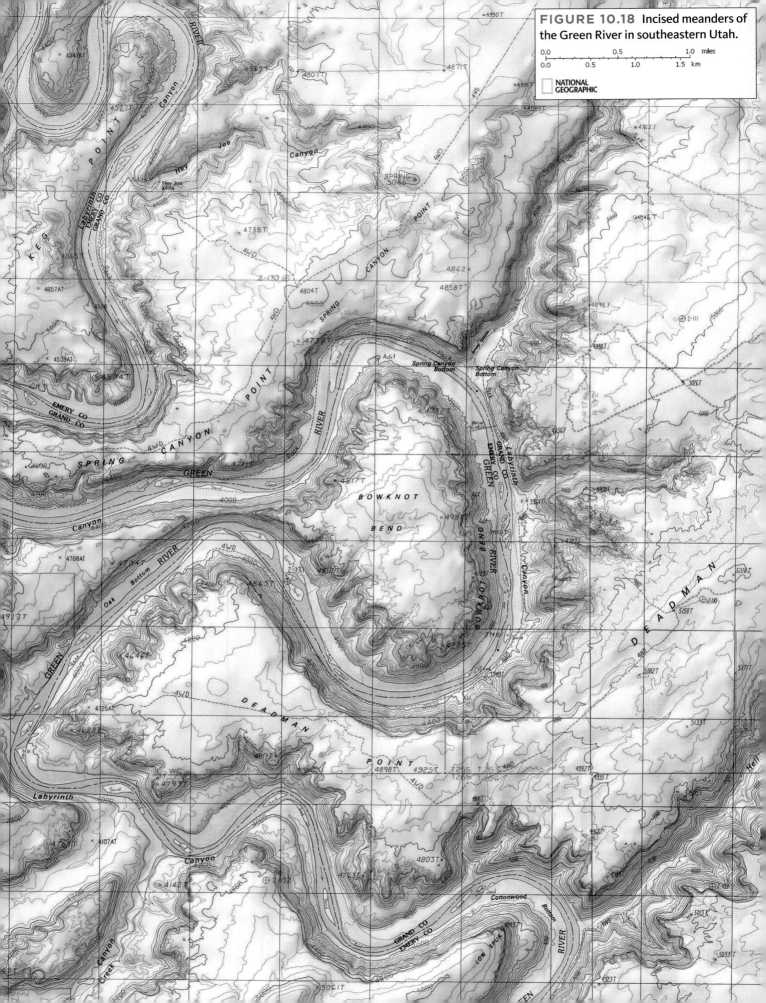

FIGURE 10.18 Incised meanders of the Green River in southeastern Utah.

10.8 When There's Too Much Water: Floods

A flood occurs when more water enters a stream than its channel can hold. Many floods are seasonal, caused by heavy spring rains or melting of thick winter snow. Others, called **flash floods**, follow storms that can deliver a foot or more of rain in a few hours. **Figure 10.19** is a map compiled by FEMA showing the estimated flood potential in the United States, based on the number of square miles that would be inundated. It might appear that states lightly shaded in Figure 10.19 would have little flood damage, but that would be an incorrect reading of the map because two of the worst river floods in U.S. history occurred in Rapid City, South Dakota, and Johnstown, Pennsylvania. Indeed, these two cities have been flooded many times. The *area* of potential flooding may not be as large as in some other states, but the *conditions* for flooding may occur frequently. South Dakota's Rapid Creek has flooded more than thirty times since the late 1800s, including a disastrous flash flood on June 9, 1972, triggered by 15 inches of rain in 6 hours. The creek overflowed or destroyed several dams, ruined more than 1,300 homes and 5,000 cars, and killed more than 200 people in Rapid City (**Fig. 10.20**).

FIGURE 10.19 Flood risk in the United States.

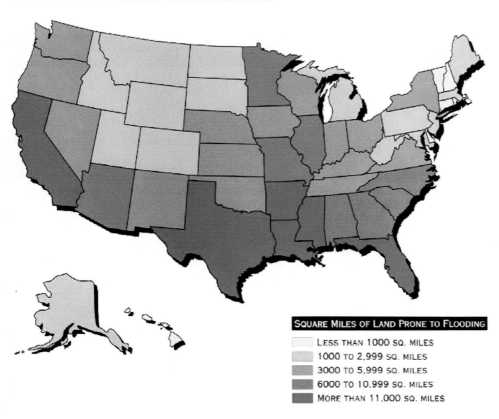

SQUARE MILES OF LAND PRONE TO FLOODING

LESS THAN 1000 SQ. MILES
1000 TO 2,999 SQ. MILES
3000 TO 5,999 SQ. MILES
6000 TO 10,999 SQ. MILES
MORE THAN 11,000 SQ. MILES

FIGURE 10.20 Effects of the June 9, 1972, flood in Rapid City, South Dakota.

(a) Rapid Creek has a narrow floodplain where it flows through Dark Canyon, 1 mile upstream of Rapid City. Floodwaters filled the entire floodplain. Concrete slabs (arrows) are all that is left of the homes built in the floodplain.

(b) The spillway of the Canyon Creek Dam (arrow) was clogged by debris carried by the floodwater, causing water to flow over the dam, which then failed completely.

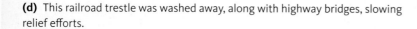

(c) Not the usual lineup of cars for gasoline.

(d) This railroad trestle was washed away, along with highway bridges, slowing relief efforts.

Name: _____   **Section:** _____
Course: _____   **Date:** _____

Streams shown on maps earlier in the chapter are all subject to flooding, but the potential problems for cities along their banks are not the same.

(a) For which streams would flooding not be a problem? Explain.

_____

_____

_____

(c) Would students at the State University of New York campus at Geneseo on the east bank of the Genesee River (Fig. 10.5) have to be evacuated if water rose 20 feet?

(b) What would be the effect of a flash flood that raised the level of the Bighorn River (Fig. 10.4) 20 feet above its banks? Shade the area that would be affected.

(d) Shade the area that would be inundated if floodwaters of the Neches River (Fig. 10.14) crested 15 feet above normal level.

FIGURE 10.21 Aerial image of the floodplain of the Missouri River near Jefferson City, Missouri (compare with Fig. 10.11).

continued

Name: _____   **Section:** _____

Course: _____   **Date:** _____

(e) Compare the map of the Missouri River in Figure 10.11 with **Figure 10.21**, a satellite image of the same area. What information does the satellite image add?

_____

_____

_____

(f) What things have been built in the floodplain that could be destroyed or made unusable in a major flood? How would the loss of these affect relief efforts?

_____

_____

_____

CHAPTER 11

GLACIAL LANDSCAPES

PURPOSE

- Learn how glaciers erode and deposit material.
- Become familiar with landforms carved and deposited by glaciers.
- Distinguish glaciated landscapes from those formed by other agents of erosion.

MATERIALS NEEDED

- Colored pencils
- Ruler and paper for constructing topographic profiles

11.1 Introduction

Glaciers hold more than 21,000 times the amount of water in all streams and account for about 75% of Earth's freshwater. Today, huge continental ice sheets cover most of one continent (Antarctica) and nearly all of Greenland. Smaller glaciers carve valleys on the slopes of high mountains, causing a characteristically sharp, jagged topography (**Fig. 11.1**). Distinctive landforms show that continental glaciers were even more widespread during the Pleistocene Epoch—the so-called ice ages—covering much of northern Europe and North America.

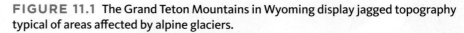

FIGURE 11.1 The Grand Teton Mountains in Wyoming display jagged topography typical of areas affected by alpine glaciers.

Today, glaciers are receding at rates unprecedented in human history. As they shrink, the balance of the water cycle changes among glaciers, oceans, streams, and groundwater, altering our water supply and modifying ecologic systems worldwide. Glaciated landscapes contain clues to past climate changes and help us understand what is happening now and plan for the future. We know, for example, that continental glaciers advanced and retreated several times during the Pleistocene and we know how fast those changes were. This enables us to measure human impact on the rates of these processes.

11.1.1 Types of Glaciers

Some small glaciers form on mountains and flow downhill, carrying out their erosion and depositional work in the valleys they carve. These are called, appropriately, **mountain, valley,** or **alpine glaciers** (**Fig. 11.2a**). Although they flow in valleys like streams, mountain glaciers work differently and form distinctly different landscapes. The Antarctic and Greenland **continental ice sheets** described earlier are not confined to valleys. They flow across the countryside, as masses of ice thousands of feet thick, dwarfing hills and burying all but the tallest peaks (**Fig. 11.2b**). The North American and European ice sheets did the same during the Pleistocene. Exercise 11.1 compares the ways in which glaciers and water do their geologic work.

FIGURE 11.2 Mountain and continental glaciers.

(a) The Crowfoot Glacier in Banff National Park, Canada.

(b) Continental ice sheet, Antarctica.

EXERCISE 11.1 **Comparison of Glaciers and Streams**

Name: _____   Section: _____

Course: _____   Date: _____

Glaciers behave differently from stream water, even those that occupy valleys. Based on what you know about streams and have learned about glaciers, complete the following table.

		Streams	Continental glaciers	Valley glaciers
State of matter		Liquid		
Composition		H_2O		
Areal distribution		In channels with tributaries flowing into larger streams		
Rate of flow		Fast or slow, depending on gradient		
Erodes by . . .		1. Abrasion using sediment load 2. Dissolving soluble minerals, rocks		
Maximum depth to which erosion can occur		Base level: sea level for streams that flow into the ocean; master stream for tributaries		
Deposits when . . .		Stream loses kinetic energy either by slowing down or evaporating		
Type(s) of sediment	**Maximum clast size range**	Small boulders to mud		
	Sorting	Generally well sorted		
	Clast shape	Moderately to well rounded, depending on distance transported		

11.1.2 How Glaciers Create Landforms

Glaciers erode destructional landforms and deposit constructional landforms. A glacier can act as a bulldozer, scraping the regolith from an area and plowing it ahead. When it encounters bedrock, a glacier uses its sediment like grit in sandpaper to abrade the rock and carve unique landforms. Glacial erosion occurs wherever ice is flowing, whether the glacier front is advancing or retreating.

Depositional landforms are equally distinctive—some are ridges tens or hundreds of miles long, others are broad blankets of sediment. All form when the ice in which the sediment is carried melts, generally when the glacier front is retreating. Some sediment, called **till**, is deposited directly from melting ice, but some is carried away by meltwater and is called **outwash** (or *glaciofluviatile* [glacial + stream] *sediment*). Till and outwash particles differ in size, sorting, and shape (**Fig. 11.3**) because of the different properties of the water and ice that deposited them.

FIGURE 11.3 Till and outwash.

(a) Till with characteristic poor sorting (small boulders through sand, silt, and mud) and angular rock fragments.

(b) Layers of outwash. Each layer is very well sorted; most contain well-rounded, sand-sized grains and some show cross bedding. Note the coarser, pebbly layer (arrow) resulting from an increase in meltwater discharge.

11.2 Landscapes Produced by Continental Glaciation

Landscapes formed by continental glaciation differ from those formed by streams in every imaginable way: the shapes of hills and valleys are different and both continental and mountain glaciers produce unique landforms not found in fluvial landscapes. **Figure 11.4** shows an area affected by continental glaciation for comparison with fluvial landscapes, and Exercise 11.2 explores the differences more fully.

Continental glaciers scour the ground unevenly as they flow, producing an irregular topography filled with isolated basins that often become lakes when the ice melts. Many of these post-Pleistocene lakes are still not incorporated into normal networks of streams and tributaries, thousands of years after the ice melted. Glaciers also deposit material unevenly, clogging old streams and choking new ones with more sediment than they can carry. *Poorly integrated drainage, with many swampy areas, is thus characteristic of recently glaciated areas.* In addition, continental glaciers carve bedrock into characteristic streamlined shapes that indicate the direction the ice flowed.

EXERCISE 11.2   **Comparison of Landscapes Formed by Glaciers and Streams**

Name: _____   Section: _____
Course: _____   Date: _____

(a) Contrast and compare the glaciated and stream-carved valleys in Figure 11.4. The U-shaped valley in Figure 11.4a is typical of glacially carved landscapes, in contrast to the V-shaped valley in Figure 11.4b, which is typical of high-gradient stream valleys.

_____

_____

FIGURE 11.4 Comparison of valleys carved by glaciers and streams.

(a) An unnamed valley in Glacier National Park, Montana.

(b) Yellowstone River valley in Wyoming.

continued

Name: _____   Section: _____
Course: _____   Date: _____

(b) Suggest a reason for the difference in shape. *Hint:* Use a colored pencil to fill in the portion of both valleys in which active erosion occurred. (We will return to this issue below.)

_____

_____

(c) Now compare an extensively glaciated area (**Fig. 11.5**) with the photographs, maps, and digital elevation models of fluvial landscapes (in Chapter 10). Describe the differences without worrying about technical terms for the glacial features.

_____

_____

_____

FIGURE 11.5 Aerial view of part of the Canadian Shield after extensive continental glaciation.

11.2.1 Erosional Landscapes

When a continental glacier passes through an area, it carves the bedrock into characteristic shapes called *sculptured bedrock*. The DEM in **Figure 11.6** shows the topography of an area eroded by continental glaciers in southern New York State. The hills are underlain by high-grade gneisses that are resistant to erosion and the valleys by less resistant marbles and schists.

11.2.2 Depositional Landscapes

Landscapes formed by continental glaciers contain several depositional landforms. Those composed of till deposited directly from the melting ice are called **moraines** or **drumlins**; those made of outwash carried by meltwater streams include **outwash plains** and **eskers**.

Name: _____   **Section:** _____
Course: _____   **Date:** _____

(a) Describe the topography in Figure 11.6 in your own words, paying particular attention to the shapes of the bedrock hills. Look for any "topographic grain" that might be related to glacial activity.

_____

_____

_____

Now look at the topographic map (**Fig. 11.7**) for more detail.

(b) Construct a topographic profile from St. John's Church to Lake Rippowam (Line A–B) on the topographic map in Figure 11.7, using 10× vertical exaggeration. Are the hills symmetrical or asymmetrical?

_____

(c) Are the steep and gentle slopes distributed randomly or systematically? Describe the distribution.

_____

_____

(d) What is controlling the steep slopes of these asymmetric hills?

_____

_____

(e) Based on your description and observations, draw arrows on Figures 11.6 and 11.7 showing the direction in which the glacier moved across the map area. Explain your reasoning.

_____

_____

(f) Fill in the blanks to complete the following sentence that describes how sculptured rock records glacial movement: The direction of ice flow indicated by sculptured rock is *from* the _____ side of the hill *toward* the _____ side.

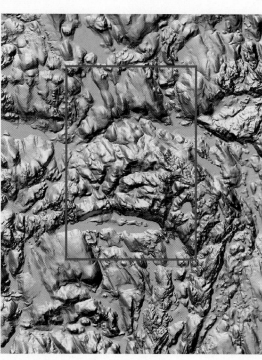

FIGURE 11.6 DEM of the Peach Lake quadrangle in New York (area in rectangle shown in Fig. 11.7).

continued

FIGURE 11.7 Portion of the Peach Lake, New York, 7.5' quadrangle showing sculptured bedrock.

Moraines: Moraines form when melting ice drops the boulders, cobbles, sand, silt, and mud that it has been carrying. Some moraines are irregular ridges; others are broad carpets pockmarked by numerous pits separated by isolated hills. The type of moraine—ridge or carpet—depends on whether the front (**terminus**) of the glacier is *retreating* (i.e., the ice is melting faster than it is being replaced) or *stagnating* (i.e., the terminus is in a state of dynamic equilibrium in which new ice is replenishing what is melting so that the position of the terminus doesn't change).

Till deposited from a retreating glacier forms an irregular carpet of debris called a **ground moraine** that may be hundreds of feet thick. Large blocks of ice are isolated as a glacier retreats and some may be buried by outwash or till. When a block melts, the sediment subsides and forms a depression in the ground moraine called a **kettle hole**. For this reason, the irregular surface of a ground moraine is often referred to as "knob and kettle" topography.

The terminus of a stagnating glacier may remain in the same place for hundreds or thousands of years, and the till piles up in a ridge that outlines the terminus. This ridge is called a **terminal moraine** and shows the maximum extent of the glacier.

Figure 11.8 shows depositional features created at the termini of continental glaciers. Use these models to identify landforms in the following DEMs and maps. A glossary of glacial depositional and erosional features can be found in **Table 11.1** at the end of this chapter.

Drumlins: Drumlins are streamlined, asymmetric hills made of till that occur in four large swarms in Nova Scotia, southern Ontario/New York State, southeastern New England (most famously Bunker Hill), and the Midwestern states of Wisconsin, Iowa, and Minnesota. Their origin is debated, but their shapes clearly reflect the dynamics of ice flowing through till, and their shape and orientation are good indicators of glacial direction.

Outwash Plains: A melting continental glacier generates an enormous amount of water, which flows away from the terminus carrying all the clasts the water is powerful enough to carry, usually sand, gravel, and small boulders. Meltwater streams round and sort this sediment and eventually deposit it beyond the terminal moraine as a generally flat outwash plain (Fig. 11.8).

Eskers: Tunnels may form at the base of a continental glacier and act as pipelines to carry meltwater even as the glacier ice surrounding the tunnels continues to flow. Well-sorted sediment carried by these subglacial streams may build up to form ridges fifty feet high or more. These ridges, called *eskers*, trace the courses of meandering subglacial streams and stand above the unsorted till of the adjacent ground moraine (Fig. 11.8).

FIGURE 11.8 Formation of depositional landforms after continental glaciation.

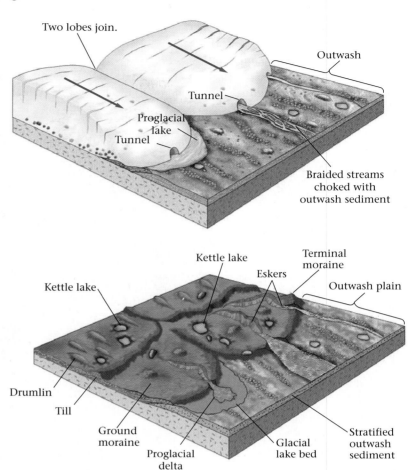

Name: _____   Section: _____

Course: _____   Date: _____

Figure 11.9 is a DEM of southeastern New York and southwestern Connecticut showing the irregular glacial topography at the southernmost extent of glaciers in the United States. Long Island is composed largely of till and outwash deposited during the advance and retreat of *two* continental ice sheets. This makes the area more complex than that in Figure 11.8, but landforms that formed at the two glacial termini are recognizable.

FIGURE 11.9 DEM of Long Island, New York, showing two terminal moraines.

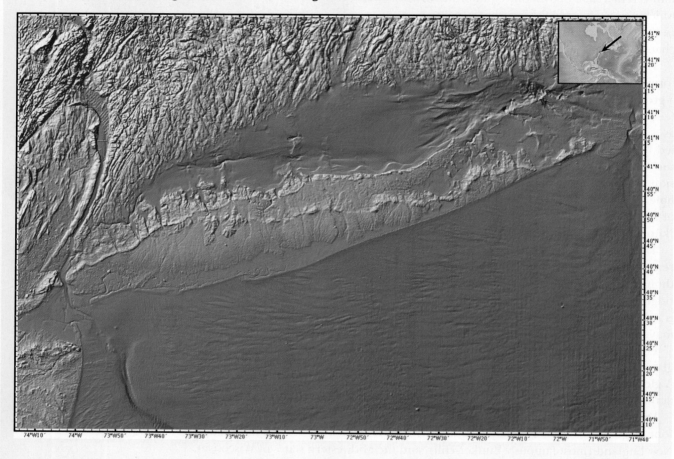

(a) Outline or otherwise identify the two terminal moraines, the southernmost outwash plain, and the outwash and glacial deltas deposited between the terminal moraines during a time of glacial retreat. Use different colors to separate features deposited during the two different glacial advances.

(b) Which of the two moraines is younger? Explain your reasoning.

_____

_____

continued

Name: _____   Section: _____
Course: _____   Date: _____

Figure 11.10 is a map of the Arnott moraine in Wisconsin, formed approximately 14,000 years ago. It shows the classic irregular "knob and kettle" topography of terminal moraines. Compare this morainal area with stream-produced landscapes from Chapter 10.

(a) Are the stream networks and divides as well developed and clearly defined in this area?

_____

(b) Suggest an explanation for the absence of a well-integrated drainage system.

_____

_____

_____

(c) Suggest an explanation for the numerous swampy areas.

_____

_____

_____

(d) Compare the sizes and shapes of the hills ("knobs") in this area with those in the fluvial landscapes.

_____

_____

(e) Suggest an origin for the several small lakes in the moraine. What will be their eventual fate?

_____

_____

_____

(f) There are several gravel pits in the area. Why might gravel be more easily and profitably mined from these specific areas than from others?

_____

_____

(g) What difficulties do you think might be involved in farming in ground moraine?

_____

_____

continued

FIGURE 11.10 Glaciated terrain in Wisconsin (Amherst and Arnott 7.5' quadrangles).

Contour interval = 10 feet

Name: _____   **Section:** _____

Course: _____   **Date:** _____

Not all ridges in glaciated areas are moraines—some may be eskers—and not all hills are knobs in a ground moraine—some may be drumlins. Eskers are quite long. **Figure 11.11** shows part of an esker that extends for more than 40 miles. You can't tell from a map or DEM that eskers are made of outwash rather than till, but they are much narrower than terminal moraines and lack the kettle holes associated with melting ice. The asymmetric shape of drumlins makes them easy to recoginze.

Examine Figures 11.9 and 11.10 and suggest other ways to diffentiate eskers from moraines.

continued

FIGURE 11.11 Part of the Passadumkeag, Maine, 7.5' quadrangle showing the Enfield Horseback esker (the DEM inset shows the esker in its regional setting).

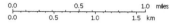

NATIONAL GEOGRAPHIC

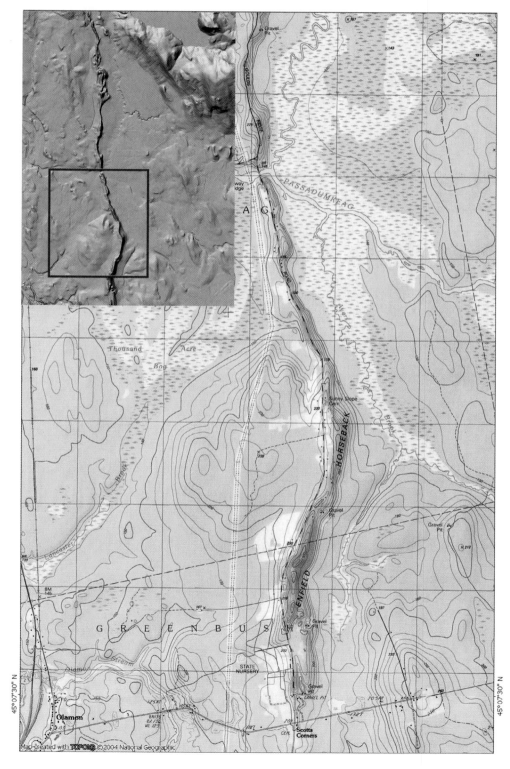

Name: _____   **Section:** _____
Course: _____   **Date:** _____

(a) Examine the DEM and map of the Weedsport, New York, area (**Fig. 11.12a, b**) and describe the topography.

_____

_____

_____

_____

(b) Sketch the long axis of one of the hills on the DEM. What is its orientation? _____ Do the same for the others. What was the direction of glacial movement in this area?_____ Show this with a large arrow on the map.

(c) Sketch or construct profiles of the five hills numbered on Figure 11.12b. Are the hills symmetrical or asymmetrical? Describe the distribution of their steep and gentle slopes.

_____

_____

_____

(d) The direction of ice flow indicated by a drumlin is *from* the _____ side of the drumlin *toward* the _____ side.

FIGURE 11.12 (a) DEM of the Weedsport, New York, 7.5′ quadrangle, showing the southern end of the Ontario/New York State drumlin field (area in rectangle shown in Fig. 11.12b).

continued

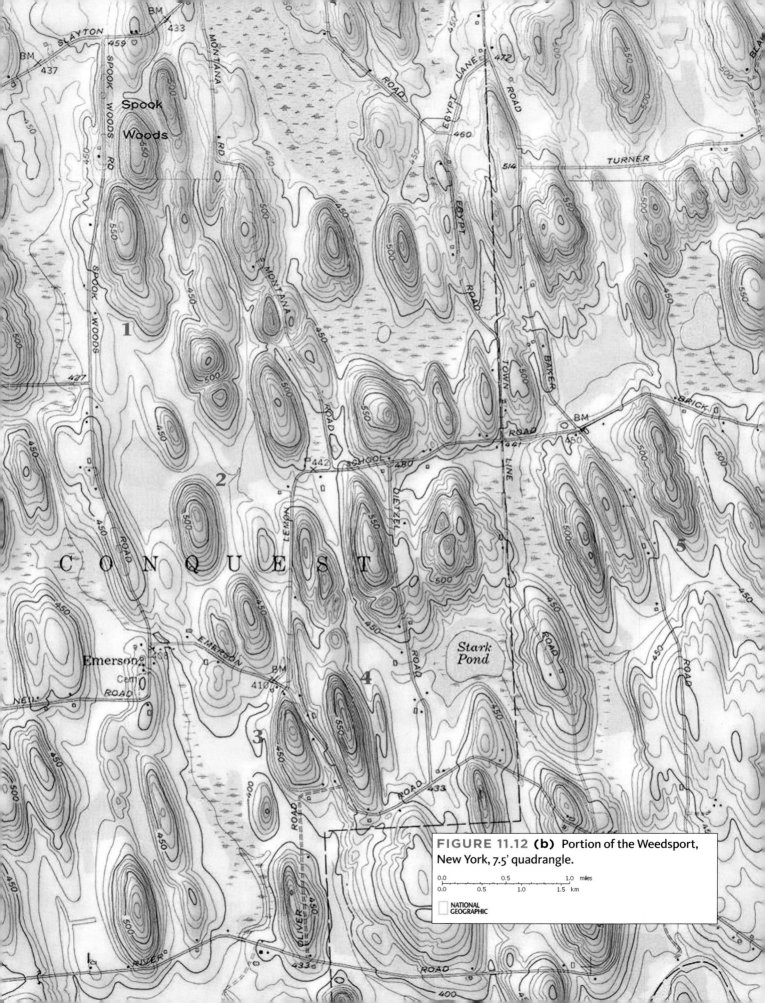

FIGURE 11.12 (b) Portion of the Weedsport, New York, 7.5' quadrangle.

11.3 Landscapes Produced by Mountain Glaciation

Alpine, Himalayan, Rocky Mountain, or Sierra Nevadan vistas with their jagged peaks, steep-walled valleys separated by knife-sharp divides, and spectacular waterfalls are the result of valley glacier erosion. Small glaciers form in depressions high up in the mountains and flow downhill, generally following existing stream valleys. The depressions are deepened and expanded headward by *plucking*, which eats into the mountainside, and the ice flows downhill, filling the former stream valleys and modifying them by abrasion (**Fig. 11.13**). When the ice melts, the expanded depressions become large bowl-shaped amphitheaters (**cirques**); formerly rounded stream divides are replaced by knife-sharp ridges (**arêtes**); and the headward convergence of several cirques leaves a sharp, pyramidal peak (**horn peak,** or **horn**, named after the Matterhorn in the Alps).

FIGURE 11.13 Erosional features formed by valley glaciers.

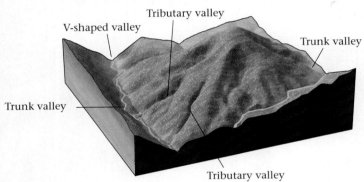

(a) Before valley glaciation the area has a normal fluvial topography.

(b) The extent of the glaciers during valley glaciation.

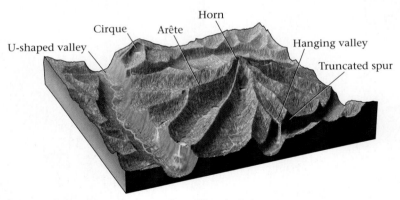

(c) The area's erosional features after valley glaciation.

Valley glaciers transform stream valleys in unmistakable ways, including their cross-sectional and longitudinal profiles, and the way that streams flow into one another. Figure 11.13c shows the characteristic U-shaped cross-sectional profile. Why are stream valleys V-shaped whereas glaciated valleys are U-shaped? Let's see if your ideas in Exercise 11.2 were right.

The answer is in how the two agents of erosion operate (**Fig. 11.14**). A stream actively erodes only a small part of its valley at any given time, cutting its channel downward or laterally (on the left in Fig. 11.14). Mass wasting gentles the valley

FIGURE 11.14 Glacial modification of a fluvial valley.

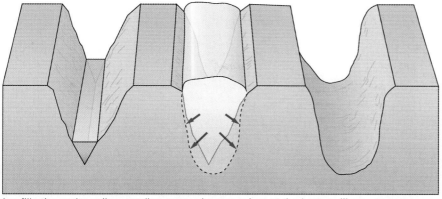

Ice fills the entire valley, eroding everywhere, not just at the bottom like a stream. As it erodes, the valley is deepened and widened everywhere, resulting in the U-shape on the right.

FIGURE 11.15 Longitudinal profiles of stream and glacial valleys.

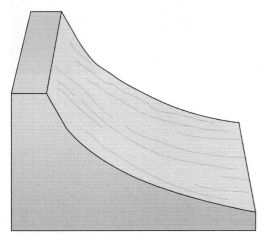

(a) Smooth, concave stream profile.

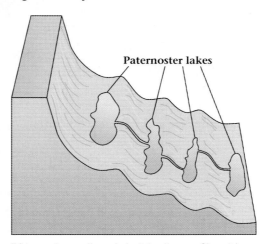

Paternoster lakes

(b) Irregular, scalloped glacial valley profile, with paternoster lakes filling the basins.

walls, creating the V shape. In contrast, a valley glacier fills the entire valley, abrading the walls not only at the bottom but everywhere it is in contact with the bedrock (center). When the ice melts, it leaves a U-shaped valley (right).

Most streams have smoothly concave **longitudinal profiles**. Valley glaciers, being solid, can overcome gravity locally and flow uphill for short distances. Their longitudinal profile is more irregular than a stream's, commonly containing a series of shallow scooped-out basins. Water can collect in some of these basins, producing a chain of lakes called **paternoster** (or **cat-step**) **lakes** (Fig. 11.15).

The base level for a tributary stream is the elevation of the stream, lake, or ocean into which it flows, and most streams flow smoothly into the water body at their base level. In contrast, glaciers have no base level. Large glaciers can carve downward more rapidly than smaller ones, so previously existing main stream valleys are deepened more than tributary valleys. When the ice melts, the result is a **hanging valley** such as the one shown in Figure 11.16. Today, Bridal Veil Creek falls almost 200 m to the floor of Yosemite Valley. The mouth of the creek and the floor of the valley were once at the same elevation, but the small valley glacier that filled Bridal Veil Creek could not keep pace as the main glacier eroded downward more rapidly.

FIGURE 11.16 The hanging valley at Bridal Veil Falls, Yosemite National Park, California.

Recognizing Features of Valley Glaciation

Name: _____ **Section:** _____

Course: _____ **Date:** _____

Figure 11.17 is a DEM showing part of Glacier National Park and several valley glacier erosional features. Label examples of horn peaks, arêtes, cirques, and U-shaped valleys, and look also for flat places (uniform color) that could be filled with lakes.

FIGURE 11.17 DEM (false color) of a portion of Glacier National Park showing classic features of valley glaciation.

Name: _____   **Section:** _____
Course: _____   **Date:** _____

Mt. Katahdin is the highest peak in Maine, reaching almost a mile above sea level. At first glance, the glacial history of the area seems to be straightforward. A closer look suggests it is a bit more complicated. Examine the shaded relief map of the Mt. Katahdin area (**Fig. 11.18**).

(a) What glacial landform is represented by North Basin, South Basin, and the basin near the northwest corner of the map in which Klondike Pond is located? _____

(b) What glacial feature is represented by the Knife Edge, Hamlin Ridge, and Keep Ridge?

_____

(c) What kind of glacial feature is Mt. Katahdin? _____

(d) What type of glacier produced these features? _____

The regional setting of Mt. Katahdin is shown in **Figure 11.19**. The area of northern New England, Quebec, and New Brunswick shown in this map was covered by continental ice sheets that pushed as far south as Long Island. The Passadumkeag esker in Figure 11.11 is located just a bit south of Mt. Katahdin, as shown by the green dot in Figure 11.19.

Suggest an explanation for the origin of the alpine glacial features on Mt. Katahdin and surrounding peaks in the midst of an area affected by continental glaciation. There is more than one possible answer—how many can you think of?

_____

_____

_____

continued

FIGURE 11.18 Shaded relief topographic map of Mt. Katahdin in Maine.

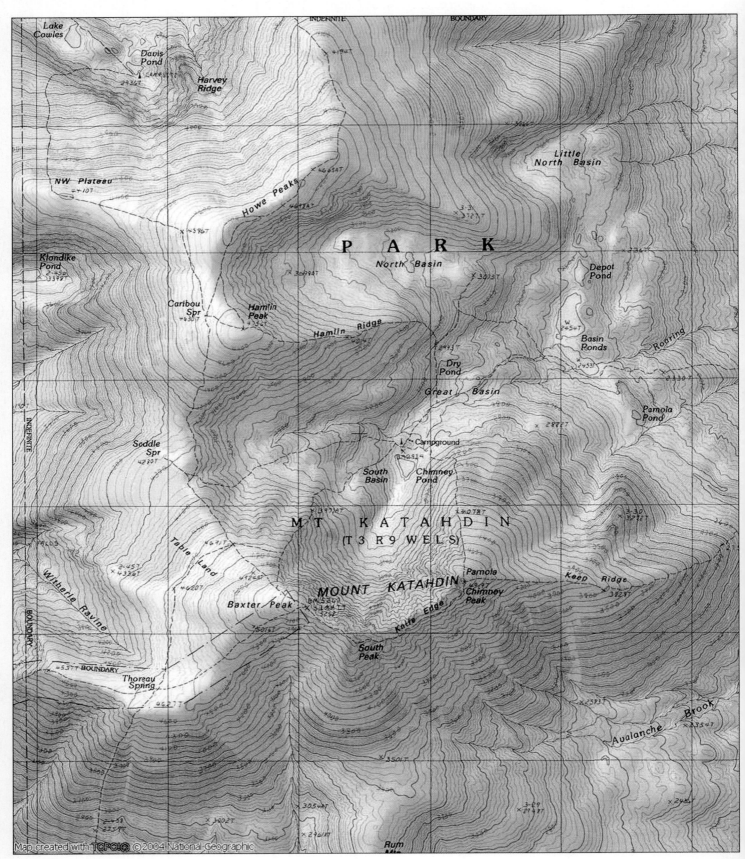

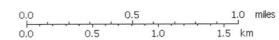

0.0   0.5   1.0   miles

0.0   0.5   1.0   1.5   km

MN ⭐ TΓ

18°

04/07/09

FIGURE 11.19 Location of Mt. Katahdin in northern New England (red dot) and the Passadumkeag esker (green dot) shown in Figure 11.11.

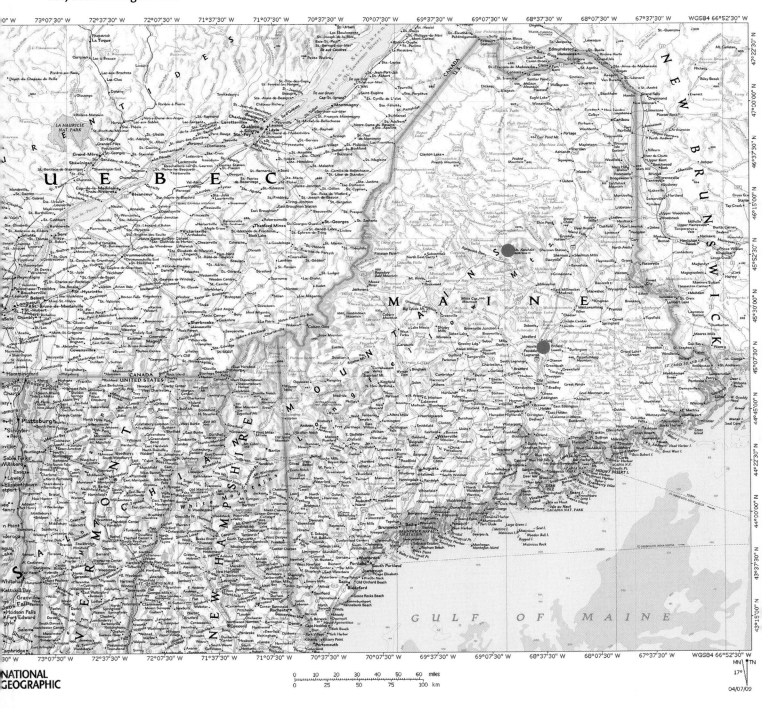

TABLE 11.1 Glossary of glacial landforms.

Erosional Landforms		
Landform	**Type of glacier**	**Description**
Arête	Valley	Sharp ridge separating cirques or U-shaped valleys
Cirque	Valley	Bowl-shaped depression on mountainside; site of snowfield that was source of valley glacier
Hanging valley	Valley	Valley with mouth high above main stream valley; formed by tributary glacier that could not erode down as deeply as larger main glacier
Horn peak	Valley	Steep, pyramid-shaped peak; erosional remnant formed by several cirques
Sculptured bedrock	Continental	Asymmetric bedrock with *steep* side facing the direction toward which the ice flowed; also called roche moutonnée
U-shaped valley	Valley	Steep-sided valley eroded by a glacier that once filled the valley
Depositional Landforms		
Landform	**Type of glacier**	**Description**
Drumlin	Continental	Streamlined (elliptical), asymmetric hill composed of till; the long axis parallels glacial direction, the gentle side faces the direction toward which the ice flowed
Esker	Continental	Narrow sinuous ridge made of outwash deposited by a subglacial stream in a tunnel at the base of a glacier
Ground moraine	Valley and continental	Sheet of till deposited by a retreating glacier
Kettle hole	Valley and continental	Depression left when a block of ice that had been isolated from a melting glacier is covered with sediment and melts
Lateral moraine	Valley	Till deposited along the walls of a valley when a valley glacier retreats
Outwash plain	Valley and continental	Outwash deposited by meltwater beyond the terminus of a glacier
Proglacial delta	Continental	Outwash deposited into a proglacial lake
Terminal moraine	Valley and continental	Wall of till deposited when the terminus of a glacier remains in one place for a long time
Lakes Associated with Glacial Landscapes		
Landform	**Type of glacier**	**Description**
Ice-marginal lake	Valley and continental	Lake formed by disruption of local drainage by a glacier
Kettle lake	Valley and continental	Lake filling a kettle hole
Paternoster lake	Valley	Series of lakes filling basins in a glaciated valley
Tarn	Valley	Lake occupying part of a cirque

## CHAPTER 12

GROUNDWATER AS A LANDSCAPE FORMER AND RESOURCE

PURPOSE

- Understand how groundwater infiltrates and flows through Earth materials.
- Explain why groundwater erodes and deposits differently from streams and glaciers.
- Learn to recognize landscapes formed by groundwater and to interpret groundwater flow direction from the topographic features.
- Understand how geologists carry out groundwater resource and pollution studies.

MATERIALS NEEDED

- Specimens of four different materials for the infiltration exercise
- Tracing paper and colored pencils.

12.1 Introduction

Some rain and snow that falls on the land runs off into streams, some evaporates into the air, and some is absorbed by plants. The remainder sinks into the ground and is called **groundwater**. Groundwater moves much more slowly than water in streams because it has to drip from one pore space to another. It therefore has much lower kinetic energy than stream water and cannot carry the particles with which streams erode bedrock by abrasion. As a result, groundwater can only erode *chemically*, and groundwater erosion is most effective in soluble rocks—usually limestone, dolostone, or marble containing carbonate minerals like calcite and dolomite.

If groundwater erodes underground, how can it form landscapes at the surface? As groundwater erodes rocks from below, it undermines their support of the surface above, and the land may collapse to produce very distinctive landscapes (**Fig. 12.1a**). Landscapes in areas of extreme groundwater erosion are among the most striking in the world, with narrow, steep-sided towers unlike anything produced by streams or glaciers (**Fig. 12.1b**). Groundwater-eroded landscape is called **karst topography,** after the area in Yugoslavia where geologists first studied it in detail, but spectacular examples are also found in Indiana, Kentucky, Florida, southern China, Puerto Rico, and Jamaica.

Groundwater is also a vital resource, used throughout the world for drinking, washing, and irrigating crops. Strict rules govern its use in many places because it flows so slowly that renewal of the groundwater supply takes a long time. In addition, its slow flow and underground location make it difficult to purify once it has been polluted. We examine both roles of groundwater in this chapter: landscape former and resource.

FIGURE 12.1 Landscapes carved by groundwater.

(a) Karst topography characterized by sinkholes and underground caves in southern Indiana.

(b) Karst towers in the Guilin area of southern China.

12.2 Aquifers and Aquitards

Materials that transmit water readily are called **aquifers** (from the Latin, meaning "to carry water"), and those that prevent water from infiltrating are called **aquitards** or **aquicludes** (they re*tard* or ex*clude* water). Groundwater flows through pores and

EXERCISE 12.1 Factors Affecting Infiltration and Groundwater Flow

Name: ____________________ **Section:** __________

Course: ____________________ **Date:** __________

Let's look first at factors that control the flow of water underground. Your instructor will provide four containers screened at the bottom and filled with materials that have different grain shapes and sizes (**Fig. 12.2**).

(a) Which materials do you think will transmit water best? Rank the four materials in decreasing order of transmissivity and explain your prediction.

Most transmissive *Least transmissive*

Because:

FIGURE 12.2 Just passing through.

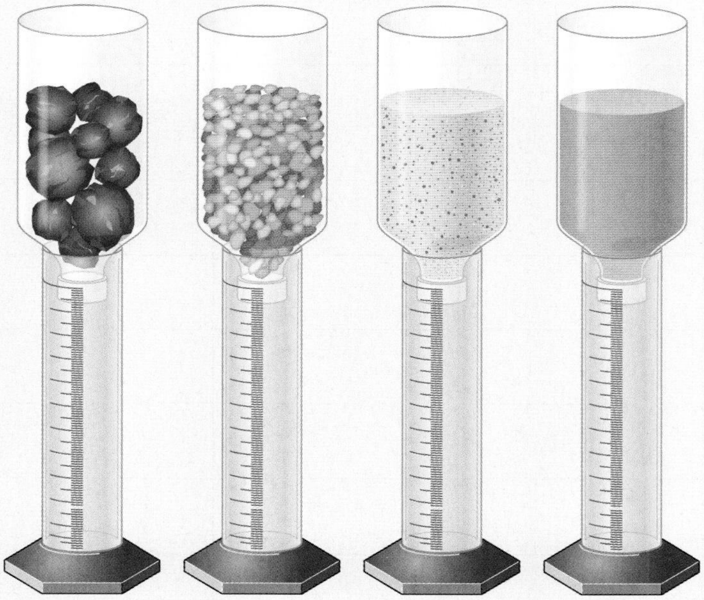

(b) Which material will transmit water fastest? Slowest? Explain.

Fastest water transmission *Slowest water transmission*

Because:

continued

Name: _____   **Section:** _____
Course: _____   **Date:** _____

Now let's see how good your predictions are. Pour equal amounts of water into the four bottles and measure the amount of water that passes through and its rate of flow. Record your observations in the following data table.

	Observation	Container A	Container B	Container C	Container D
Properties of material that might affect water flow	Grain size				
	Sorting				
	Grain shape				
	Porosity				
	Permeability				
Volume of water transmitted (ml)	30 seconds				
	60 seconds				
	90 seconds				
	120 seconds				
	150 seconds				
	180 seconds				
	210 seconds				
	240 seconds				
Other observations	Water color				
	Amount of water retained				

continued

Name: _____   Section: _____
Course: _____   Date: _____

(c) Which materials transmitted the most water? Which retained the most water?

_____

_____

_____

(d) What properties of the materials are correlated with good water transmission? Which are correlated with water retardation and retention?

_____

_____

_____

_____

fractures in bedrock as well as through unconsolidated sediment. Some rocks, like a poorly cemented sandstone, contain pores when the rock forms (**primary porosity**), but others that have no pores at all when they form (e.g., granite, obsidian) become porous and permeable when broken by faulting or fracturing (**secondary porosity**).

Being porous is clearly not enough for a material to be a good aquifer. Pumice and scoria are very porous but their pores are not connected. Pore spaces must be connected for water to move from one to another—a property called **permeability**. The materials in Exercise 12.1 that transmitted water easily had to be both porous (there was room for the water between grains) *and* permeable (the water could move through the material). *An aquifer must be both porous and permeable.*

Water's relatively slow pore-to-pore movement promotes chemical erosion because the longer the water is in contact with minerals, the more it can dissolve them. Some dissolved ions are poisonous if consumed directly or used for agriculture. For example, widespread arsenic poisoning occurred in Bangladesh when groundwater wells were drilled to replace surface water supplies contaminated with cholera and typhoid bacteria. An international geologic team is searching for the source of the arsenic and for ways to resolve this health issue confronting millions of people.

EXERCISE 12.2   The Difference between Porosity and Permeability

Name: _____   Section: _____
Course: _____   Date: _____

Are all porous rocks aquifers? Hold pieces of highly porous pumice and scoria above two beakers or rest them on the rims as shown in **Figure 12.3**. Slowly drop or sprinkle water onto the rocks and observe what happens.

(a) Are pumice and scoria porous? Permeable? Explain.

_____

_____

_____

continued

Name: _____   **Section:** _____
Course: _____   **Date:** _____

(b) Is pumice an aquifer or aquitard? Explain.

_____

_____

_____

(c) Is scoria an aquifer or aquitard? Explain.

_____

_____

_____

FIGURE 12.3 Porosity and permeability in pumice and scoria.

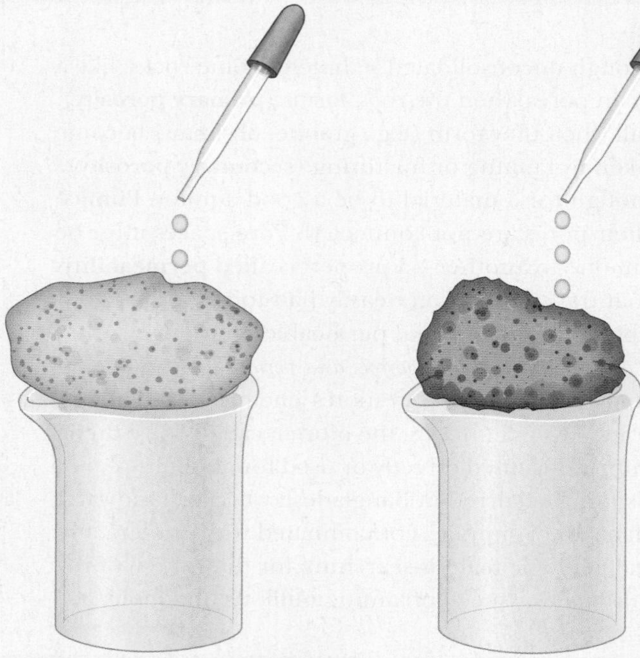

12.3 Landscapes Produced by Groundwater

Groundwater is the only agent of erosion that creates landforms from *below* the surface of the Earth, and all karst topographic features are destructional. **Sinkholes** form when underground cavities dissolved by groundwater grow so large that there is not enough rock to support the ground surface (**Fig. 12.4a**). **Karst towers** are what remains when the rock around them has been dissolved (**Fig. 12.4b**). **Karst valleys** form when several sinkholes develop along an elongate fracture (**Fig. 12.4c**). Caves and caverns, the largest groundwater erosional features, are underground, unseen, and often unsuspected at the surface. Groundwater depositional features—stalactites, stalagmites, and columns (**Fig. 12.5**)—are found in caverns and are therefore not surface landforms.

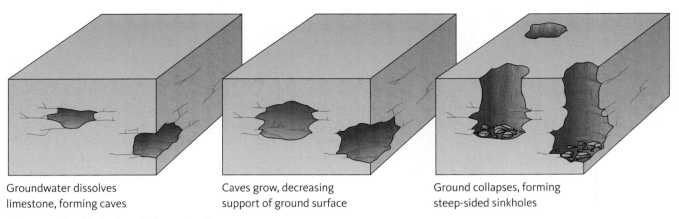

Groundwater dissolves limestone, forming caves

Caves grow, decreasing support of ground surface

Ground collapses, forming steep-sided sinkholes

(a) Origin of sinkholes by solution and collapse

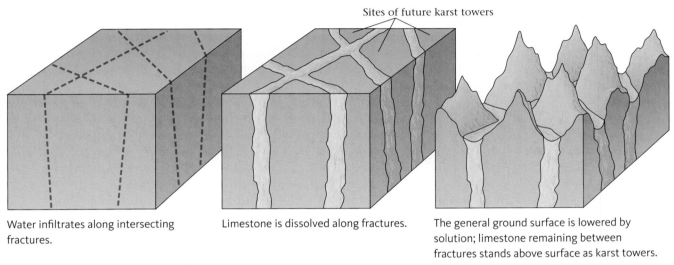

Sites of future karst towers

Water infiltrates along intersecting fractures.

Limestone is dissolved along fractures.

The general ground surface is lowered by solution; limestone remaining between fractures stands above surface as karst towers.

(b) Origin of karst towers as erosional remnants

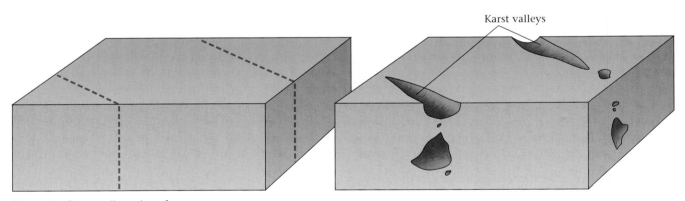

Karst valleys

(c) Origin of karst valleys along fractures

FIGURE 12.5 Groundwater deposition in caverns.

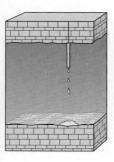
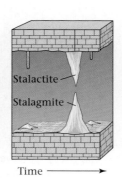
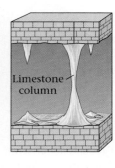

Time ⟶

Stalactites grow *downward* from the roof; stalagmites grow *upward* from the floor. Columns form when stalactites and stalagmites merge.

EXERCISE 12.3 **Karst Topography**

Name: _____ Section: _____

Course: _____ Date: _____

This exercise examines two classic karst areas in the United States to help you become familiar with groundwater erosional landscapes—one in Kentucky and one in Florida.

Compare the map of the Mammoth Cave area (**Fig. 12.6**) with the diagrams showing how karst features develop (Fig. 12.4).

(a) This area receives a moderate amount of rainfall, yet there are very few streams. Suggest an explanation for this phenomenon. (This is a characteristic of nearly all karst regions.)

_____

_____

_____

(b) Look for changes in topographic grain or texture across the map area. Draw lines separating places with different topographic textures.

(c) Describe the topography in the southernmost part of the area. What karst features are present?

_____

_____

_____

_____

(d) Some of the circular depressions in this part of the map are partially filled with water. Suggest an origin for these lakes (we will return to this point later).

_____

_____

_____

continued

NATIONAL GEOGRAPHIC

Name: _____   Section: _____
Course: _____   Date: _____

(e) Describe the topography in the northern two-thirds of the map. What karst feature is represented by The Knobs? The Woolsey and Owens valleys?

_____

_____

_____

_____

(f) Karst features are found throughout the map but have a different appearance in the south compared to the north. Describe the differences and suggest a possible explanation.

_____

_____

_____

_____

(g) Compare this area with Mammoth Cave. In what ways are the landscapes similar? How do they differ?

_____

_____

_____

_____

(h) Some of the sinkholes contain lakes, but others are dry. Suggest an explanation for the difference. *Hint:* Determine the elevation of lakes in the sinkholes and write it in each lake. Compare the elevation at the bottoms of sinkholes that do not have lakes.

_____

_____

_____

_____

12.4 The Water Table

The level of water in the lakes in Figure 12.8 is controlled by a feature called the **water table**. Gravity pulls groundwater downward until it reaches an aquitard that it cannot penetrate, and the water begins to fill pores in the aquifer just above the aquitard. More water percolating downward finds pores at the bottom of the aquifer already filled, and so saturates those even higher. The water table is the boundary between the **zone of saturation** (where all pore spaces are filled with water) and the **zone of aeration** (where some pores are partly filled with air (**Fig. 12.7**). You can model the water table by adding water to a beaker filled with sand and watching the level of saturation change.

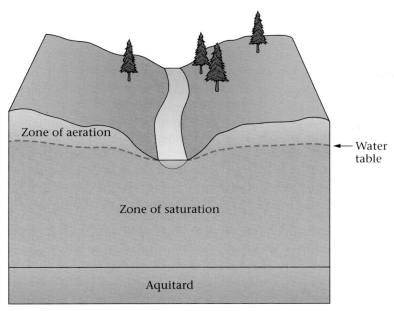

A stream or lake forms where the water table intersects the land surface.

12.5 Groundwater Resources and Environmental Problems

Groundwater is vital for human existence, both for drinking and to irrigate crops. We drill wells to depths below the water table and use pumps to extract the water from an aquifer. But groundwater flows so slowly that it can rarely replenish itself as rapidly as our needs dictate. Indeed, some water that we drink today entered its aquifer system hundreds of years ago. If we take too much groundwater, our wells run dry, a serious problem when many farms between the Rocky Mountains and the Mississippi River depend on the same aquifer for their water.

When we pump water from an aquifer, it comes from the saturated pore spaces below the water table, and water must flow downward to replace it from other parts of the aquifer (**Fig. 12.10**). This lowers the water table, especially in the area around the well. Each well creates a **cone of depression** in the water table as shown in Figure 12.10. You model this phenomenon every time you order a thick milkshake from your local fast food restaurant. Next time, take the top off the shake and insert the straw just below the surface of the shake. Sip slowly and look at what happens to the surface. Because the shake flows sluggishly (not unlike groundwater), the area below the straw is drawn down more than the area near the edges of the cup. You've just made a small cone of depression.

Two types of problems are associated with groundwater resources: issues of *quantity* (whether there is enough groundwater to meet our needs) and *quality* (whether the groundwater is pure enough for us to drink or whether it has become contaminated through natural or human causes). Two case histories, both based on real situations, explore these issues.

Name: _____   **Section:** _____
Course: _____   **Date:** _____

Interlachen area, Florida

Now look at the map of the Interlachen area in Florida (**Fig. 12.8**), another karst landscape underlain by highly soluble limestone. In addition to showing another karst region, this map illustrates other important concepts about groundwater, its value as a resource and environmental and other problems associated with living in karst topography.

The water table generally mimics the surface topography in a slightly subdued shape (Fig. 12.7). Groundwater flows downhill from higher parts of the water table to lower parts.

(a) Construct a profile of the Interlachen area from Point A to Point B in Figure 12.8. From your data of lake level elevations, sketch the water table on this profile.

(b) Draw an arrow on the map to indicate the direction of groundwater flow. The case histories at the end of this chapter illustrate the importance of knowing which way groundwater is flowing.

(c) The city of Interlachen has grown dramatically in the past 20 years, and part of its suburban street grid can be seen in Figure 12.8. What construction problems do developers face in an area like this?

_____
_____
_____
_____

(d) What environmental and other problems do the new residents face?

_____
_____
_____
_____

FIGURE 12.8 Karst topography in the Interlachen area of Florida.

NATIONAL GEOGRAPHIC

WGS84 81°52'30" W

Carlsbad Caverns National Park in New Mexico (**Fig. 12.9**) contains spectacular caverns with stalactites, stalagmites, and columns, yet the area in which it is located does not exhibit the karst features found at Mammoth Cave and southern Florida. Suggest an explanation.

_____

_____

_____

_____

FIGURE 12.9 Topography of the Calsbad Caverns area of New Mexico.

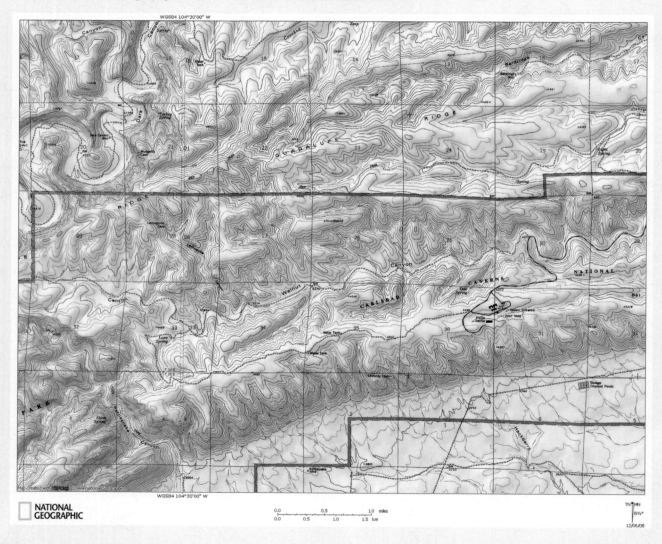

FIGURE 12.10 Formation of a cone of depression by removal of water from an aquifer.

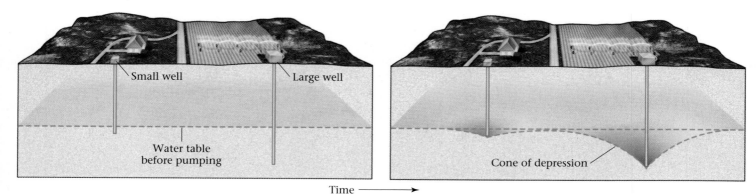

Small well   Large well

Water table
before pumping

Cone of depression

Time ⟶

EXERCISE 12.6   Environmental Issue 1: Where Has My Water Gone?

Name: _____   Section: _____

Course: _____   Date: _____

This exercise presents a real life problem, although details have been changed to protect the privacy of the parties involved.

Figure 12.11 is a map showing an area in which farmers have relied on groundwater for generations for drinking and to irrigate their crops. In an attempt to diversify the local economy, the town selectmen offered tax incentives to bring two new industries into the area: Grandma's Soup Company and the Queens Sand and Gravel Quarry. Trouble arose when Farmer Jones (see location on map) found that his well had run dry despite the fact that it was the rainy season. Having taken Geology 101 at the State University several years ago, he suspected that pumping by the new companies had lowered the water table. He hired a lawyer and sued Grandma's Soup Company to prevent it from pumping "his" water. Grandma's lawyers responded that the company had nothing to do with his problem.

The judge, who has never taken a geology class, has appointed *you* to resolve the problem. Your task is to (1) find out why Farmer Jones has lost his water, and (2) figure out how to keep everyone happy. After all, Jones and his friends have been benefiting from the fact that Grandma's soup factory has been paying an enormous amount of taxes. Your assistants have already gathered a lot of useful data.

- A survey shows that the land is as flat as a pancake, with an elevation of 965 feet.

- Records show the depth to the water table in wells *before* the companies opened and *after* the two companies had been pumping for a year (**Table 12.1**).

Analyze the data and make a formal recommendation to the court. *Hint:* Use your contouring skills to make topographic maps of the water table *before* (Fig. 12.11) and *after* (**Fig. 12.12**) the companies started pumping.

continued

FIGURE 12.11 Contour map of water table *before* commerical pumping.

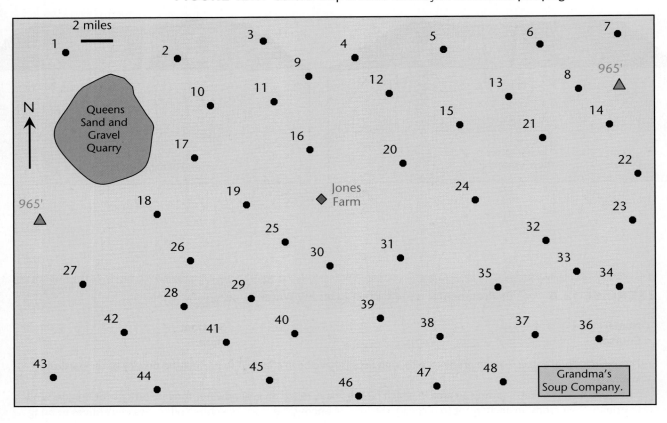

FIGURE 12.12 Contour map of water table *after* commercial pumping.

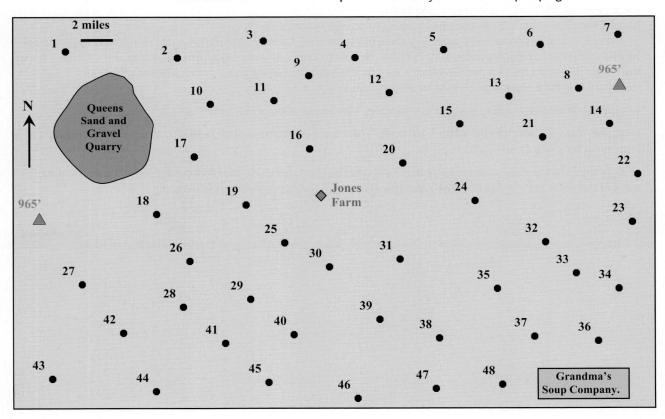

Name: _____ Section: _____

Course: _____ Date: _____

TABLE 12.1 Water table data for the *area* surrounding Jones Farm.

Before the companies began withdrawing water				After the companies had been pumping for a year			
Location	Depth to water table (feet)	Location	Depth to water table (feet)	Location	Depth to water table (feet)	Location	Depth to water table (feet)
Jones Farm	32	25	40	Jones Farm	146	25	103
1	78	26	52	1	132	26	118
2	68	27	57	2	118	27	124
3	62	28	50	3	97	28	111
4	39	29	46	4	81	29	102
5	40	30	30	5	72	30	97
6	48	31	49	6	82	31	82
7	60	32	61	7	80	32	99
8	57	33	70	8	107	33	108
9	51	34	77	9	123	34	118
10	63	35	62	10	113	35	100
11	56	36	81	11	92	36	120
12	27	37	71	12	71	37	111
13	52	38	59	13	89	38	98
14	64	39	50	14	90	39	79
15	47	40	44	15	107	40	93
16	49	41	43	16	128	41	102
17	62	42	51	17	133	42	116
18	64	43	58	18	116	43	118
19	53	44	42	19	89	44	106
20	41	45	48	20	88	45	91
21	59	46	59	21	100	46	80
22	70	47	64	22	107	47	99
23	71	48	74	23	84	48	117
24	51			24	No data		

Groundwater *quality* is becoming a critical issue as growing populations place stress on the environment. Commercial pollution or leakage from residential septic tanks may make water unsuitable for drinking. The following case history happens all too commonly.

The Smiths bought a home in a suburban development where the water supply was a shallow groundwater aquifer. A few months ago, they noticed a strange smell when they took showers, and the drinking water has begun to taste foul. Last week, while washing dishes, the Smiths thought they smelled gasoline in their kitchen. The local TV news channel ran an investigative report on leakage from storage tanks at gas stations and the Smiths suspected that this might be the cause of their problem. They hired a groundwater consulting company to solve the problem and you have been put in charge of the project.

Figure 12.13 is a map of the Smiths's neighborhood, showing the location of their home and two local gas stations the Smiths suspect to be the cause of their water problem. You have surveyed the neighborhood using state-of-the-art detection instruments to measure the amount of semivolatile compounds in the soil around the Smiths's home. These chemicals

FIGURE 12.13 Concentrations (in parts per million) of semivolatile compounds in the area surrounding the Smith home (nd = not detected).

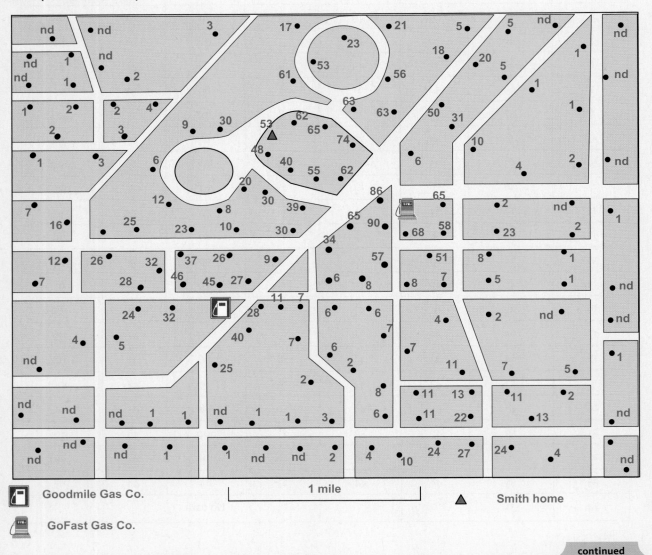

Goodmile Gas Co.

GoFast Gas Co.

1 mile

Smith home

continued

Name: _____   **Section:** _____
Course: _____   **Date:** _____

are released into the ground when corroded gas tanks leak and are indicators of soil and groundwater contamination. Concentrations greater than 50 parts per million (ppm) are considered dangerous. The data are shown in red numbers on Figure 12.13.

(a) Does either gas station have a leakage problem? Explain your conclusion.

_____

_____

_____

(b) In what direction does the local groundwater flow? How do you know?

_____

_____

(c) Which homes will be the next to feel the effects of the gasoline leakage?

_____

_____

(d) What additional problems has your research discovered?

_____

_____

_____

Hint: Make a contour map of the soil contamination level using a 10 ppm contour interval. This is not a map showing elevation, but the rules are very much the same. Color code the map to show levels of potential danger: red for values above the danger threshold (50 ppm); a second color for probable future threat (between 30 and 50 ppm), and a third color for possible future threat (between 10 and 30 ppm).

## CHAPTER 13

PROCESSES AND LANDFORMS IN ARID ENVIRONMENTS

PURPOSE

- Understand how erosion and deposition in arid regions differ from those in humid regions.
- Recognize how these differences make arid landscapes unlike those of humid environments.
- Become familiar with landforms and landscapes of arid regions.

MATERIALS NEEDED

- Colored pencils

13.1 Introduction

Arid regions are defined as those that receive less than 25 cm (10 in) of rain per year, which provides very little water for stream and groundwater erosion or for chemical and physical weathering. Weathering, mass wasting, and wind and stream erosion are still the major factors in arid landscape formation, but because they are applied in very different proportions than in humid regions, the results are strikingly different from temperate and tropical landscapes (Fig. 13.1).

FIGURE 13.1 Arid landforms from the southwestern United States.

(a) Erosional remnants in Monument Valley, Arizona.

(b) Hoodoo panorama in Bryce Canyon, Utah.

(c) Delicate Arch in Arches National Park, Utah.

Movies and television portray arid regions with camels resting at oases amid mountainous sand dunes. Some of these places, like parts of the Sahara and Mojave deserts, do look like that (minus camels in the Mojave), but some have more rock than sand, others lie next to shorelines where there is no shortage of water, and the North and South Poles are among Earth's most arid regions (Fig. 13.2).

FIGURE 13.2 Types of arid regions.

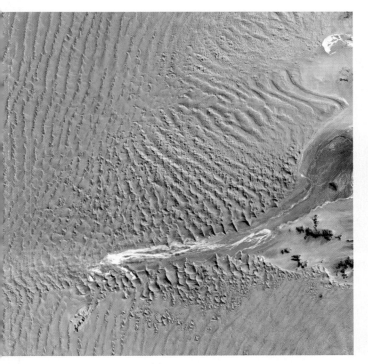

(a) Sandy desert, Namibia.

(b) Polar desert, Norway.

(c) Rocky desert, Utah.

EXERCISE 13.1 Comparing Arid and Humid Landscapes

Name: ______________________ **Section:** ________

Course: ______________________ **Date:** ________

Compare arid landscapes in a part of Canyonlands National Park and the Grand Canyon, located in Utah and Arizona, respectively (**Fig. 13.3a,b**), with humid landscapes of the Delaware Water Gap at the New Jersey/Pennsylvania border (**Fig. 13.3c**) and an area in Utah (**Fig. 13.3d**).

FIGURE 13.3 Comparison of arid and humid landscapes.

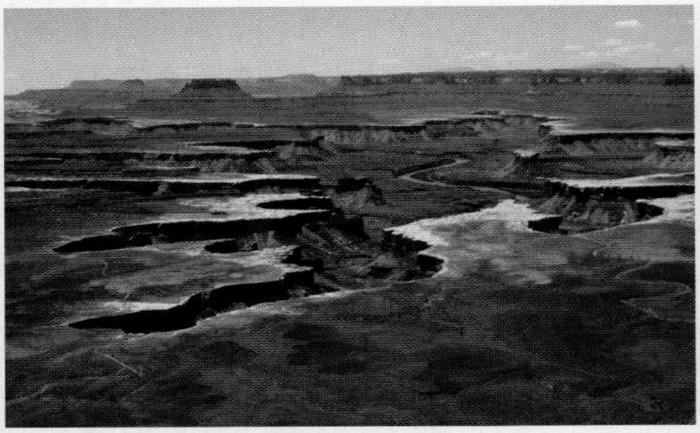

(a) Canyonlands National Park, Utah.

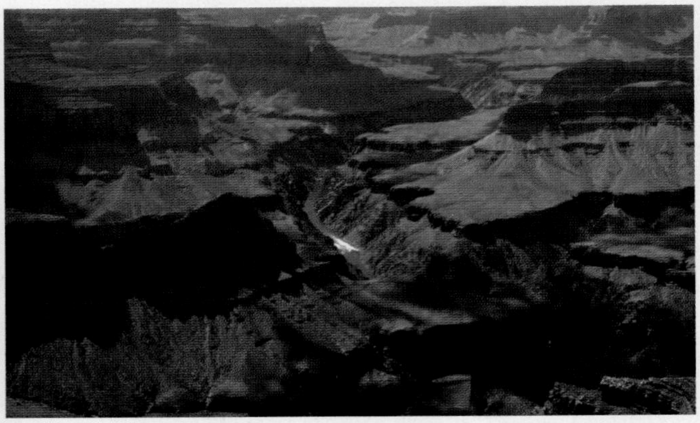

(b) Grand Canyon, Arizona.

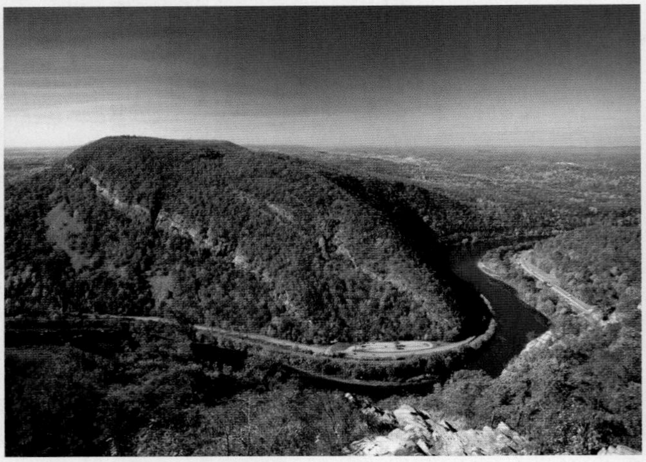

(c) Delaware Water Gap, New Jersey/Pennsylvania.

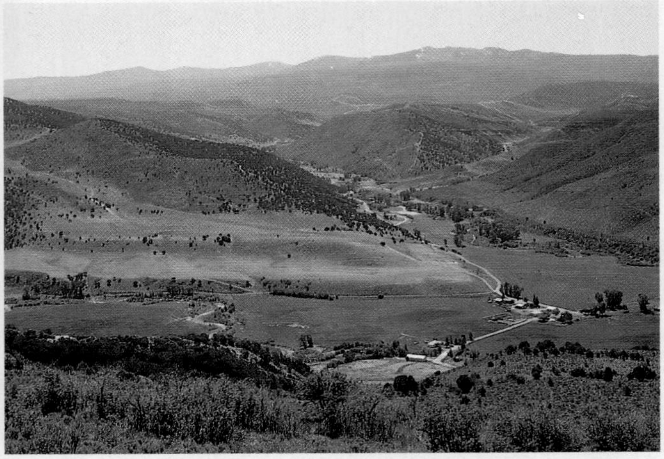

(d) Humid region in Utah.

(a) In which areas are the landforms more angular? In which are they more rounded? Suggest an explanation for the difference.

____

____

____

____

continued

Name: _____   Section: _____
Course: _____   Date: _____

(b) In which areas do you think landscape changes are more rapid? Are slower? Explain your answer.

_____

_____

_____

_____

(c) What other differences do you see between the arid and humid regions?

_____

_____

_____

_____

(d) What type of erosion appears to be dominant in the arid regions? In the humid regions? How is this possible?

_____

_____

_____

_____

13.2 Processes in Arid Regions

Now let's examine the underlying causes for the different evolution of landscapes in arid and humid areas. Features tend to survive longer in arid areas than in humid ones because weathering, erosion, and deposition are slowed by the scarcity of water needed to abrade, dissolve, and carry debris away. In the absence of water, soluble rocks like limestone, dolostone, and marble—easily weathered and eroded in humid regions to form valleys—become ridge formers.

Both physical and chemical weathering occur in arid regions. The dominant processes of physical weathering are usually different from those in humid areas—for example, where do you think root wedging is more common? In areas where rain is rare, the major source of water for chemical weathering may be the thin film of dew formed on rocks each morning. The scarcity of water limits chemical weathering severely and soil formation is slow.

13.3 Progressive Evolution of Arid Landscapes

The evolution of arid landscapes differs from that in humid regions for all the reasons that you just outlined. **Figure 13.4** shows the stages in the development of an arid landscape, starting with a block made of resistant and nonresistant rocks that has been uplifted by faulting (Fig. 13.4a). Physical weathering takes advantage of fractures, causing cliffs to retreat and forming a rubble-strewn plain (Fig. 13.4b).

Name: _____   Section: _____

Course: _____   Date: _____

How are the following processes different in arid regions and humid regions? Consider the nature and intensity of the processes and the role that water plays in each. What are the dominant processes in each category in the two different areas?

	Arid regions	Humid regions
Physical weathering		
Chemical weathering		
Mass wasting		
Soil formation		
Stream erosion		
Wind erosion		
Groundwater activity		

Streams—perhaps flowing after only a few storms during rainy season but still very effective—erode materials from the highlands and redeposit them as **alluvial fans** that begin to fill in the lowlands (Fig. 13.4c). With further deposition, alluvial fans coalesce to form a broad sand and gravel deposit (a **bajada**) flanking the uplifted block (Fig. 13.4c).

Where several fault blocks and valleys are present, bajada sediment from neighboring blocks eventually fills in the intervening valley, burying all but a few eroded remnants of the original blocks (Fig. 13.4d). These remnants stand out as isolated peaks, called **inselbergs**, above the sediment fill. Streams flowing into the basin from surrounding blocks have no way to leave and so form **playa lakes**, which may evaporate to form **salt flats**. Monument Valley (Fig. 13.4e) is a classic example of remnants of resistant rock isolated from their source layers by arid erosion.

Arid landscapes are typically more angular than humid landscapes. Compare, for example, Figures 13.3b and 13.3c. In the Grand Canyon, there is little soil. Rockfall is the dominant process of mass wasting, and sharp, angular slope breaks indicate contacts between resistant and nonresistant rocks.

FIGURE 13.4 Stages in the evolution of an arid landscape.

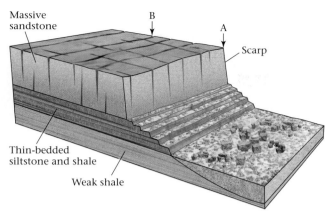

(a) An uplifted block of sandstones and shales.

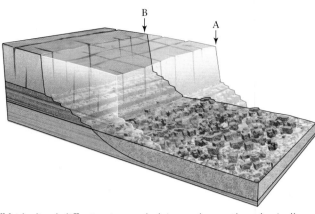

(b) Idealized cliff retreat as underlying rocks weather physically, causing rockfall.

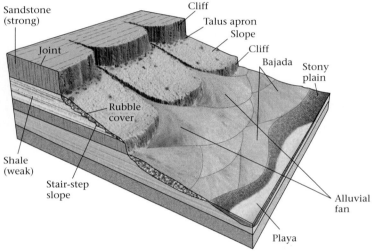

(c) A debris apron covers the lowlands and protects the bedrock slope from further erosion.

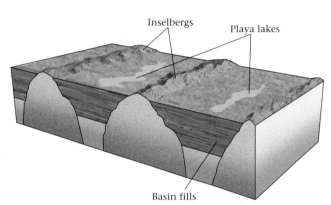

(d) A late stage of landscape development involving several uplifted blocks.

(e) Buttes in Monument Valley, Arizona.

Name: _____   **Section:** _____
Course: _____   **Date:** _____

(a) Why hasn't soil creep smoothed the slopes of the Grand Canyon (Fig. 13.3b)?

_____

_____

_____

Death Valley, California/Nevada

Examine the topographic overview (**Fig. 13.5**) of the Death Valley area. This is one of the driest places in North America and, with an elevation of more than 250 feet *below* sea level, one of the lowest spots on the continent. It also contains classic examples of arid landscapes.

(b) Compare this region with Figure 13.4. Is this a single fault block like Figure 13.4a or several like 13.4d? Explain your reasoning.

_____

_____

_____

(c) Sketch in the border fault(s) with a colored pencil on both the map view and the profile in Figure 13.5.

(d) Why is Death Valley lower in elevation than the other major valleys shown in the topographic profile in Figure 13.5?

_____

_____

_____

Now examine Death Valley more closely, starting with the Stovepipe Wells area (**Fig. 13.6**).

(e) Locate this map on Figure 13.5 using the latitude/longitude grids on both maps.

(f) What are the broad, rounded landforms that project into Death Valley from the mountains adjacent on the west and southeast? How did they form?

_____

_____

_____

(g) Several streams flow into Death Valley from these mountains. Where do they go? What is their base level?

_____

_____

_____

(h) A stippled pattern identifies areas of rapidly moving sand dunes. Where did this sand come from?

_____

_____

_____

continued

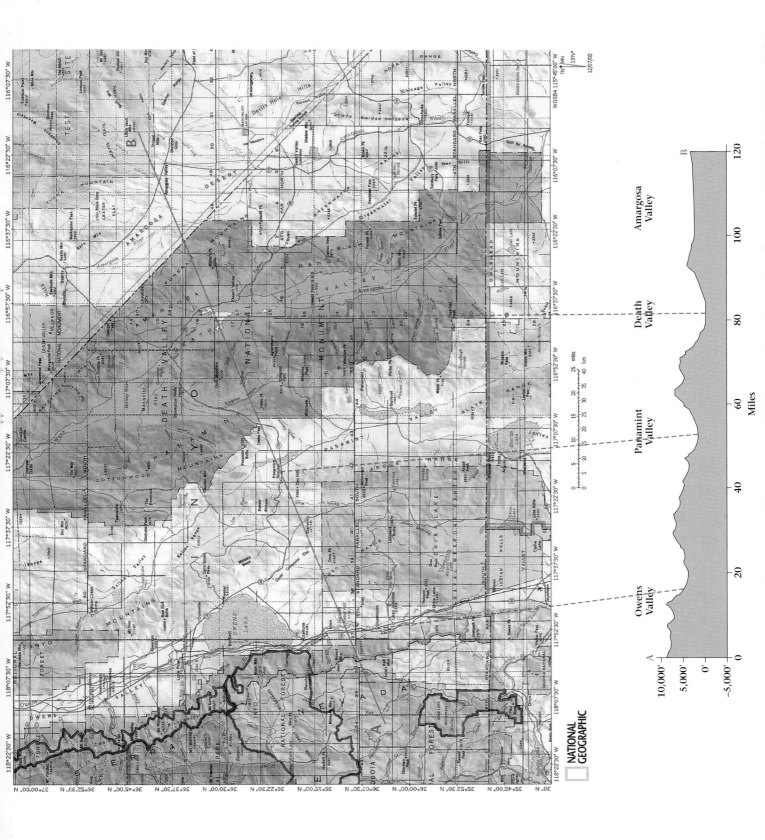

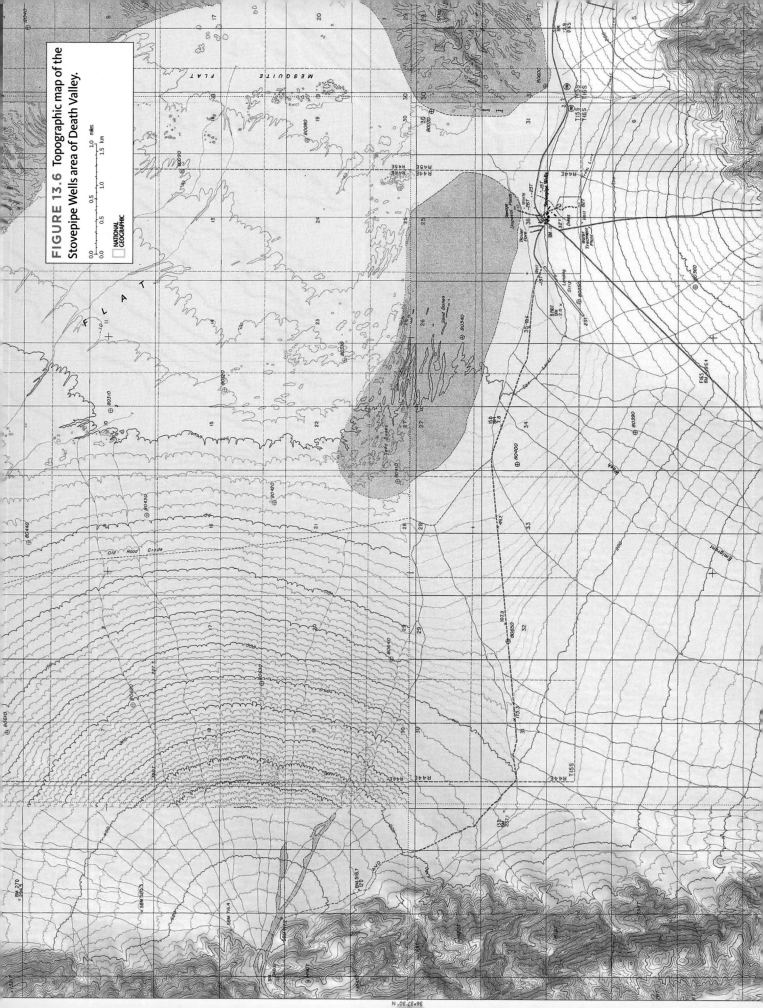

FIGURE 13.6 Topographic map of the Stovepipe Wells area of Death Valley.

Name: _____   **Section:** _____
Course: _____   **Date:** _____

(i) A more detailed view of the sand areas would show individual dunes, but these would not be accurate representations of the current land surface. Why not?

_____

_____

Examine the map of the area surrounding the town of Death Valley, including the Death Valley Airport and Furnace Creek Inn (**Fig. 13.7**).

(j) What arid environment depositional landforms are present? Label these on the map.

(k) Two lakes are shown on the map, one in Cotton Ball Basin, the other in Middle Basin. What are they called and how did they form?

_____

_____

_____

(l) What evidence is there on the map that the water in the streams and lakes is unsafe to drink? Explain why the water is not drinkable.

_____

_____

_____

(m) Is all of the water in these lakes brought in by the streams shown on the map? Explain your reasoning.

_____

_____

_____

Compare Figures 13.5, 13.6, and 13.7 with Figure 13.4. How far advanced in the arid erosion cycle is the Death Valley area? What changes can we expect to take place in the future?

_____

_____

_____

_____

Buckeye Hills, Arizona
The Buckeye Hills southwest of Phoenix are in an arid area similar to Death Valley, but this area differs in some ways. A topographic overview of the Buckeye Hills is shown in **Figure 13.8**.

continued

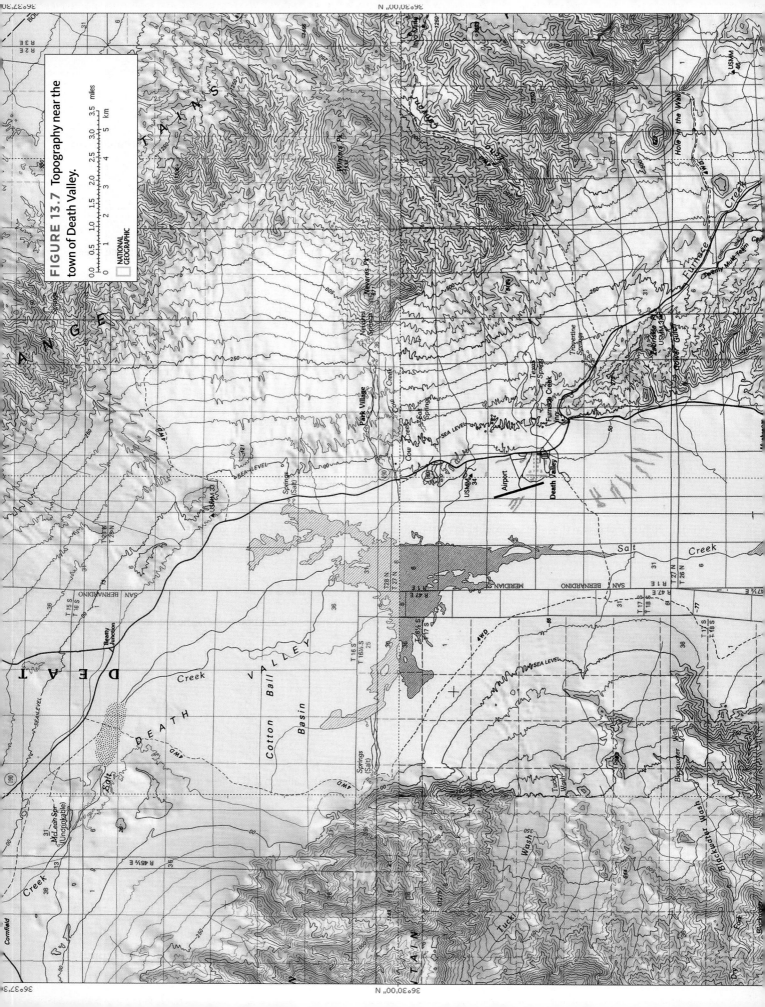

FIGURE 13.7 Topography near the town of Death Valley.

FIGURE 13.8 Topographic overview of the Buckeye Hills area of Arizona (area in rectangle shown in Figure 13.9).

Name: _____   **Section:** _____

Course: _____   **Date:** _____

(n) Describe the landscape shown in this overview.

_____

_____

_____

_____

(o) How are the Buckeye Hills similar to the part of Death Valley shown in Figure 13.5?

_____

_____

_____

(p) How are they different?

_____

_____

_____

Now look at the detailed map of the central part of the Buckeye Hills (**Fig. 13.9**).

(q) Identify as many erosional and depositional features of arid landscapes as you can. List them here and label them on the map.

_____

_____

_____

_____

(r) What is the probable origin of the small hills just east of I-80, southeast of Little Rainbow Valley?

_____

_____

_____

(s) What evidence is there of stream activity in the Buckeye Hills?

_____

_____

_____

continued

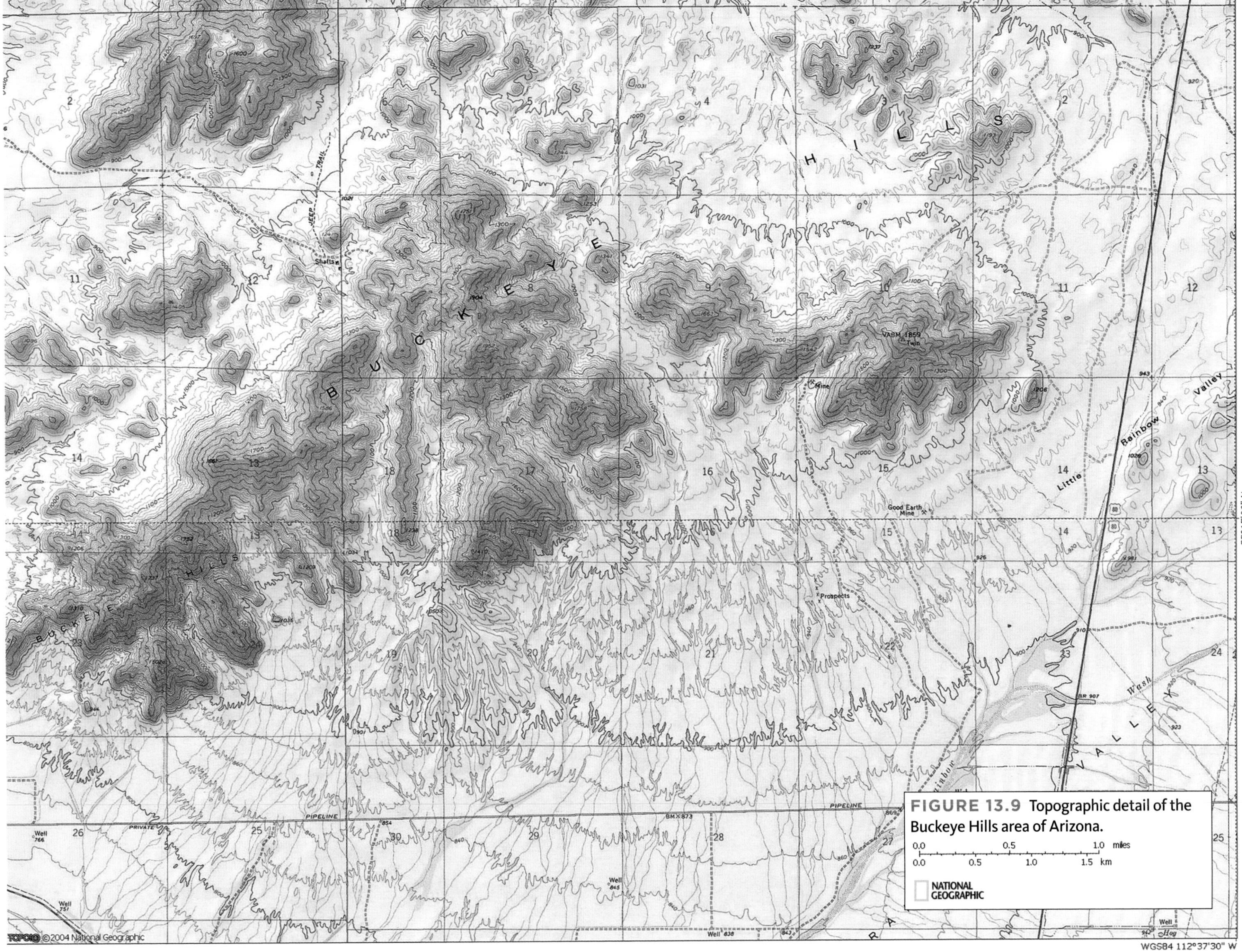

FIGURE 13.9 Topographic detail of the
Buckeye Hills area of Arizona.

Name: _____   Section: _____

Course: _____   Date: _____

(t) Refer to the diagram showing the evolution of arid landscapes (Fig. 13.4). Which area, Death Valley or the Buckeye Hills, is in a more advanced state of development? Explain your reasoning.

_____

_____

_____

(u) What evidence is there of mineral deposits in the Buckeye Hills?

_____

_____

_____

Nearly vertical slopes characterize the resistant sandstones, and more gently sloping surfaces characterize the less resistant shales. In contrast, a well-developed soil profile has formed above the bedrock in the Delaware Water Gap, partially masking erosional differences in the underlying rock. Soil creep is the dominant mass wasting process, and this smooths the topography.

13.4 Wind and Sand Dunes

Wind is more active in arid areas than humid areas because there is less cohesion between sand grains and little vegetation to hold the sand in place. This is because a small amount of moisture between grains at the surface holds the grains together weakly in humid climates, but that phenomenon is absent in arid regions. There are also fewer trees or other obstacles to block the full force of the wind. Wind erodes the way streams do—using its kinetic energy to pick up loose grains and then using this sediment load to abrade solid rock—literally sandblasting cliffs away.

Dunes are the major depositional landform associated with wind activity in deserts, but they also form along shorelines where there is an abundant source of sand. There are several kinds of dunes, depending on the amount of sand available, the strength of the wind, and how constant the wind direction is (**Fig. 13.10**). **Figure 13.11** shows the ripple-marked surface of a large transverse dune from the stippled area on the Stovepipe Wells map in Death Valley (Fig. 13.6).

FIGURE 13.10 Types of sand dunes (red arrows indicate wind direction).

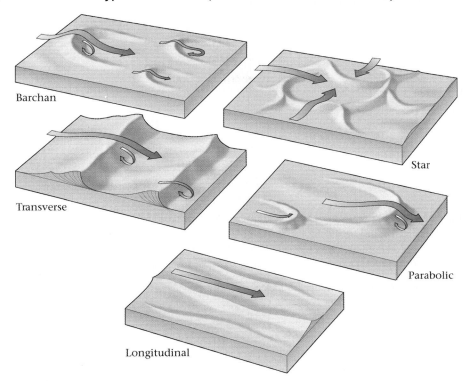

Barchan

Star

Transverse

Parabolic

Longitudinal

FIGURE 13.11 Origin and movement of transverse dunes.

(a) Transverse dunes at Stovepipe Wells in Death Valley.

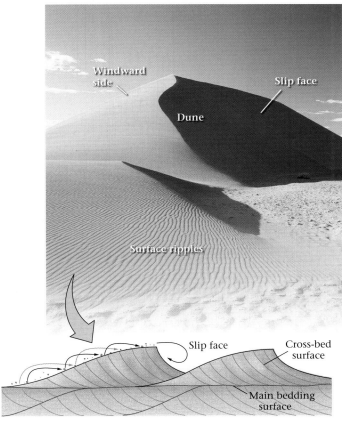

Windward side

Slip face

Dune

Surface ripples

Slip face

Cross-bed surface

Main bedding surface

(b) Movement of sand grains in a transverse dune. Sand is moved from the upwind dune face to the slip face, causing the dune to migrate downwind.

Name: _____   **Section:** _____
Course: _____   **Date:** _____

(a) Compare Figures 13.11a and 13.11b. Based on your observations, draw an arrow on the ripple-marked surface to indicate the wind direction. Explain your reasoning.

_____

_____

_____

The Nebraska Sand Hills

The Nebraska Sand Hills are the remnant of a vast mid-continent sea of sand that existed beyond the Pleistocene ice front. More humid conditions today permit grasses, trees, and shrubs to stabilize the dunes so that they no longer move across the countryside like those in Death Valley or the Sahara, but the hills preserve evidence of the more arid conditions under which they formed. Examine the overview of the Nebraska Sand Hills (**Fig. 13.12**).

(b) What evidence is there that this is still a relatively arid region?

_____

_____

_____

_____

(c) Describe the general topographic grain of the area including the size and shape of the ridges and valleys present.

_____

_____

_____

_____

(d) These ridges are the remnants of the Pleistocene sand sea. What features do they represent?

_____

_____

_____

Examine the map detail (**Fig. 13.13**) of the area outlined in red in Figure 13.12.

(e) Describe the shape of the ridges. Draw a topographic profile along the line indicated to help answer the next question.

_____

_____

(f) What does the shape of the dunes tell you about the wind during the Pleistocene?

_____

_____

continued

FIGURE 13.12 Topographic overview of part of the Nebraska Sand Hills (area in rectangle shown in Fig. 13.13).

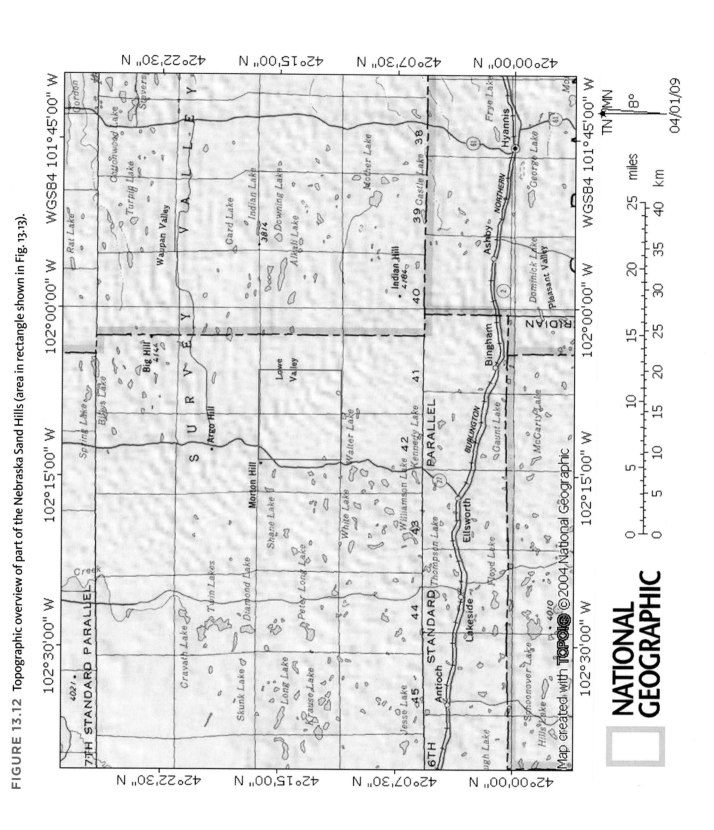

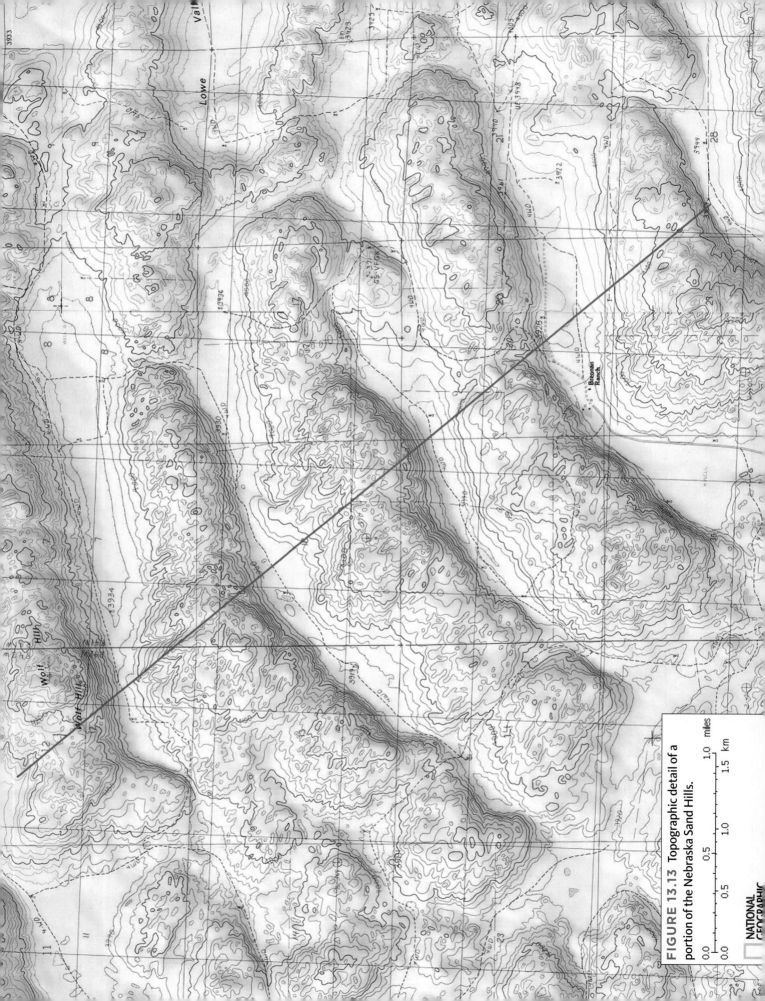

FIGURE 13.13 Topographic detail of a portion of the Nebraska Sand Hills.

CHAPTER 14

SHORELINE LANDSCAPES

PURPOSE

- Understand processes of shoreline erosion and deposition.
- Recognize different types of shoreline landforms and landscapes.
- Interpret active shoreline processes and identify possible hazards.

MATERIALS NEEDED

- Colored pencils
- Magnifying glass or hand lens to help read close-spaced contour lines

14.1 Introduction

Europeans coming to North America first settled the Pacific, Atlantic, Gulf of Mexico, and Great Lakes shores. These areas are still home to much of the continent's population and millions more vacation at their beaches, but shorelines are the most dynamic places on Earth, capable of changing in just hours or days. Shores are unique places where the rapidly moving atmosphere and hydrosphere interact with the more stable geosphere. The results of these interactions depend on factors involving all parts of the Earth System.

14.2 Factors Controlling Shoreline Erosion and Deposition

Factors shaping shoreline landscapes include rock or sediment type; wind direction and strength; interaction with mass-wasting; wind, stream, and other erosional agents; and tectonic activity. A few basic principles explain how these interact to produce shoreline erosion and deposition:

- Waves are the dominant erosional and depositional agent at shorelines.
- Waves are generated by the interaction of wind and the surfaces of oceans or large lakes.
- The kinetic energy of waves causes erosion and redeposition of unconsolidated material.
- Wind also moves sediment by itself, forming coastal sand dunes.
- Shoreline currents redistribute sediment to produce barrier islands, spits, and other landforms.
- Like streams and glaciers, waves use loose sediment to abrade the bases of solid bedrock cliffs. When support at the base of cliffs is undermined, the cliffs collapse by rockfall and slump. Waves erode the rubble, exposing the new cliff to wave erosion, and the cycle repeats. In this way, shoreline cliffs gradually retreat inland.
- Shorelines are in a constant state of conflict between the erosional processes of waves and wind and the depositional processes of sediment, lava flows, reef building, and so on.

14.3 Evolution of Coastal Landforms

The evolution of coastal landforms depends on the complex combination of factors listed above and on long-term climate change. The following sections and Exercise 14.1 help explain the role each of these factors plays.

14.3.1 Waves

Waves form by friction generated when wind blows across the surface of an ocean or lake. The symmetrical shape of waves offshore shows how the kinetic energy of the wind is transferred to the water. Offshore, water molecules move in circular paths, each molecule passing some energy on to those it contacts (**Fig. 14.1**). Loss of energy in these contacts limits the depth of wave action to approximately half of the wavelength, a depth referred to as the **wave base**.

FIGURE 14.1 Mechanics of wave action offshore and near the shoreline.

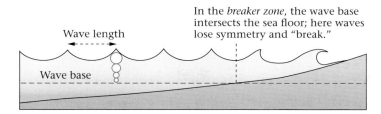

When water is shallower than the wave base, as on the right in Figure 14.1, orbiting water molecules interfere with the sea floor or lake bottom, causing waves to lose symmetry and "break." The pileup of water increases the waves' kinetic energy, enhancing their ability to erode the shoreline.

Figure 14.2 shows the breaker zone associated with waves in California (Fig. 14.2a) and Hawaii (Fig. 14.2b). In both cases, the wind that generated the waves blew for thousands of miles uninterrupted by land or trees, which is the most favorable condition for wave formation. The arrow in Figure 14.2b shows where the waves have cut a notch in the lava cliffs of Kauai.

FIGURE 14.2 Waves striking shorelines.

(a) Waves at the California coastline.

(b) Waves at the island of Kauai in Hawaii.

14.3.2 Coastal Materials

Figure 14.3 shows shorelines made of different materials ranging from loose sand to solid bedrock. Other shorelines may contain large amounts of organic materials, as in mangrove swamps, coral reefs, and marshy river deltas. Results of shoreline erosion differ greatly depending on which combination of these is present.

14.3.3 Climate Change

Long-term climate change has a profound effect on shorelines. For example, enormous amounts of water are locked up on land in the form of glaciers during an ice age. Most of that water would normally have been in the oceans. The opposite phenomenon is a major issue facing us today as global warming melts the Greenland and Antarctic ice caps. Use your common sense and your knowledge of Earth Systems to predict the result on the world's shorelines.

Name: _____   **Section:** _____
Course: _____   **Date:** _____

(a) Indicate with arrows the wind directions that generated the waves shown in Figure 14.2. Explain your reasoning.

_____

_____

_____

(b) In which area does the wind appear to be stronger? Explain your reasoning.

_____

_____

_____

_____

(c) In which of the two areas would you expect wave erosion to be most effective? Explain.

_____

_____

_____

_____

(d) What factors other than wind velocity must be considered?

_____

_____

_____

_____

(e) How does erosion of the materials in Figure 14.3 differ?

_____

_____

_____

_____

(f) How can waves erode the rock cliffs shown in Figures 14.3a and 14.3b?

_____

_____

_____

(g) What factors determine how resistant these bedrock cliffs are to erosion?

_____

_____

_____

(h) Which of the shorelines in Figure 14.3 would be the most difficult to erode? The easiest? Explain.

_____

_____

_____

_____

FIGURE 14.3 Shoreline materials.

(a) Rocky shoreline at West Quoddy Head in Maine.

(b) Sandy and rocky shoreline in Hawaii.

(c) Black sand beach in Hawaii.

(d) Boulder and cobble beach at West Quoddy Head, Maine.

(e) Coral reef along the Hawaiian shoreline.

(f) Salt marsh along the coast of Cape Cod in Massachusetts.

(g) Mangrove swamp in northeastern Brazil.

Effects of Climate Change on Shorelines

Name: _____ **Section:** _____
Course: _____ **Date:** _____

(a) What would be the effect on the oceans of an episode of worldwide glacial advance? Explain.

_____

_____

_____

(b) What would be the effect on the world's shorelines?

_____

_____

_____

(c) In what way was the location of the world's shorelines at the Pleistocene glacial maximum different from that of today's shorelines? Explain your reasoning.

_____

_____

_____

14.3.4 Tectonic Activity

Global warming and cooling change sea level by adding or removing water from the ocean basins. Tectonic activity can have the same effect by (1) widening an ocean basin by sea-floor spreading or shrinking it by subduction, or (2) locally uplifting or downdropping the coast relative to the ocean basin.

EXERCISE 14.3   **Effects of Plate Tectonic Processes on Shorelines**

Name: _____ **Section:** _____
Course: _____ **Date:** _____

(a) What effect will continued sea-floor spreading in the Atlantic Ocean have on east and Gulf coast sea level? Explain your reasoning.

_____

_____

_____

(b) What effect would partial closing of the Atlantic Ocean have on sea level? Explain.

_____

_____

_____

14.4 Emergent and Submergent Shorelines

Short-term changes in sea level occur in the tidal cycle and during major storms. Tides change sea level twice daily, by only a few feet in most places but by more than 50 feet in places like the Bay of Fundy. Hurricane storm surges tens of feet high bring the ocean much farther inland than normal tides, as demonstrated all too well by storms in the past few years. *Long-term* sea level changes are caused by interaction of tectonic and climatic factors, and one way to classify shorelines is whether they are **emergent** (recently uplifted relative to sea level) or **submergent** (lowered so that former land areas are below sea level) (**Fig. 14.4**).

FIGURE 14.4 Uplift revealed along the California coastline.

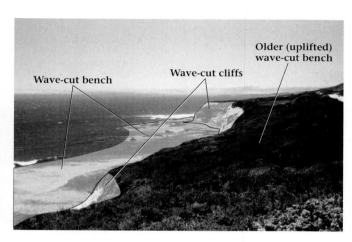

Waves move sediment back and forth across the tidal zone, abrading a flat **wave-cut bench** and carving wave-cut notches (Fig. 14.2b), which undermine the cliffs and cause rockfalls and slumping. The cliff gradually retreats and the bench widens. If the land is uplifted or sea level drops, the bench may be raised far above sea level (Fig. 14.4a) and a new bench forms at the new level of the sea (Fig. 14.4b). Figure 14.4b shows a single uplifted bench in addition to the current one. In tectonically more active coastal areas, there may be several uplifted benches; dating them enables geologists to estimate the rate of uplift.

EXERCISE 14.4   **Recognizing Emergent and Submergent Shorelines**

Name: _____   Section: _____

Course: _____   Date: _____

Emergent shorelines look very different from submergent shorelines. Examine **Figure 14.5**, part of the Atlantic coast in Maine, and **Figure 14.6**, part of the Pacific coast in California. One of these is a typical submergent shoreline, the other a classic emergent shoreline. Apply your geologic reasoning to tell which is which.

(a) Compare and contrast the shapes of the two shorelines.

_____

_____

_____

continued

Name: _____   Section: _____

Course: _____   Date: _____

(b) Which of these shorelines is emergent and which is submergent? Explain your reasoning.

_____

_____

_____

FIGURE 14.5 Maine coastline near Boothbay Harbor.

NATIONAL
GEOGRAPHIC

0.0 0.5 1.0 1.5 2.0 2.5 3.0 3.5 miles

0   1   2   3   4   5 km

FIGURE 14.6 California coastline south of Half Moon Bay.

Name: _____   **Section:** _____

Course: _____   **Date:** _____

The Pacific Coast, California

Figure 14.7 is a map showing details of the California coast not far from where Figure 14.4 was photographed.

FIGURE 14.7 Portion of the Dos Pueblos Canyon quadrangle of California.

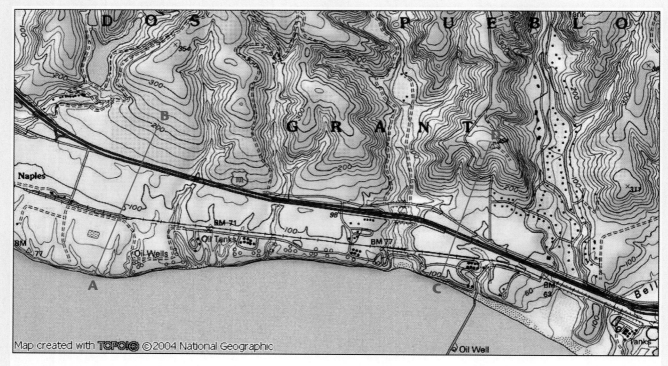

 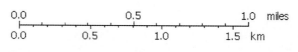

(a) Describe the shoreline in your own words.

_____

_____

_____

(b) Sketch profiles along lines A–B and C–D.

continued

Name: _____   Section: _____

Course: _____   Date: _____

(c) What evidence shows that a change in sea level has taken place?

_____

_____

_____

(d) Based on the map and profile, is this an emergent or submergent shoreline? By how much has sea level changed? Explain your reasoning.

_____

_____

_____

(e) Is the map or photograph more useful for measuring the amount of change? Explain.

_____

_____

_____

Lake Erie, Ohio

The shorelines of many lakes that formed shortly after the retreat of Pleistocene glaciers from North America have changed markedly in the last few thousand years. Some glacial lakes have shrunk to a fraction of their former size (such as Glacial Lake Bonneville, which is now the Great Salt Lake in Utah) or disappeared entirely (Glacial Lake Hitchcock in Massachusetts). The Great Lakes have adjusted to post-glacial conditions and their shorelines reveal those changes. **Figure 14.8** shows an area in Ohio just south of Lake Erie.

(f) Examine the spacing of the contour lines. What do they suggest about the evolution of Lake Erie?

_____

_____

_____

(g) Draw a profile along line A–B.

_____

_____

_____

_____

_____

_____

(h) How are Sugar, Chestnut, and Butternut ridges related to the post-glacial history of Lake Erie?

_____

_____

_____

continued

FIGURE 14.8 Area just south of Lake Erie near Elyria, Ohio.

Contour interval = 10 feet

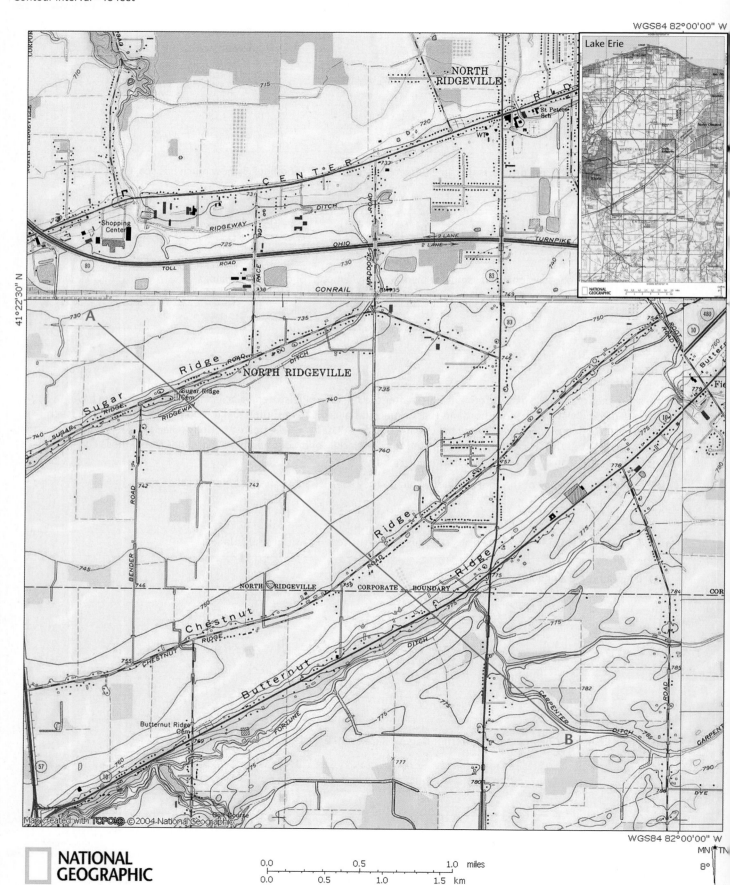

Name: _____　　　Section: _____

Course: _____　　　Date: _____

(i) Label the previous shoreline position(s).

(j) The current elevation of Lake Erie is 174 feet. How much has lake level changed, based on the evidence on this map?

_____

(k) Was the change continuous or did it take place in sporadic episodes? Explain.

_____

_____

_____

(l) Assuming the retreat of continental glaciers took place about 10,000 years ago, calculate the rate at which lake level dropped if the change had been continuous.

_____

(m) If the change was episodic, suggest a way to estimate the relative amount of time associated with each "still-stand" of lake level. What assumptions must you make?

_____

_____

_____

14.5 Erosional and Depositional Shoreline Features

14.5.1 Erosional Features

You have already seen wave-cut notches, benches, and cliffs, and below you will see sea arches and sea stacks, all of which are the most common erosional features of bedrock shorelines. Bedrock shorelines do not erode at the same rate in all places, and the differences are the result of rock type, extent of fracturing, and location on promontories (more rapid erosion) or embayments (slower). **Figure 14.9** shows erosion along four different bedrock shorelines.

FIGURE 14.9 Erosional features of bedrock shorelines.

(a) A wave-cut notch along the Hawaiian coast.

(b) A wave-cut bench at the foot of the cliffs at Etrétat, France.

FIGURE 14.9 Erosional features of bedrock shorelines. (*con't.*)

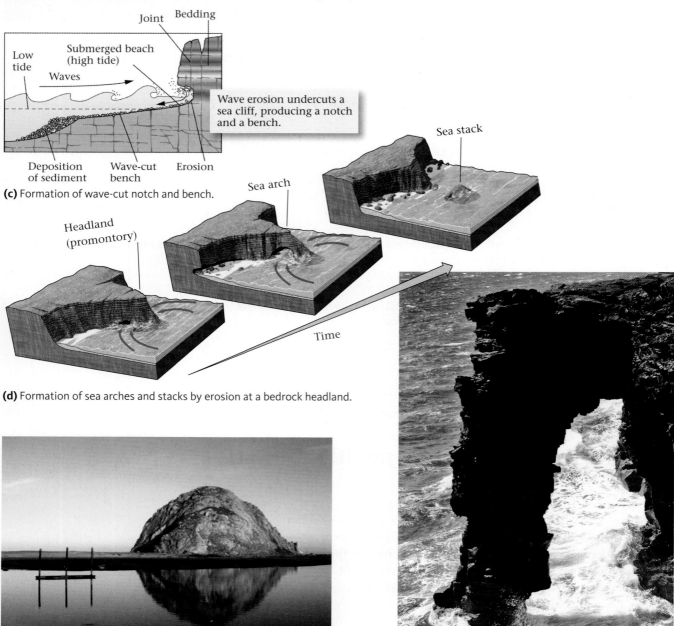

(c) Formation of wave-cut notch and bench.

Wave erosion undercuts a sea cliff, producing a notch and a bench.

(d) Formation of sea arches and stacks by erosion at a bedrock headland.

(e) Morro Rock in California, a classic sea stack.

(f) A sea arch on the coast of Hawaii.

EXERCISE 14.6　　**Erosional Processes and Shoreline Landforms**

Name: _____　　Section: _____
Course: _____　　Date: _____

(a) How did the large blocks in Figure 14.9a get into the surf zone? What is their eventual fate?

_____

_____

continued

Name: _____   **Section:** _____
Course: _____   **Date:** _____

(b) Suggest a reason for the preferential erosion of the center of a sea arch.

_____

_____

(c) Suggest a reason for the scalloped shape of the Hawaiian coastline in (a).

_____

_____

(d) How did Morro Rock become isolated from the bedrock shoreline?

_____

_____

(e) How do sea stacks form?

_____

_____

(f) What is the eventual fate of the sea arch in (f)?

_____

_____

Examine the shoreline in **Figure 14.10**. Point Sur and False Sur are the same kind of feature and record a multistage development for this part of the California shoreline.

(g) Based on its size and steepness, Point Sur is made of what material? Explain.

_____

_____

(h) Of what material is the area between Point Sur lighthouse and the California Sea Otter Game Refuge made? Explain your reasoning.

_____

_____

(i) Suggest a sequence of events by which the Point Sur shoreline could have formed.

_____

_____

_____

continued

FIGURE 14.10 Shoreline at Point Sur, California.

Contour interval = 40 feet

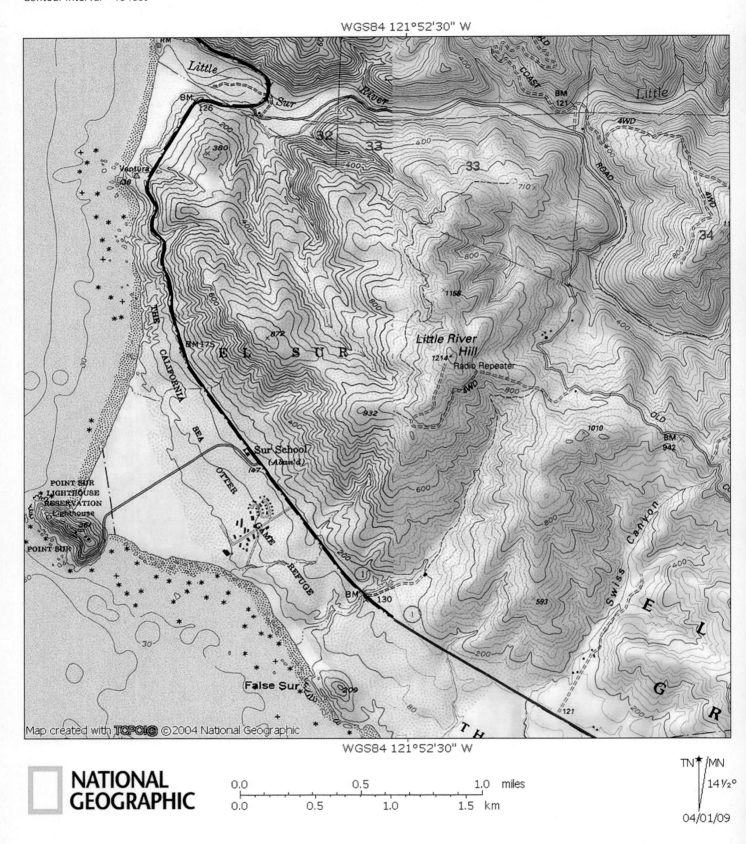

WGS84 121°52'30" W

WGS84 121°52'30" W

NATIONAL GEOGRAPHIC

0.0   0.5   1.0 miles

0.0   0.5   1.0   1.5   km

TN MN

14½°

04/01/09

Cliffs of poorly consolidated coastal plain sediments and unconsolidated glacial debris are common along the Atlantic coastline of North America. Waves undermine these cliffs, causing slumping and rapid cliff retreat. **Figure 14.11** captures the process in a dramatic fashion. Storm-generated waves undermined the bluff beneath this Florida house, causing partial collapse of the building.

FIGURE 14.11 Coastal bluff erosion caused parts of this house in Destin, Florida to collapse. The erosion was exacerbated by the high waves during Hurricane Dennis in July of 2005.

14.5.2 Depositional Features

Prominent depositional features develop along shorelines that have an abundant supply of sand, ranging from continuous features that extend for miles along the coastline to small isolated beaches. The Gulf and Atlantic coastal plains are underlain by unconsolidated sediments and display numerous landforms characteristic of such environments.

Beaches are the most common landforms in these areas. Long, narrow sand deposits formed by the interaction of waves and longshore currents are also abundant (**Fig. 14.12**). Others lie offshore and are called **barrier islands** because they block direct access from the sea to the continent. These islands may be barriers to explorers, fishermen, and recreational sailors, but they are our first line of defense from hurricanes and other storms. Some of these elongate sand bars (called **spits**) are attached to the mainland, and some spits, like Cape Cod, are sharply curved (**hooks**).

FIGURE 14.12 Barrier islands along the Texas Gulf coast, southwest of Galveston.

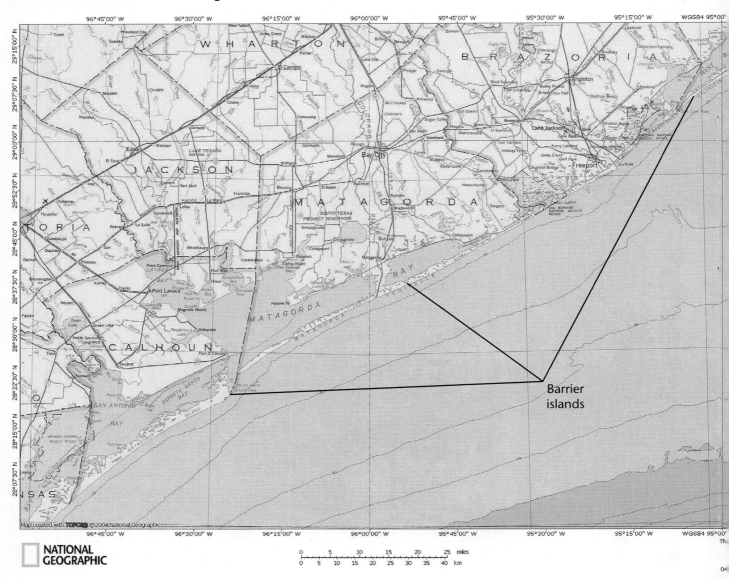

NATIONAL GEOGRAPHIC

EXERCISE 14.8   Interpreting Depositional Shoreline Processes

Name: _____   Section: _____

Course: _____   Date: _____

(a) Does the shoreline in Figure 14.12 appear to be emergent or submergent? Explain.

_____

_____

_____

(b) Identify and label as many depositional landforms as you can in Figure 14.12.

continued

Name: _____   **Section:** _____

Course: _____   **Date:** _____

(c) What evidence is there that sediment redistribution is taking place _landward_ of the barrier island as well as on the barrier islands and spits that protect these areas?

_____

_____

_____

(d) The rapid movement of sand by longshore drift could block access to the mainland by closing gaps in barrier islands and between spits. What steps can be taken to prevent this in the Galveston area?

_____

_____

_____

(e) Indicate the dominant direction of longshore drift. Explain your reasoning.

_____

_____

_____

(f) Outline the wetlands that could protect the mainland from the next hurricane.

Marshy wetlands are common between barrier islands and the shore and are important parts of the food chain, supporting a diverse group of organisms and providing breeding areas for fish and birds. They are also part of our natural storm-protection system. If a storm surge manages to overflow the barrier island (as happend in Galveston, Texas in 2008, for example), the wetlands act as sponges, soaking up the water and lessening damage to the more densely inhabited mainland.

EXERCISE 14.9   **Depositional Processes and Shoreline Landforms**

Name: _____   **Section:** _____

Course: _____   **Date:** _____

Figures 14.13a and **14.13b** are topographic maps of the northern end of Cape Cod made more than 100 years apart. As mentioned earlier, a terminal moraine here provides an abundant source of sand and gravel for shoreline processes.

continued

Name: _____   **Section:** _____

Course: _____   **Date:** _____

These two maps help illustrate how much change can occur along a coastline in a (geologically) short period. They also show how population pressures affect our use of limited, and therefore very valuable, shoreline space.

FIGURE 14.13 Shoreline changes at Cape Cod, Massachusetts.

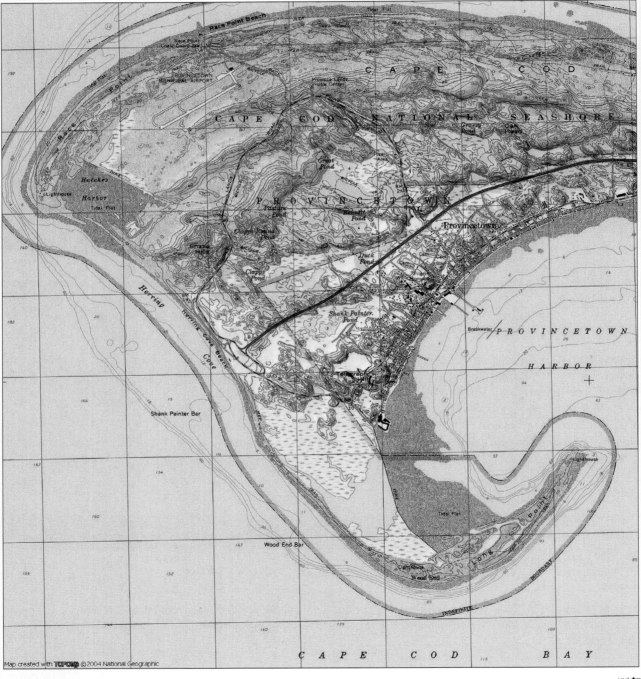

(a) The shoreline of Cape Cod, MA in 1983.

Name: _____   **Section:** _____
Course: _____   **Date:** _____

FIGURE 14.13 Shoreline changes at Cape Cod, Massachusetts. (*con't.*)

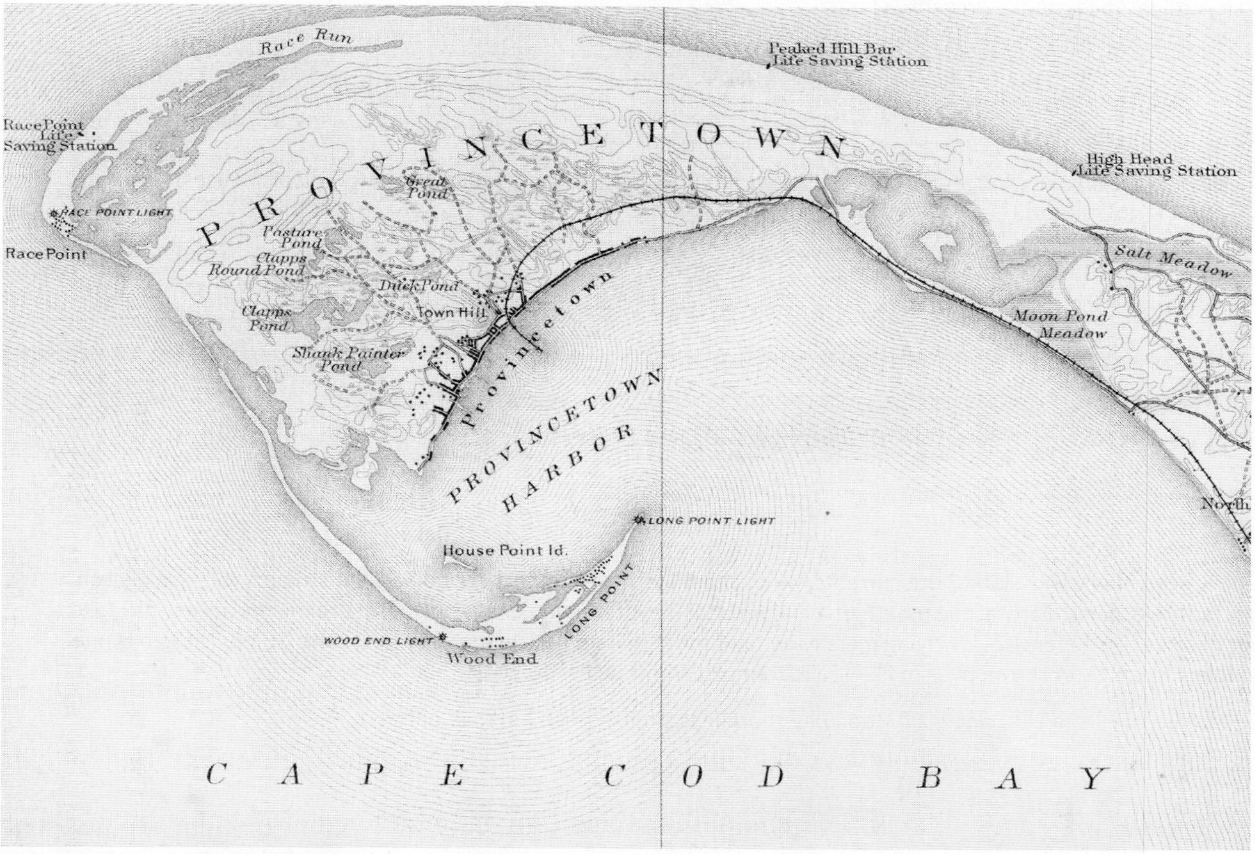

(b) The shoreline of Cape Cod, MA in 1985.

(a) What kind of landform is Cape Cod? _____

(b) Draw arrows to indicate the direction of longshore currents in this area.

(c) Describe the *geologic* changes that have occurred in the century separating the compilation of these two maps.

_____

_____

_____

(d) How have Cape Codders tried to prevent Provincetown Bay from changing?

_____

_____

_____

(e) Based on what happened in the 107 years recorded by these two maps, which landforms might disappear? Which might change shape drastically? Explain your reasoning.

_____

_____

_____

continued

Name: _____   **Section:** _____
Course: _____   **Date:** _____

(f) Which of these changes would be beneficial to people living or vacationing in Provincetown? Which would be negative? How might the latter be prevented?

_____
_____
_____

(g) Describe the *human* changes that have affected this area (e.g., transportation, housing, and other uses).

_____
_____
_____

EXERCISE 14.10   **Interaction of Shoreline and Land Processes**

Name: _____   **Section:** _____
Course: _____   **Date:** _____

We have seen how waves, longshore currents, wind, and mass wasting affect shorelines, but streams can also play an important role, particularly at the mouth of a major river. The interaction of stream and shoreline processes may take several forms. The extensive, low-lying marshy delta of the Mississippi River extending into the Gulf of Mexico is one example. **Figure 14.14** shows two smaller scale examples from the Pacific coast.

(a) How are the Oceanside and Half Moon Bay shorelines similar? How are they different?

_____
_____
_____

(b) What shoreline depositional landforms are found in these areas?

_____
_____

(c) What is the probable source of sediment for these landforms?

_____
_____

(d) How does the relationship between stream and shoreline processes in the Oceanside area differ from that of Half Moon Bay? Suggest a reason for the difference.

_____
_____
_____

(e) Compare these Californian shorelines with the Galveston area. Why are they different?

_____
_____
_____

continued

FIGURE 14.14 Interaction of shoreline and stream processes.

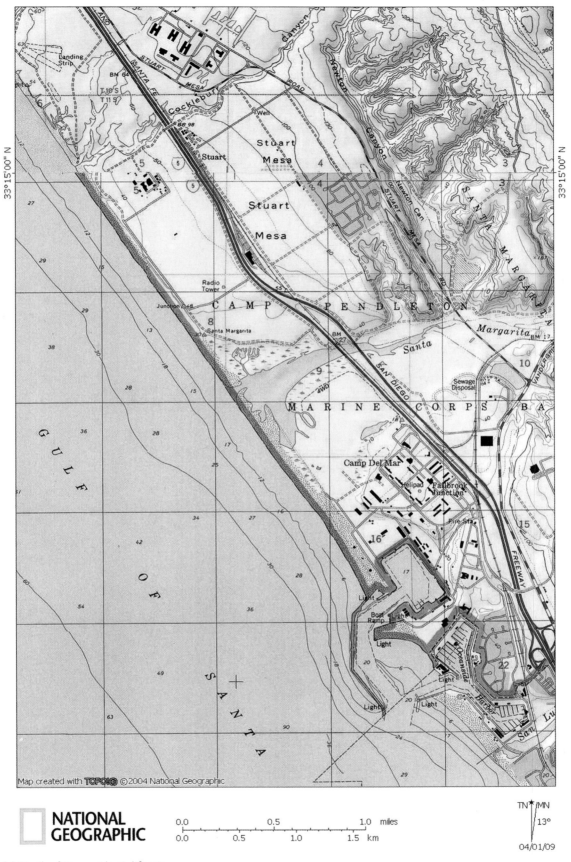

NATIONAL GEOGRAPHIC

0.0 ————— 0.5 ————— 1.0 miles
0.0 — 0.5 — 1.0 — 1.5 km

TN ✱ ∤MN
13°
04/01/09

(a) North of Oceanside, California

FIGURE 14.14 (*con't.*)

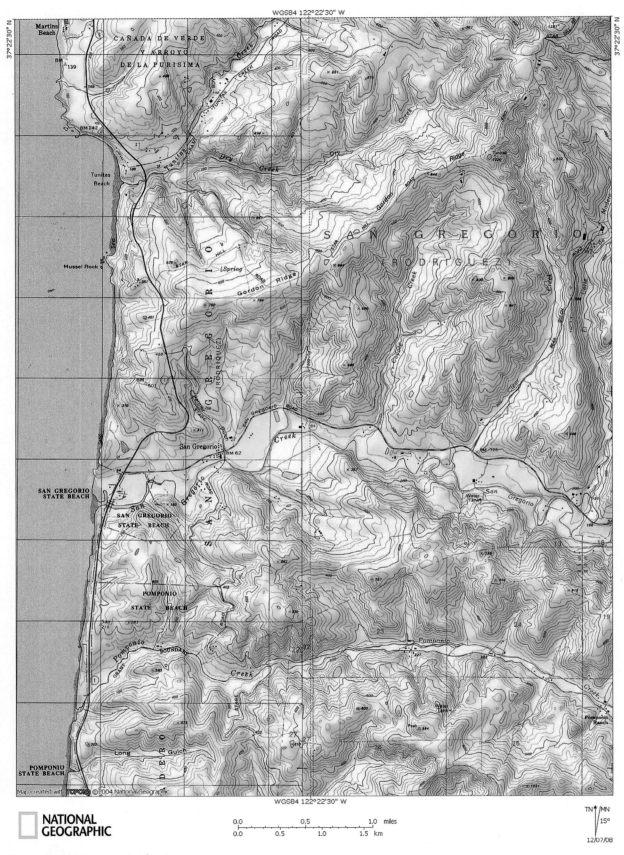

(b) Near Half Moon Bay, California

14.6 When Shorelines Become Dangerous

Events in December 2004 and August 2005 imprinted the dangers of shoreline processes indelibly on our minds. First, earthquake-generated tsunamis wreaked havoc along the margins of the Indian Ocean. Then hurricane Katrina—called the worst natural disaster in U.S. history—struck the Atlantic and Gulf coasts. In contrast, the rise in sea level caused by melting glaciers worldwide is not violent—indeed it is almost imperceptible—but it will eventually disrupt society more than any single hurricane or tsunami.

14.6.1 Coastal Storms

When the normal force of wind is multiplied fivefold in winter "nor'easters" in New England or tenfold in hurricanes, wave-related processes are amplified catastrophically. Typical coastal storms come from a single direction and are easier to prepare for than hurricanes, whose wind velocities can be above 150 miles per hour and whose direction can shift rapidly.

Hurricanes and typhoons, their Pacific Ocean relatives, are enormous storm systems that develop when an atmospheric disturbance passes over warm ocean water (over 80°F). The storm absorbs energy from the ocean water and forms a low-pressure system as hot air rises. Low-level winds flow toward the center of the system, with a swirling counterclockwise pattern caused by the Coriolis effect (**Fig. 14.15**).

FIGURE 14.15 Hurricane Katrina striking the Gulf Coast on August 29, 2005.

Note the spiral form with a well-developed eye and swirling rain bands outlining the counterclockwise wind circulation pattern.

Heat from the ocean adds energy and moisture, and water vapor condensing at high altitudes adds even more energy. If winds in the upper atmosphere are weak, they cannot prevent the storm from intensifying to hurricane levels (wind velocities greater than 74 miles per hour). While most wind associated with hurricanes is horizontal, winds near the center are a vertical downdraft of warm air that surrounds the eye, a generally clear and ominously calm sector of the storm. Most hurricanes are around 300 miles wide, including the eye, which is typically 20 to 40 miles across.

Hurricanes are pushed slowly by *steering currents* in the atmosphere, generally at 10 to 15 miles per hour. Winds throughout the lower atmosphere, ocean temperature, and interaction with landmasses can make it difficult to predict the path of a hurricane. Hurricanes weaken after making landfall because they are no longer nourished by the warm ocean water, but they may strengthen if they cross over water again. This happened to Katrina, which was only a category 1 hurricane when it hit Florida but grew to category a 5 as it headed across the Gulf of Mexico toward New Orleans.

14.6.2 Hurricane Hazards

A common misconception about coastal storms is that wind is the major hazard. Winds of 100 to 175 miles per hour are truly dangerous (**Fig. 14.16**) but the **storm surge**, a wall of water driven onshore by the hurricane, is much more hazardous. Katrina's storm surge was estimated at 30 feet above normal sea level; it carried boats, houses, and other debris inland, causing widespread destruction and wiping out entire communities (**Fig. 14.17a,b**).

FIGURE 14.16 Windows in a New Orleans hotel shattered by wind from Hurricane Katrina.

Coastal flooding often accompanies hurricanes, which is not surprising because some storms deliver as much as 15 inches of rain. After Katrina, water from Lake Pontchartrain broke through levees north of New Orleans, adding to the rain from the storm. Thankfully, the levees protecting New Orleans from the Mississippi River remained intact. Floodwaters damage buildings, block relief efforts (**Fig. 14.17c,d**), and can lead to serious health problems if water treatment plants are overwhelmed by storm runoff and dump millions of gallons of sewage into the flooded areas. Exposure to toxic materials in the floodwater, like gasoline from damaged gas stations and cars, is a long-term health problem for flood victims.

FIGURE 14.17 Storm surge damage and flooding associated with Hurricane Katrina.

(a) A neighborhood complex flattened by the storm surge in Biloxi, Mississippi.

(b) Storm surge debris and damage.

(c) Approximately 80% of New Orleans was flooded when levees that protected the city from Lake Pontchartrain failed.

(d) Breached levees in New Orleans had to be repaired before floodwaters could be pumped out.

EXERCISE 14.11   **Planning for Hurricanes**

Name: _____   Section: _____
Course: _____   Date: _____

Emergency planners must consider all the factors that determine how dangerous a storm will be: shoreline topography and composition, population density, type of building construction, as well as the strength and path of the storm. These conditions can vary widely over short distances and change rapidly if a storm changes direction.

Landforms

Floridians are used to hurricanes, but the topography of the state's east and west coasts offers different degrees of protection from wind and storm surge (**Fig. 14.18**).

(a) Which shoreline offers more protection from a storm—that of Naples or Port St. Lucie? Explain.

_____

_____

continued

Name: _____   Section: _____
Course: _____   Date: _____

FIGURE 14.18 A tale of two coasts.

(a) Naples on Florida's west coast.

(b) Port St. Lucie on Florida's east coast.

Storm Path

The path of a hurricane determines whether an area is spared or severely damaged, because the path determines the effective wind velocity at any point along the coast. The effective velocity of hurricane winds is a combination of the wind speed in the hurricane and that of the steering winds. In **Fig. 14.19**, Point A, in the direct path of the hurricane, will receive winds of 125 mph because the velocity of the steering winds adds to that of the hurricane itself.

(b) What will be the effective wind velocity of the hurricane at Point B? Explain.

_____

_____

FIGURE 14.19 Effect of storm path on effective wind velocity.

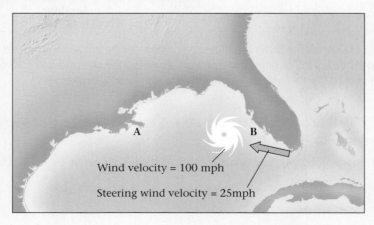

continued

Name: _____   **Section:** _____
Course: _____   **Date:** _____

In general, the *right side of a hurricane* (viewed looking toward the direction in which it is moving) is the most dangerous because of the addition of steering and hurricane wind velocities.

The direction from which hurricane winds strike an area also depends on the precise storm path. Remember that hurricane winds flow in a counterclockwise direction and answer the following questions referring to **Fig. 14.20**:

(c) Sketch paths by which the first wind from the hurricane [to the left] would strike each of the indicated locations (A through F) from the *south*.

(d) From the *north*.

(e) From the *east*.

(f) From the *west*.

(g) As a hurricane passes, the direction of its winds shifts. Explain how a single storm could cause winds from opposite directions to affect an area.

_____

_____

_____

(h) The deep estuaries indicated by the arrows are highly vulnerable to storm surge because their funnel shapes concentrate water to heights well above typical storm surge. Sketch the path that would cause the greatest storm surge into each estuary—the worst case scenario.

FIGURE 14.20 Relationship between wind direction and hurricane path.

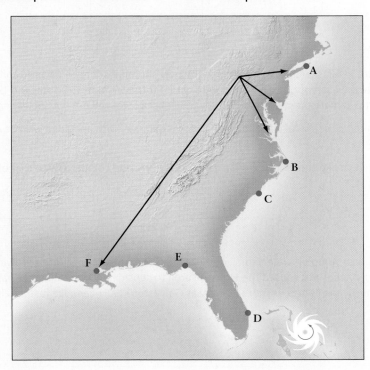

Name: _____   Section: _____
Course: _____   Date: _____

The accelerated melting of the polar ice caps and of alpine glaciers on every continent is well documented, and the resulting worldwide rise of sea level will be catastrophic for millions of people. Estimates for the maximum change if the ice caps melt range up to several hundred feet. Let's first see the effect of a smaller sea level rise on the infrastructure of Washington, D.C. (**Fig. 14.21**). Look at the contour lines in the area including the Capitol, White House, National Mall, and Washington Monument.

(a) How much would sea level have to rise to flood the White House basement? _____ feet

(b) The base of the Washington Monument? _____ feet

(c) The tracks at Union Station? _____ feet

(d) What would happen if sea level rose 100 feet, the low end of the estimated change.

_____

_____

_____

How will our coastlines be affected? **Fig. 14.22** shows topographic profiles for selected coastal cities. From these profiles, indicate how far inland the shoreline would move if sea level rose 50, 100, or 150 feet.

Sea level rise	Distance inland				
	Amityville	**Morehead City**	**Savannah**	**Houston**	**San Diego**
50 feet					
100 feet					
150 feet					

(e) Using the scale in Figure 14.22, sketch the positions of the new shorelines for the Atlantic and Gulf coasts assuming that they would parallel the current one. Use a separate color for the 50-foot, 100-foot, and 150-foot sea level rises. Now you understand the earlier comment that the slow rise of sea level will disrupt society enormously.

Now consider the combination of sea level rise and coastal storm damage, with an added wrinkle caused by increasing use of natural resources.

(f) How can a rise in sea level make the danger from a category 2 hurricane comparable to that of a category 3 or 4?

_____

_____

_____

(g) Over the past few decades, pumping water from southeast Texas to support the state's growing population and to extract oil for our energy needs has caused the Houston area to subside, by several feet. What effect will this have on damage if a major hurricane comes onshore directly at Galveston? Hurricane Ike just barely missed in 2008!

_____

_____

_____

continued

FIGURE 14.21 Topographic map of the National Mall area of Washington, D.C.

Contour interval = 10 feet

Name: _____   Section: _____

Course: _____   Date: _____

FIGURE 14.22   Coastal profiles for Exercise 14.12.

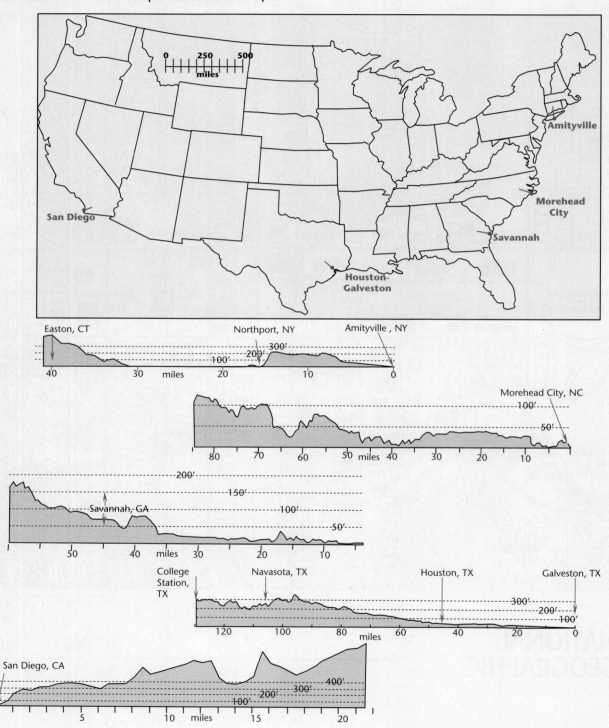

INTERPRETING GEOLOGIC STRUCTURES ON BLOCK DIAGRAMS, GEOLOGIC MAPS, AND CROSS SECTIONS

PURPOSE

- Become familiar with common geologic structures, such as folds and faults.
- Visualize structures in three dimensions, using block diagrams, maps, and cross sections.
- Recognize the presence of folded and faulted rock from landscape features.
- Interpret the geologic structure of an area from a geologic map.
- Learn how to read a geologic map of a region.

MATERIALS NEEDED

- Colored pencils
- A fine-tipped black pen
- Tracing paper
- A pair of scissors and tape
- A protractor
- A straight edge

15.1 Introduction

The Earth is a dynamic place! Over time, lithosphere plates move relative to one another: at convergent boundaries, one plate sinks into the mantle beneath another; at rifts, a continental plate stretches and may break apart; at a mid-ocean ridge, two oceanic plates move away from each other; at a collision zone, continents press together; and at a transform boundary, two plates slip sideways past each other. All these processes generate stress that acts on the rocks in the crust. In familiar terms, *stress* refers to any of the following (**Fig. 15.1**): **pressure**, which is squeeezing equally from all sides; **compression**, which is squeezing or squashing in a specific direction (indicated by the inward-pointing arrows); **tension**, which is stretching or pulling apart (indicated by the outward-pointing arrows); **shear**, which happens when one part of a material moves relative to another part in a direction parallel to the boundary between the parts (indicated by adjacent arrows pointing in opposite directions).

FIGURE 15.1 Kinds of stress.

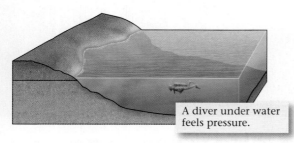

A diver under water feels pressure.

(a) Pressure.

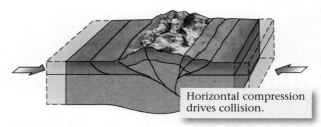

Horizontal compression drives collision.

(b) Horizontal compression.

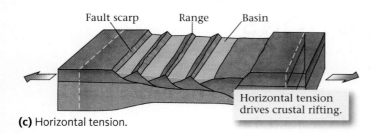

Fault scarp   Range   Basin

Horizontal tension drives crustal rifting.

(c) Horizontal tension.

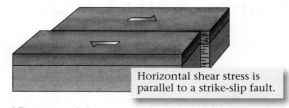

Horizontal shear stress is parallel to a strike-slip fault.

(d) Horizontal shear.

The application of stress to rock produces **deformation**, which includes many phenomena, such as the displacement of rocks on sliding surfaces called **faults**, the bending or warping of layers to produce arch-like or trough-like shapes called **folds**, or the overall change in the shape of a rock body by thickening or thinning. Under certain conditions, a change in the shape of a rock body produces **foliation**, a fabric caused by the alignment of platy or elongate minerals.

The products of deformation, such as faults, folds, and foliations, are called **geologic structures**. Some geologic structures are very small and can be seen in their entirety within a single hand specimen. Typically, however, geologic structures in the Earth's crust are large enough that they affect the orientation and geometry of rock layers, which in turn may control the pattern of erosion and, therefore, the shape of the land surface.

Name: _____   **Section:** _____
Course: _____   **Date:** _____

For each of the phenomena described below, name the stress state involved.

(a) You spread frosting on a cake with a knife. The frosting starts out as a thick wad, then smears into a thin sheet.

_____

(b) You step on a filled balloon until the balloon flattens into a disk shape. _____

(c) You pull a big rubber band between your fingers so that it becomes twice its original length.

_____

(d) A diver takes an empty plastic milk jug, with the lid screwed on tightly, down to the bottom of a lake. The jug collapses inward from all sides. _____

Geologists represent the shapes and configurations of geologic structures in the crust with the aid of three kinds of diagrams. A **block diagram** is a three-dimensional representation of a region of the crust that depicts the configuration of structures on the ground (the map surface) as well as on one or two vertical slices into the ground (cross-section surfaces). A **geologic map** represents the Earth's surface as it would appear looking straight down from above, showing the boundaries between rock units and where structures intersect the Earth's surface. A **cross section** represents the configuration of structures as seen in a vertical slice through the Earth. Figure 15.2 shows how these different representations depict Sheep Mountain in Wyoming.

The purpose of this chapter is to help you understand the various geometries of geologic structures, and to develop the ability to visualize structures and other geologic features by examining block diagrams, geologic maps, and cross sections. In addition, this chapter will help you to see how the distribution of rock units, as controlled by geologic structures, influences topography, as depicted on topographic maps and digital elevation maps (DEMs). Geologic structures can be very complex, and in this chapter we can only work with the simplest examples. Again, our main goal here is to help you develop the skill of visualizing geologic features in three dimensions.

15.2 Beginning with the Basics: Contacts and Attitude

15.2.1 Geologic Contacts and Geologic Formations

When you looked at Figure 15.2d, you saw patterns of lines. What do these lines represent? Each line is the trace of a **contact**, the boundary between two geologic units. In this context, a "trace" is simply the line representing the intersection of a planar feature with the plane of a map or cross section, a "unit" may be either a **stratigraphic formation** (a sequence of sedimentary and/or volcanic layers that has a definable age and can be identified over a broad region), an igneous intrusion, or an interval of a specified type of metamorphic rock. Geologists recognize several types of contacts: (1) an **intrusive contact** is the boundary surface of an intrusive igneous body; (2) a **conformable contact** is the boundary between successive beds, sedimentary formations,

FIGURE 15.2 Geology of Sheep Mountain in Wyoming.

(a) Oblique air photo.

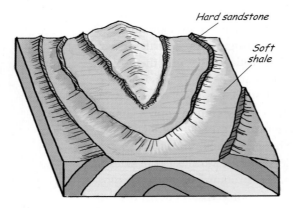

Hard sandstone

Soft shale

(b) Block diagram.

(c) Cross section.

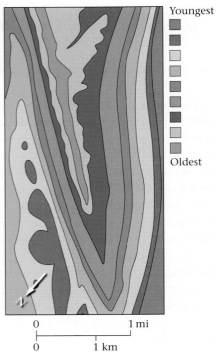

Youngest

Oldest

0   1 mi

0   1 km

(d) Geologic map. The color bands represent rock units. The map is oriented to correspond with the photo of part (a).

or volcanic extrusions in a continuous stratigraphic sequence; (3) an **unconformable contact** (or **unconformity**) occurs where a period of erosion and/or deposition has interrupted deposition; and (4) a **fault contact** is where two units are juxtaposed across a fault. Throughout this chapter, you'll gain experience interpreting contacts, but to be sure you understand the definitions from the start, complete Exercise 15.2.

15.2.2 Describing the Orientation of Layers: Strike and Dip

You can efficiently convey information about the orientation, or **attitude**, of any planar geologic feature, such as a bed or a contact, by providing two numbers. The first number, called the **strike**, is the compass direction of a horizontal line drawn on the surface of the feature (**Fig. 15.4a**). You can think of the strike as the intersection between a horizontal surface (e.g., the flat surface of a lake) and the surface of the feature. We can give an approximate indication of strike by saying "the bed strikes

Name: _____   **Section:** _____
Course: _____   **Date:** _____

In **Figure 15.3**, each arrow points to one of the four basic kinds of contacts. Add the labels.

FIGURE 15.3 Contacts.

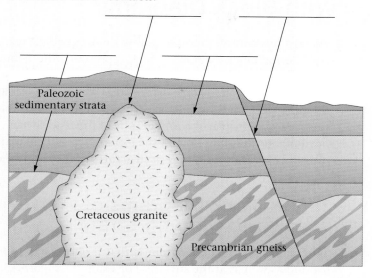

northeast," or we can be very exact by saying "the bed has a strike of N45°E," meaning that there is a 45° angle between the strike line and due north, as measured in a horizontal plane. The second number, called the **dip**, is the angle of tilt or the angle of slope of the bed. A horizontal bed has a dip of 00°, and a vertical bed has a dip of 90°. A bed dipping 15° has a gentle dip, and a bed dipping 60° has a steep dip. The direction of dip is perpendicular to the direction of strike.

Geologists use a shorthand notation for giving the strike and dip of a bed. We always write the strike as a three-digit number, for we divide the compass dial into 360° (**Fig. 15.4b**). A strike of 000° (or 360°) means the bed strikes due north; a strike

FIGURE 15.4 Strike and dip show the orientation of planar structures.

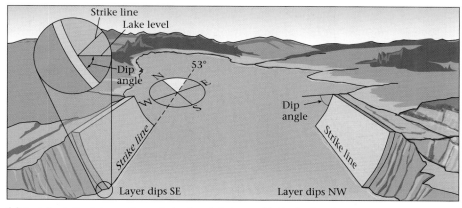

(a) A strike is the intersection of a horizontal plane with the bed surface.

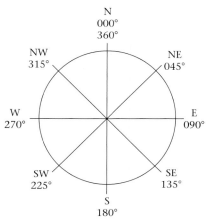

(b) Strikes are specified by azimuths (compass angles between 000° and 360°).

of 045° means that the strike line trends 45° east of north (i.e., northeast); a strike of 090° is 90° east of north (i.e., due east); and a strike of 180° is due south. We write the dip as a two-digit number (an angle between 00° and 90°) followed by a general direction. Let's consider an example: if a bed has an attitude of 045°/60°NW, we mean that it strikes northeast and dips steeply northwest. A bed with an attitude of 053°/72°SE strikes *approximately* northeast and dips steeply to the southeast.

15.3 Working with Block Diagrams

We start our consideration of how to depict geologic features on a sheet of paper by considering block diagrams, which represent a three-dimensional chunk of Earth's crust by utilizing the artist's concept of perspective (**Fig. 15.5a**). Typically, geologists draw blocks so that the top surface and two side surfaces are visible. The top is called the **map view**, and the side is a **cross-section view**. In the real world, the map-view surface would display the topography of the land surface, but for the sake of simplification, our drawings portray the top surface as a flat plane. In the next sections, we introduce a variety of different structures as they appear on block diagrams.

15.3.1 Block Diagrams of Flat-Lying and Dipping Strata

The magic of a block diagram is that it allows you to visualize rock units underground as well as at the surface. For example, **Figure 15.5b** shows three horizontal layers of strata. If the surface of the block is smooth and parallel to the layers, you can see only the top layer in the map view; the layers underground are visible only in the cross-section views. But if a canyon erodes into the strata, you can see the strata on the walls of the canyon, too (**Fig. 15.5c**).

Now, imagine what happens if the layers are tilted during deformation so that they have a dip. **Figure 15.5d** shows the result if the layers dip to the east. In the

FIGURE 15.5 Block diagrams.

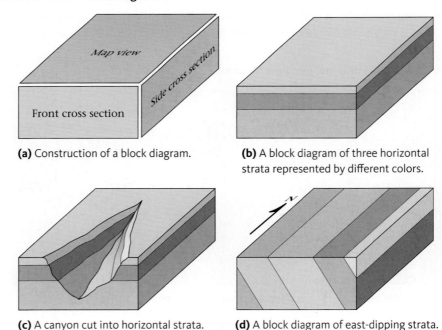

(a) Construction of a block diagram.

(b) A block diagram of three horizontal strata represented by different colors.

(c) A canyon cut into horizontal strata.

(d) A block diagram of east-dipping strata.

front cross-section face, we can see the dip. Because of the dip, the layers intersect the map-view surface, so the contacts between layers now appear as lines (the traces of the contacts) on the map-view surface. Note that, in this example, the beds strike due north, so their traces on the map surface trend due north. Also note that, in the case of tilted strata, the true dip angle appears in a cross section face only if the face is orientated perpendicular to the strike. On the right-side face in Figure 15.5d, the beds look horizontal because the face is parallel to the strike. (On a randomly oriented cross-section face, the beds have a tilt somewhere between 0° and the true dip.)

EXERCISE 15.3   **Portraying Tilted Strata on a Block Diagram**

Name: _____   Section: _____
Course: _____   Date: _____

(a) On the block diagram template below, sketch what a sequence of three layers would look like if their contacts had north-south-trending traces on the map view and dipped to the west at about 45°.

(b) On the block template below, sketch what a sequence of three layers would look like if the traces had an east-west trend and dipped south at about 45°.

(c) On the block template below, sketch what three layers would look like if the traces trend northeast-southwest and the layers dip to the southeast at about 45°. (*Hint:* This is a bit trickier, because tilts appear in both cross-section faces.)

15.3.2 Block Diagrams of Simple Folds

When rocks are deformed and structures develop, the geometry of layers depicted on a block diagram become more complicated. If deformation causes rock layers to bend and have a curved trace, we say that a **fold** has developed. Geologists distinguish between two general shapes of folds: an **anticline** is an arch-like fold whose layers dip away from the crest, whereas a **syncline** is a trough-like fold whose layers dip toward the base of the trough (**Fig. 15.6a, b**). Anticlines arch layers of rocks upward; synclines do the opposite—their strata bow *downward*. For the sake of discussion, the side of a fold is a **fold limb**, and the line that separates the two limbs (i.e., the line along which curvature is greatest) is the **fold hinge**. We can represent the hinge with a line and associated arrows on the map—the arrows point outward on an anticline and inward on a syncline. On a block diagram of the folds, we see several layers exposed (**Fig. 15.6c**). Note that the same set of layers appear on both sides of the hinge. (Exercise 15.4 allows you to discover that the age relations of layers, as seen on the map view, indicate whether a given fold is an anticline or syncline.)

FIGURE 15.6 The basic types of folds.

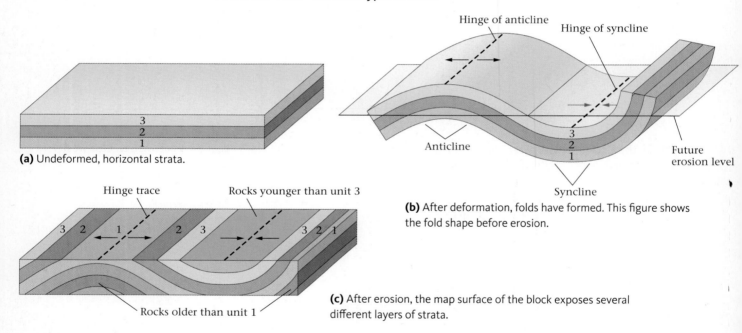

(a) Undeformed, horizontal strata.

Hinge trace — Rocks younger than unit 3

Rocks older than unit 1

Hinge of anticline — Hinge of syncline

Anticline

Syncline

Future erosion level

(b) After deformation, folds have formed. This figure shows the fold shape before erosion.

(c) After erosion, the map surface of the block exposes several different layers of strata.

The hinge of a fold may be horizontal, producing a **nonplunging fold** (**Fig. 15.7a**), or it may have a tilt, or "plunge," producing a **plunging fold** (**Fig. 15.7b**); an arrowhead on the hinge line indicates the direction of plunge. Note that if the fold is nonplunging, the contacts are parallel to the hinge trace, whereas if the fold is plunging, the contacts curve around the hinge—this portion of a fold on the map surface is informally called the fold "nose." Note that anticlines plunge *toward* their noses, synclines *away from* their noses.

Name: _____   **Section:** _____
Course: _____   **Date:** _____

Refer to the block diagram in Figure 15.6c. When erosion bevels the land surface, the map surface is like a horizontal slice through the fold. Keeping in mind that anticlines bow strata up and synclines bow them down, answer the following:

(a) Are the strata along the hinge of the anticline younger or older than the strata on the exposed part of the limbs, as seen in the map-view surface? _____

(b) Are the strata along the hinge of the syncline younger or older than the strata on the exposed part of the limbs, as seen in the map-view surface? _____

FIGURE 15.7 Block diagrams showing the contrast between nonplunging and plunging folds.

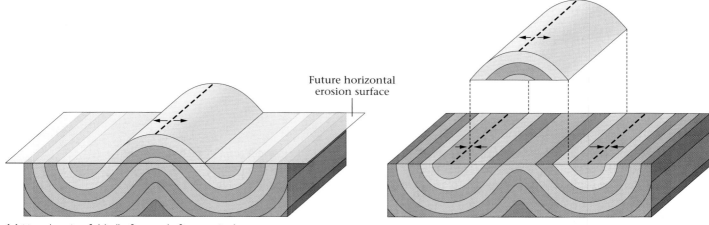

(a) Nonplunging folds (before and after erosion).

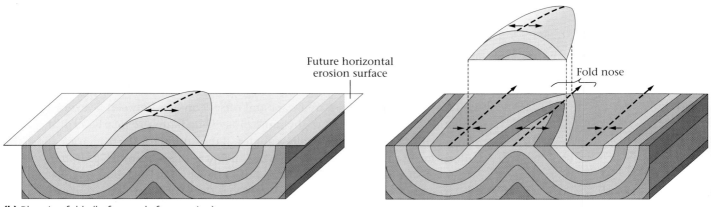

(b) Plunging folds (before and after erosion).

In some situations, the hinge of a fold is itself curved, so that the plunge direction of a fold changes along its length. In the extreme, a fold can be as wide as it is long. In the case of down-warped beds, the result is a **basin**, a bowl-shaped structure; and in the case of up-warped beds, the result is a **dome**, shaped like an overturned bowl.

Name: _____   Section: _____

Course: _____   Date: _____

(a) In **Figure 15.8**, which of the following block diagrams illustrates a basin, and which illustrates a dome? Add labels to the figure.

FIGURE 15.8   The difference between a basin and a dome.

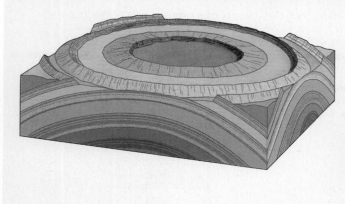 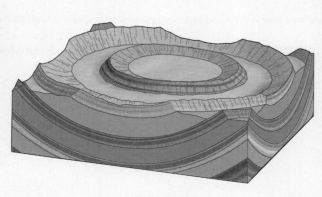

_____   _____

(b) Look at the distribution of strata in each of the blocks and circle the correct term in each statement.

(i) The center of a dome exposes (older/younger) strata, relative to its outer edge.

(ii) The center of a basin exposes (older/younger) strata relative to its outer edge.

15.3.3 Block Diagrams of Faults

As we noted earlier, a fault is a surface on which one body of rock slides past another by an amount called the **fault displacement** (Fig. 15.9) Faults come in all sizes—some have displacements of millimeters or centimeters and are contained within a single layer of rock; others are larger and offset contacts between layers or between formations. Not all faults have the same dip—some faults are nearly vertical, whereas others dip at moderate or shallow angles. If the fault is not vertical, rock above the fault surface is the **hanging wall**, and rock below is the **footwall** (Fig. 15.9a).

Geologists distinguish among different kinds of faults based on the direction of displacement. **Strike-slip faults** tend to be nearly vertical, and the displacement on them is horizontal, parallel to the *strike* of the fault (Figure 15.9b, c). On **dip-slip faults**, the displacement is parallel to the dip direction on the fault; if the hanging wall block moves up dip, it's a **reverse fault**, and if it moves down dip, it's a **normal fault** (Fig. 15.9d, e). Reverse faults form in response to compression, and normal faults form in response to tension.

You can recognize faulting, even if the fault, surface itself is not visible (due to cover by soil or vegetation), if you find a boundary along which contacts terminate abruptly (**Fig. 15.10**). The configuration that you find depends on both the attitude of the fault and the attitude of the layers.

FIGURE 15.9 Hanging wall, footwall, and the classification of faults.

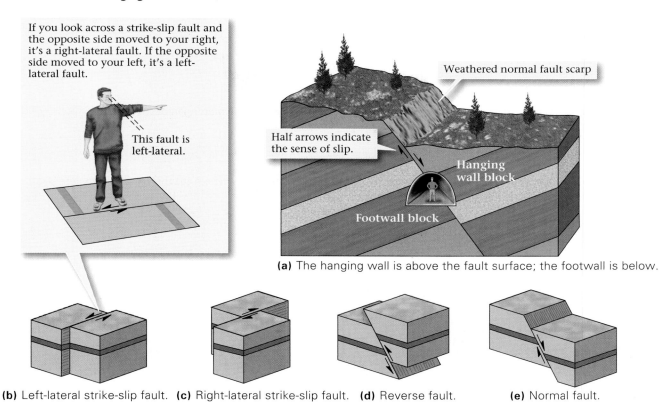

If you look across a strike-slip fault and the opposite side moved to your right, it's a right-lateral fault. If the opposite side moved to your left, it's a left-lateral fault.

This fault is left-lateral.

Weathered normal fault scarp

Half arrows indicate the sense of slip.

Hanging wall block

Footwall block

(a) The hanging wall is above the fault surface; the footwall is below.

(b) Left-lateral strike-slip fault. **(c)** Right-lateral strike-slip fault. **(d)** Reverse fault. **(e)** Normal fault.

FIGURE 15.10 Examples of the consequences of faulting on strata.

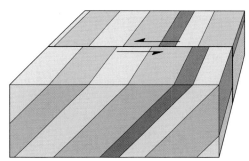

In this case, displacement on a strike-slip fault causes layers to terminate abruptly.

EXERCISE 15.6   **Faulted Strata on a Block Diagram**

Name: _____   **Section:** _____
Course: _____   **Date:** _____

The following questions refer to **Figure 15.11**.

(a) The block in (Fig. 15.11a) below shows a vertical fault cutting across a nonplunging syncline. Complete the block diagram by adding arrows to show the sense of slip across the fault, and by adding colored bands for the appropriate stratigraphic units in the blank areas. What type of fault is it? _____

continued

Name: _____   **Section:** _____

Course: _____   **Date:** _____

(b) The block in 15.11b below shows a dip-slip fault. Is this a normal or reverse fault? _____

(c) As you walk from west to east across the map surface, would you cross Layer 3 more than once? Explain.

_____

_____

(d) The red line on the front cross-section face represents a drill hole. Does the drill hole cut through the complete stratigraphic section, or do you see repetition or loss of section?

_____

_____

FIGURE 15.11 Blocks displaying offset by faults.

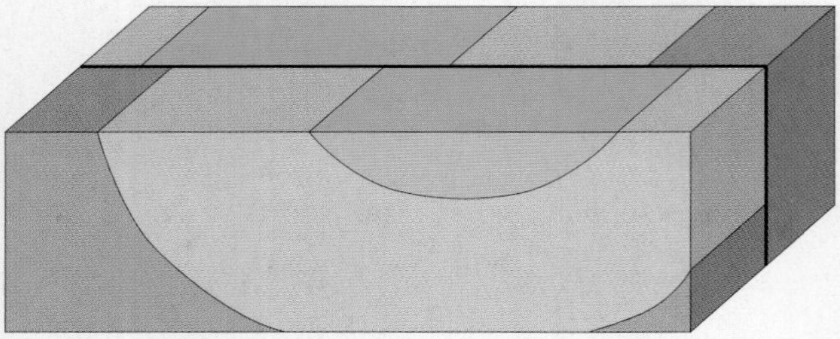

(a) Vertical fault.

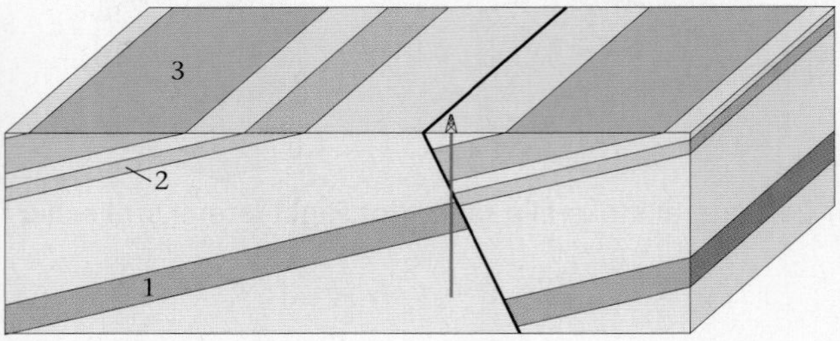

(b) Dip = slip fault.

15.3.4 Block Diagrams of Unconformities

An **unconformity** is a contact that represents a period of nondeposition and/or erosion, as we noted earlier. Geologists recognize three different kinds: (1) At a **disconformity**, bedding above and below the unconformity are parallel, but there is a significant time gap between the age of the strata below and the age of the strata above, (2) At a **nonconformity**, strata are deposited on a "basement" of intrusive igneous and/or metamorphic rock, and (3) At an **angular unconformity**, the orientation of the beds above the unconformity is not the same as that below. Exercise 15.7 gives you a chance to distinguish among these three types.

EXERCISE 15.7 Interpreting Unconformities

Name: ______________________ **Section:** __________

Course: ______________________ **Date:** __________

(a) In the space provided below each block in **Figure 15.12**, indicate what type of unconformity is shown.

FIGURE 15.12 Blocks of unconformities.

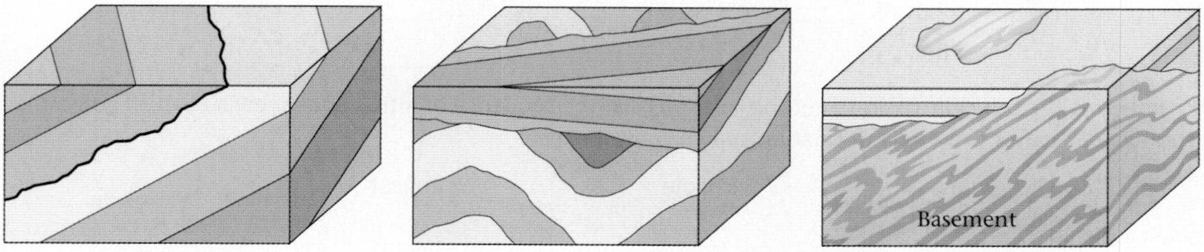

Block 1: ______________ Block 2: ______________ Block 3: ______________

(b) All the layering is tilted in Block 1. Keeping in mind the principle of original horizontality, what does this tilting mean? Was the unconformity tilted when it was first formed?

____

____

(c) In Block 2, in which direction is the post-unconformity strata dipping? ______________

(d) In Block 3, a pink blob-shaped area appears on the map surface. What geologic observation(s) could prove that the contact between the gray area and the sedimentary beds is an unconformity and not an intrusive contact?

____

____

____

____

15.3.5 Block Diagrams of Igneous Intrusions

An igneous intrusion forms when molten rock (magma) pushes into or "intrudes" preexisting rock. Geologists distinguish between two general types of igneous intrusions. (1) **Tabular intrusions** have roughly parallel margins; These include wall-like intrusions, called **dikes**, that cut across preexisting layering, and sheet-like intrusions, called **sills**, that are parallel to layering. (2) **Plutons** are irregularly shaped, blob-shaped, or bulb-shaped, intrusions. On a block diagram, you can generally distinguish among different types of intrusions based on their relationship with adjacent layering. To see how, try Exercise 15.8.

EXERCISE 15.8 Interpreting Intrusions on a Block Diagram

Name: ______________________________ **Section:** ____________

Course: ______________________________ **Date:** ____________

(a) The two blocks below (**Figure 15.13**) show sedimentary beds and intrusions. Match the name of the intrusion to the appropriate intrusion on the block.

pluton Block 1: _______ Block 2: _______

dike Block 1: _______ Block 2: _______

sill Block 1: _______ Block 2: _______

(b) Remembering the principle of cross-cutting relations, list the sequence of intrusions for each block. If you can't determine an answer from the information shown, indicate so.

	Oldest	Middle	Youngest
Block 1:	_______	_______	_______
Block 2:	_______	_______	_______

(c) Analyses indicate the unlabeled intrusion in the front cross-section face of Block 2 is part of the same body as Intrusion A exposed on the top surface. Explain why you can't *see* the connection between the map-view exposure and the subsurface cross section on this block.

____

____

Block 1 **Block 2**

FIGURE 15.13 Blocks of igneous intrusions.

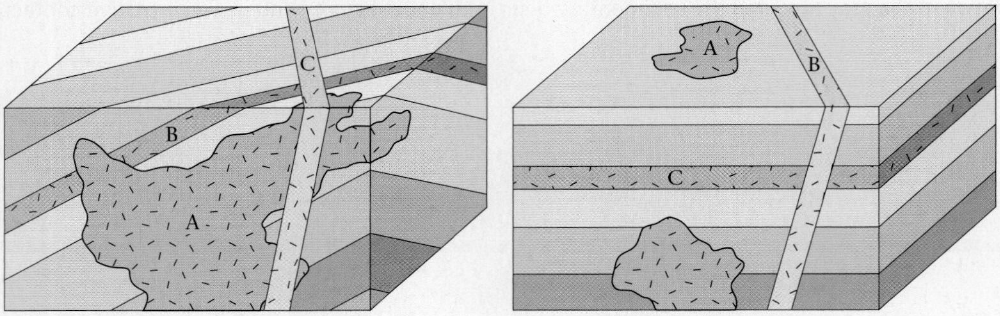

15.4 Geologic Maps

15.4.1 Introducing Geologic Maps and Map Symbols

Now that you've become comfortable reading and interpreting block diagrams, we can focus more closely on how to interpret geology on the map-view surface. A map that shows the positions of contacts, the distribution of rock units, the orientation of layers, the position of faults and folds, and other geologic data is called a **geologic map**. Contacts between rock units are shown by lines (traces) and the units themselves are highlighted by patterns and/or colors and symbols that indicate their ages. The orientation of beds, faults, and foliations, as well as the position of fold hinges, can be represented by strike and dip symbols. The map's **explanation** (or "legend")

Name: _____   **Section:** _____
Course: _____   **Date:** _____

Four cut-out block diagrams are provided at the end of the book for additional practice and to help visualize structures in three dimensions. Cut and fold the diagrams as indicated and use tape to hold the tabs together to make three-dimensional block diagrams.

(a) Complete the blank cross-section panels for Block 1, and describe the structure present. Does the block show horizontal or tilted strata? Folds? Faults?

_____

_____

_____

_____

(b) Complete the map view and blank cross-section views for Blocks 2 and 3. Compare and contrast the structures in these two blocks.

_____

_____

_____

_____

(c) Complete the cross-section panels for Block 4. Describe the nature of the folding and explain how you arrived at your conclusion. Describe the nature of the faulting. Is it dip-slip? If so, is it normal or reverse? Is it strike-slip? If so, is it left-lateral or right-lateral? Explain your reasoning.

_____

_____

_____

_____

defines all the symbols, abbreviations, and colors on the map. **Figure 15.14**, a geologic map of the Bull Creek quadrangle in Wyoming, illustrates the components of a geologic map. Note that all maps should have a scale, north arrow, and explanation.

Let's begin our discussion of geologic maps by considering the various features that can be portrayed on these maps.

• *Rock units*: Geologic maps show the different rock units in an area. These units may be bodies of intrusive igneous rock, layers of volcanic rock, sequences of sedimentary rocks, or complexes of metamorphic rock. The most common unit of sedimentary and/or volcanic rock is a stratigraphic formation, as noted earlier. These are commonly named for a place where it is well exposed. A formation may consist entirely of beds of a single rock type (e.g., the Bright Angel Shale consists only of shale), or it may contain beds of several different rock types (e.g., the Bowers Mountain Formation contains shale, sandstone, and rhyolite).

Typically, maps use patterns, shadings of gray, or colors to indicate the area in which a given unit occurs. On geologic maps produced in North America, an abbreviation for the map unit may also appear within the area occupied by the formation. This abbreviation generally has two parts: the first part represents the formation's age, in capital letters; the second part, in lowercase letters, represents the formation's name.

FIGURE 15.14 Geologic map of the Bull Creek quadrangle in Wyoming.

Geologic map of the Bull Creek quadrangle, Teton and Sublette Counties, Wyoming M.L. Schroeder, 1976

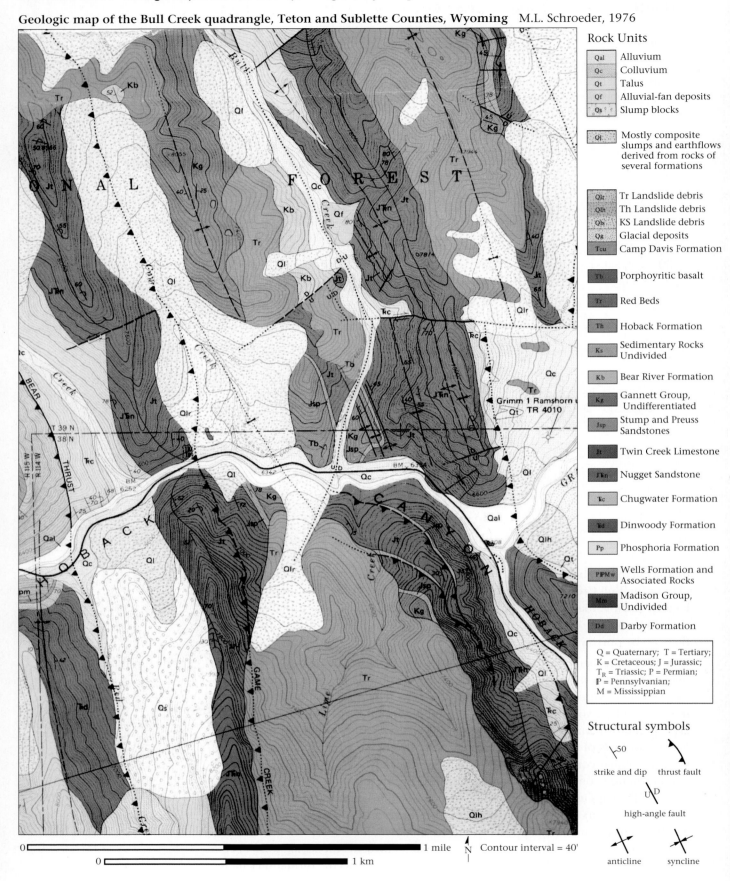

Rock Units

Qal	Alluvium
Qc	Colluvium
Qt	Talus
Qf	Alluvial-fan deposits
Qs	Slump blocks
Ql	Mostly composite slumps and earthflows derived from rocks of several formations
Qlr	Tr Landslide debris
Qlh	Th Landslide debris
Qls	KS Landslide debris
Qg	Glacial deposits
Tcu	Camp Davis Formation
Tb	Porphoyritic basalt
Tr	Red Beds
Th	Hoback Formation
Ks	Sedimentary Rocks Undivided
Kb	Bear River Formation
Kg	Gannett Group, Undifferentiated
Jsp	Stump and Preuss Sandstones
Jt	Twin Creek Limestone
Jʀn	Nugget Sandstone
Ʀc	Chugwater Formation
Ʀd	Dinwoody Formation
Pp	Phosphoria Formation
PIPMw	Wells Formation and Associated Rocks
Mm	Madison Group, Undivided
Dd	Darby Formation

Q = Quaternary; T = Tertiary;
K = Cretaceous; J = Jurassic;
T_R = Triassic; P = Permian;
IP = Pennsylvanian;
M = Mississippian

Structural symbols

strike and dip   thrust fault

high-angle fault

anticline   syncline

0 _____ 1 mile   N   Contour interval = 40'

0 _____ 1 km

For example, O indicates rocks of Ordovician age (Oce = Cape Elizabeth Formation, Osp = Spring Point Formation) and SO indicates Silurian or Ordovician age (SOb = Berwick Formation, SOe = Eliot Formation).

- *Contacts:* Different kinds of contacts are generally shown with different types of lines. For example, a conformable or intrusive contact is a thin line, a fault contact is a thicker line, and an unconformity may be a slightly jagged or wavy line. In general, a visible contact is a solid line, whereas a covered contact (buried by sediment or vegetation) is a dashed line.

- *Strike and dip:* On maps produced in North America, geologists use a symbol to represent the strike and dip of a layer. The symbol consists of a line segment drawn exactly parallel to the direction of strike, and a short tick mark drawn perpendicular to the strike and pointing in the direction of dip (**Fig. 15.15**). A number written next to the tick mark indicates the angle of dip. (It is not necessary to write a number indicating the strike angle, because that is automatically represented by the map trend of the strike line.) Different symbols are used to represent bedding and foliation; in this book, we only use bedding symbols.

- *Other structural symbols:* The explanation also includes symbols representing the traces of folds and faults.

FIGURE 15.15 Indicating strike and dip on geologic maps.

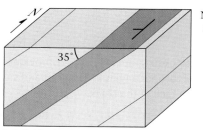

(a) A block diagram showing the dip angle.

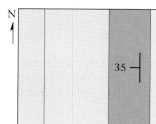

(b) A map showing a strike and dip symbol.

EXERCISE 15.10 **Using Symbols to Indicate Structures on a Geologic Map**

Name: _____   Section: _____

Course: _____   Date: _____

(a) On the blank map (**Fig. 15.16**) of a region with no hills or valleys given below, draw the appropriate strike and dip symbol next to the appropriate point. To do this, you must use a protractor and measure the angle between the north direction (the side edge of the map) and the strike angle. Then, look at the direction of dip so that you put the dip tick on the correct side of the strike symbol. These strike and dip symbols are on a contact between two formations:

A: 045°/30°SE   B: 280°/10°SW   C: 350°/25°W

(b) Based on the strike and dip symbols you show, draw a line representing the contact that passes through these points. Remember, the line needs to be parallel to the strike and dip symbol.

(c) What is the structure shown by the structure symbols? _____

FIGURE 15.16 Blank map.

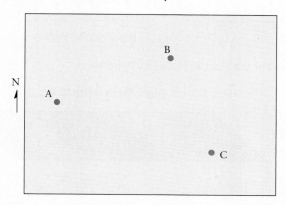

15.4.2 Constructing Cross Sections

We've seen that a cross section represents a vertical slice through the crust of the Earth. Thus, the sides of a block diagram are cross sections. If you start with a block diagram, you can construct the structure in the cross-section planes simply by drawing lines representing the contacts so that they connect to the contact traces in the map plane—the strike and dip data on the map tell you what angle the contact makes, relative to horizontal, and you use a protractor to draw the correct angle. If a fold occurs on the map surface, it generally also appears in the cross section.

EXERCISE 15.11   **Construct a Cross-Section Face View of a Block Diagram**

Name: _____   Section: _____
Course: _____   Date: _____

(a) The map surface of the block diagram in **Fig. 15.17** provides strikes and dips of the layers shown. From this information, show the layers with their proper angles in the front and side cross-section faces. Note that the strike of the layers is perpendicular to the front face of the block.

FIGURE 15.17   Block diagram for Exercise 15.10.

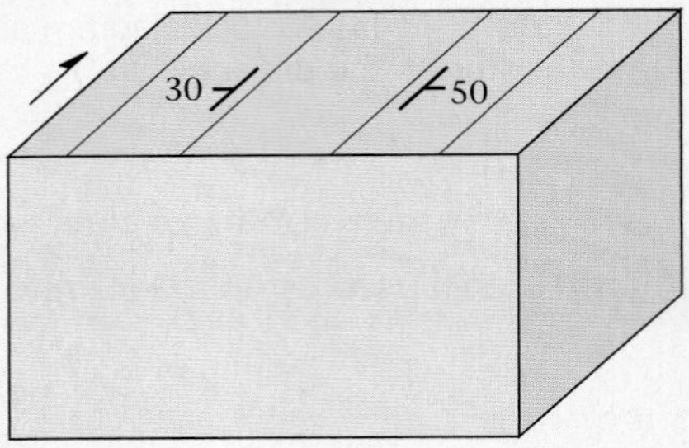

• What kind of structure is shown? _____

(b) Complete the map view and cross-section views of the block diagrams below (**Fig. 15.18**) by showing a sequence of sedimentary rocks with the indicated orientations. Show at least three layers in each block, and plot the strike and dip symbol on the top surface. Assume the bedding planes are planar.

FIGURE 15.18   Blank blocks for Exercise 15.11b.

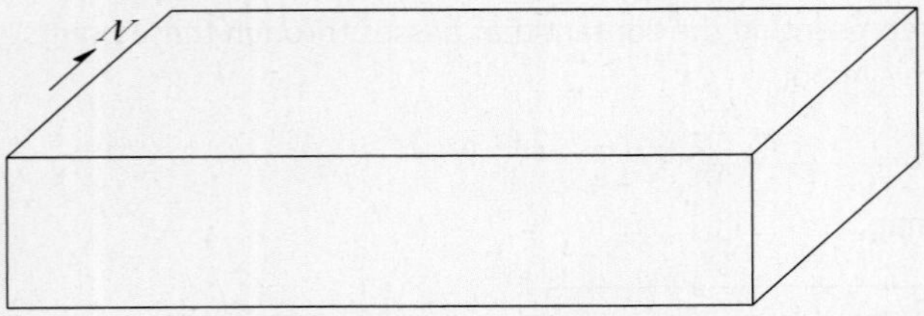

Block A: 090°/40°S 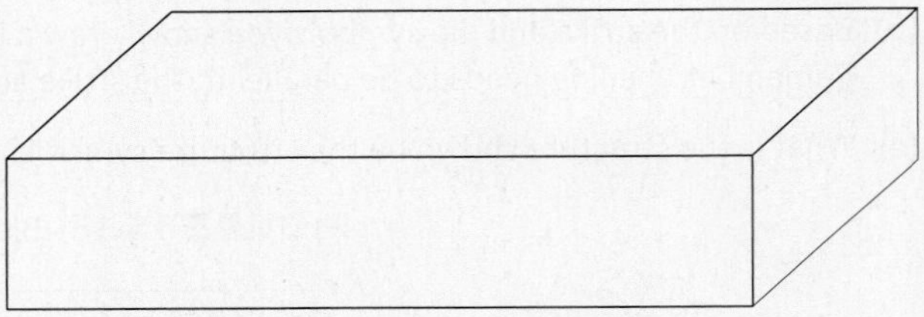   Block B: 000°/60°E

(c) Complete the cross-section views of the block in **Figure 15.19** below.

FIGURE 15.19   Block for Exercise 15.11c.

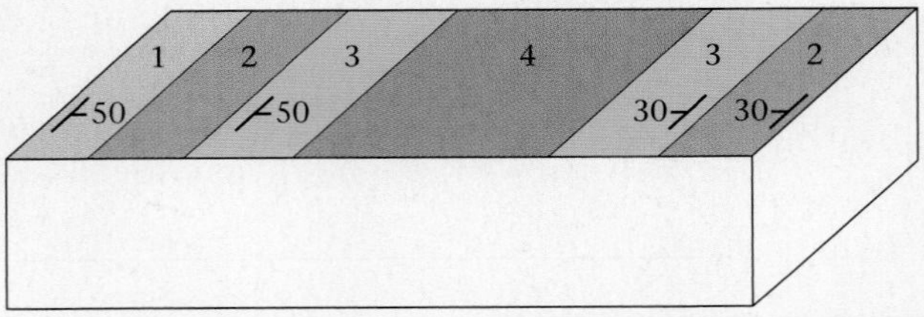

So far, we've worked with data depicted on block diagrams. Now let's consider the more common challenge of producing a cross section from a geologic map. This takes a couple of extra steps—to see how to do it, refer to **Figure 15.20**. On the left, you see a simple geologic map. The **line of section** (AA') is the line on the map view along which you want to produce the cross section. The cross section is a vertical plane inserted into the ground along the line of section. Now, take a scrap of paper and align it with your line of section. Mark off the points where contacts cross the scrap of paper. Transfer these points to the cross-section frame on the right. Using a protractor, make a little red tick mark indicating the dip of the contact; use the strike and dip symbol closest to each contact to provide this angle. Next, in the subsurface, sketch in lines that conform to the positions of the contacts and the dip angles. In this example, the contacts curve underground to define a syncline. Unless there is a reason to think otherwise, the layers should have constant thickness. Note that, because of this constraint, Layer Q appears in the lower left corner of the cross section; it would come to the surface to the west of the map area.

FIGURE 15.20 Constructing a cross section.

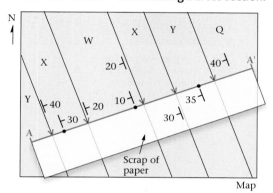

Step 1: Mark data locations on the corss-section paper.

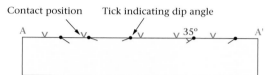

Step 2: Identify contact positions. Add dip marks at correct angles.

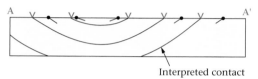

Step 3: Draw contacts so they obey location and dip data.

15.4.3 Basic Geologic Map Patterns

Geologic maps can get pretty complicated especially where structures are complicated or where topography is rugged. But, by applying what you have learned so far about block diagrams, you can start to interpret them. To make things simple, we begin with very simple maps of areas that have no topography (i.e., the ground surface is flat), as in the block diagrams that you've worked with. Exercise 15.12 challenges you to look at the map and imagine the three-dimensional structure it represents. Also keep in mind that sedimentary and *extrusive* igneous rocks are commonly deposited in horizontal layers with the youngest layer at the top of the pile and the oldest at the bottom.

EXERCISE 15.12   **Interpreting Simple Geologic Maps**

Name: _____   Section: _____

Course: _____   Date: _____

In the space below each of the following maps in **Figure 15.21**, identify the structure or geologic feature portrayed. Does the map show a fault, fold, tilted strata, dike, pluton, unconformity, or some combination? To answer these questions, you may need to refer to the block diagrams presented earlier in the chapter. Remember—think in three dimensions! With a little practice, geologists learn to recognize the basic patterns quickly.

continued

EXERCISE 15.12 continued

Name: ______________________________ Section: ____________
Course: ______________________________ Date: ____________

FIGURE 15.21 Geologic maps.

Map A

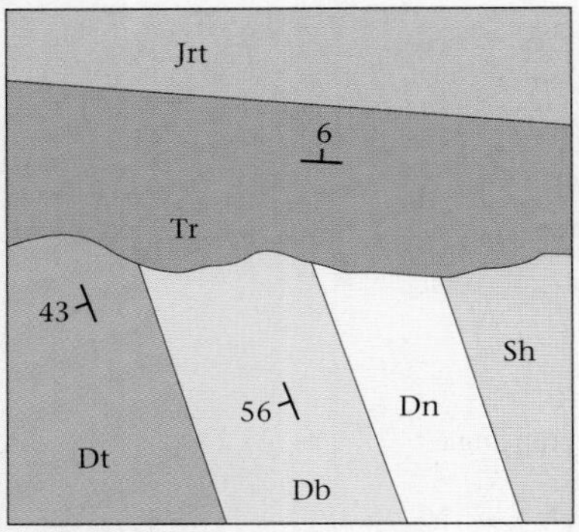

- Jrt Jurassic Tinta Fm.
- Tr Triassic Jones Sh.
- Dt Devonian Tella Fm.
- Db Devonian Bouser Fm.
- Dn Devonian Norfolk Sh.
- Sh Silurian Hallo Fm.

Map B

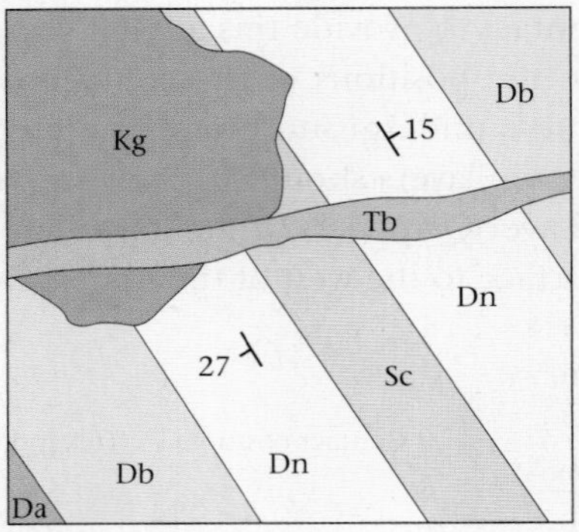

- Tb Tertiary basalt
- Kg Cretaceous granite
- Da Devonian Alsen Fm.
- Db Devonian Becraft Ls.
- Dn Devonian Norfolk Sh.
- Sc Silurian Cligfell Fm.

Map C

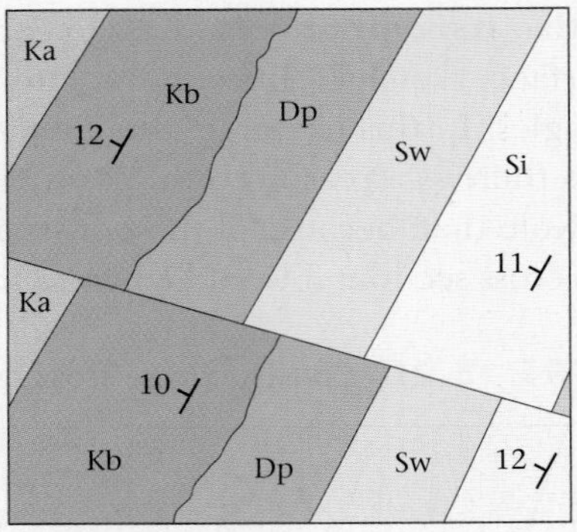

- Ka Cretaceous Altoona Fm.
- Kb Cretaceous Barrell Fm.
- Dp Devonian Potomoo Ls.
- Sw Silurian Wala Sh.
- Si Silurian Jack Fm.
- Ot Ordovician Trent Fm.

(a) Describe the geologic features of Map A.

____

____

____

(b) Describe the geologic features of Map B.

____

____

____

(c) Describe the geologic features of Map C.

____

____

____

(d) If you walk from left to right (west to east) along the southern edge of Map A, are you walking "up section" (i.e., into rocks of progressively younger age) or "down section" (i.e., into rocks of progressively older age)? ____________

Earlier, we distinguished between non-plunging folds and plunging folds, in the context of discussing block diagrams. We can recognize these folds on geologic maps simply by the pattern of color bands representing formations—on a map, formation contacts of non-plunging folds trend parallel to the hinge trace, whereas those of plunging folds curve around the hinge trace so we can see the fold nose. Furthermore, we can distinguish between anticlines and synclines by the age relationships of the color bands—strata get progressively younger away from the hinge of an anticline and progressively older away from the hinge of a syncline. If the hinge isn't shown on the map, you can draw it in where the reversal of age takes place. With this background, try Exercise 15.13.

EXERCISE 15.13   **Map Patterns of Plunging Folds**

Name: _____   **Section:** _____
Course: _____   **Date:** _____

(a) Examine the strikes and dips shown on the map below (**Fig. 15.22**). Based on these, draw the hinge traces of the folds depicted, and indicate which fold is an anticline and which is a syncline. Indicate the direction of plunge with arrow.

(b) Assume that the ground surface is flat. Keeping in mind that anticlines bow layers up and synclines bow layers down, which bed is oldest—X, Y, or Z? _____

(c) In the space provided, sketch a cross section showing what Layer Y would look like along the AA' line. This exercise is a little tricky because the cross-section line is not exactly perpendicular to the strike—for simplicity, assume that the dip angle shown in the cross section is the dip angle shown on the map.

FIGURE 15.22 Patterns of plunging folds.

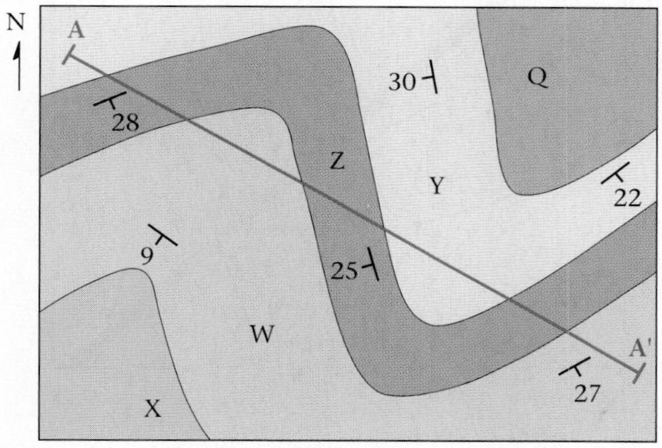

15.4.4 Geologic Maps with Contour Lines

When the map surface is not flat, geologic maps become even more challenging to interpret. That's because the trace of a contact that appears on a geologic map depends on both the slope angle and the slope direction of the land surface as well as the strike and dip of beds. We only introduce two simple situations here—the pattern of horizontal contacts and the pattern of vertical contacts. You'll address more complex situations in other geology courses. Work through Exercise 15.14 to see how they appear on a map.

Name: _____   **Section:** _____
Course: _____   **Date:** _____

(a) Look at the block diagram and the geologic map in **Figure 15.23**. What is the relationship between a horizontal contact and a contour line—parallel, perpendicular, or oblique. Keeping in mind the definition of a contour line, explain why. Geologists refer to the arrangement of valleys on this map as a *dendritic pattern,* because it resembles the veins of a leaf.

_____

_____

_____

_____

(b) The black stripe is a vertical basalt dike. Remember that when you are looking at a map, you are looking straight down from the sky. With this in mind, explain why the dike appears as a straight line on the map.

_____

_____

_____

FIGURE 15.23 Map pattern of horizontal strata in a valley.

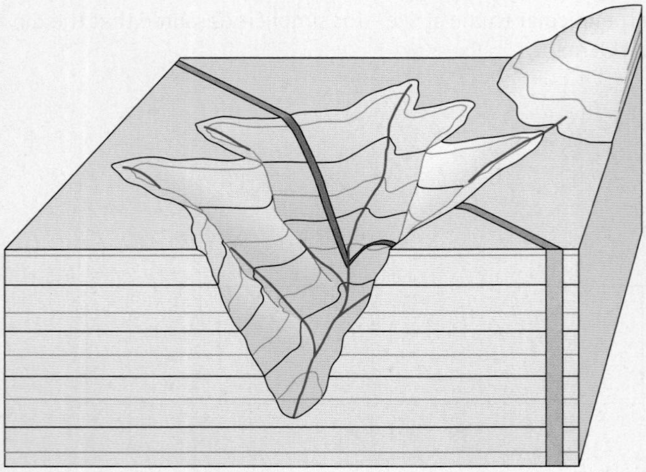

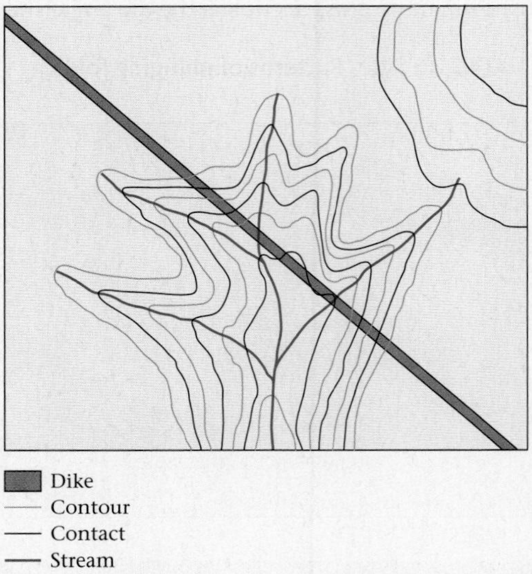

- Dike
- Contour
- Contact
- Stream

15.5 Structures Revealed in Landscapes

Unless you live in the Great Plains or along the Gulf Coast of the United States, you know that landscapes tend not to be as flat as the tops of the idealized block diagrams that we've worked with so far. In many cases, the distribution of rock units controls the details of the landscape, so erosion may cause structures to stand out in the landscape, especially in drier climates. For example, in regions of flat-lying strata, resistant rocks form cliffs, whereas nonresistant rocks form gentler slopes. Thus, a cliff exposing alternating resistant and nonresistant rocks develops a **stair-step profile**. Where strata are tilted, resistant rocks form topographic ridges, whereas nonresistant rock types tend to form valleys (**Fig 15.24a**). Generally, the ridges are asymmetric—a dip slope parallel to the bedding forms on one side, and a scarp cutting across the bedding forms on the other. In the Appalachian Mountains of Pennsylvania, ridges trace out the shape of plunging folds (**Fig. 15.24b**). A region in which the structure of bedrock strongly influences topography is called a **structurally controlled landscape**.

FIGURE 15.24   Structural control of topography.

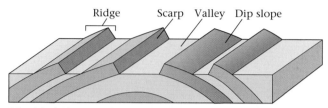

(a) A block diagram showing how resistant layers hold up ridges.

(b) A satellite photo of the Valley and Ridge Province of Pennsylvania (from *Google Earth™*).

EXERCISE 15.15   **Interpreting Structurally Controlled Landscapes**

Name: _____   Section: _____
Course: _____   Date: _____

Examine the topographic maps below (**Fig. 15.25a,b**) and suggest what type of bedrock structure is present.

(a) The structure shown in Figure 15.25a is: _____

(b) The structure shown in Figure 15.25b is : _____

continued

Name: _____   Section: _____

Course: _____   Date: _____

FIGURE 15.25a Topography of an area near Coburn, Pennsylvania.

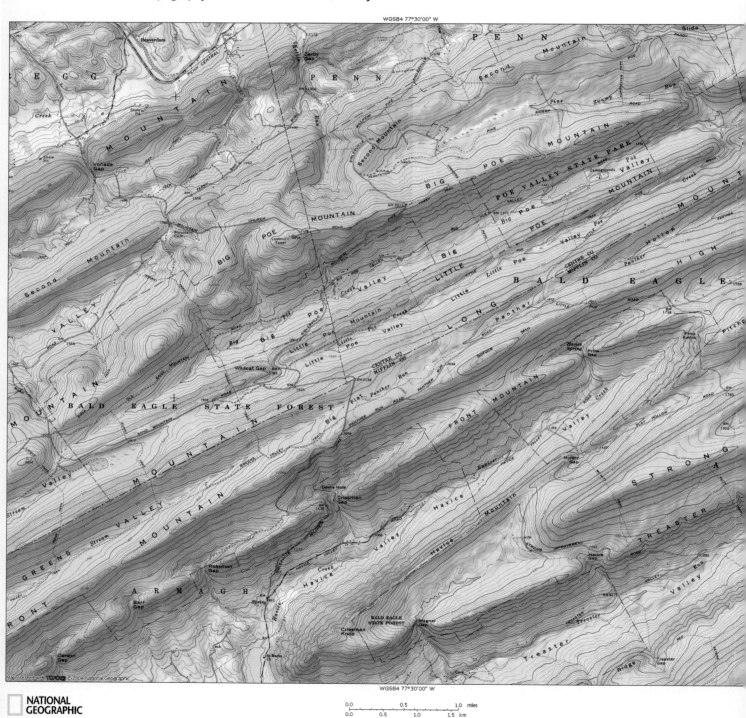

NATIONAL
GEOGRAPHIC

Name: _____   Section: _____

Course: _____   Date: _____

FIGURE 15.25b Topography of an area near Driftwood, Pennsylvania.

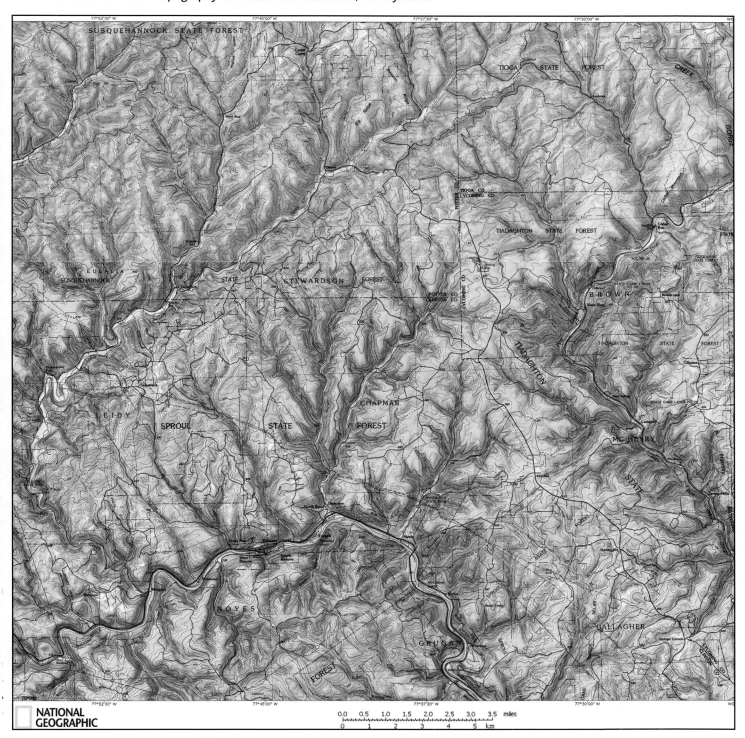

15.6 Reading Real Geologic Maps

15.6.1 Geologic Maps of Local Areas

You are now ready to apply what you've learned to interpreting the structure of selected areas of North America using excerpts from published geologic maps. These exercises give you a sense of how to see "clues" in a map that help you to picture the three-dimensional configuration of rocks underground.

EXERCISE 15.16   **The Grand Canyon in Arizona**

Name: _____   Section: _____

Course: _____   Date: _____

Examine the geologic map of part of the Grand Canyon area in Arizona (**Fig. 15.26**). There is no legend because you don't have to know the specific ages of the units to answer the questions.

(a) Noting the dendritic pattern shown on the map, what is the attitude of bedding in the Grand Canyon?

_____

FIGURE 15.26 Geologic map of part of the Grand Canyon.

Name: _____   **Section:** _____
Course: _____   **Date:** _____

Examine the Geologic Map of a portion of the Observation Peak Quadrangle in Wyoming (**Fig. 15.27**). This map area contains a number of interesting geologic features, some of which you can understand based on the work you have done earlier in this chapter. Each of the questions below refers to a specific feature on the map.

(a) In the northern part of the map, there is a large area of yellow (i.e., Quaternary deposits). What kind of contact forms the boundary between these deposits and older bedrock?

_____

(b) The bright red color represents igneous intrusions.

- What kind of intrusion is the larger round area? _____

- What kind of intrusion are the narrow bands? _____

(c) A thrust fault (a gently to moderately dipping reverse fault) is exposed in the southwest quarter of the map. The "teeth" of the thrust fault symbol lie on the hanging wall. In this locality, the fault dips about 40° in a westerly direction.

- Where Indian Creek crosses the fault, what rock unit is in the hanging wall, and what rock unit is in the footwall?

  _____

- Is the older rock in the hanging wall, or in the footwall? _____

- Thinking about the movement direction on a thrust fault, does this make sense? (Explain your answer.)

  _____

(d) In the southeastern quarter of the map, you can see the trace of a fold?

- What type of fold is it? _____

- In which direction does it plunge? _____

- In the Aspen Shale, what is the strike and dip of the strata on the western limb of the fold?

  _____

(e) In the southwestern-most corner of the map, in the area west of the north-south trending high-angle fault, the contact between the Phosphoria Formaion and the Mead Peak Shale has the shape of a 'V' with the point of the 'V' pointing in a northerly direction. (Note that there is a strike and dip measurement in the Dinwoody Formation, just east of the 'V'. The dip is only 15°, so the beds are very gently dipping.)

continued

Name: _____   Section: _____

Course: _____   Date: _____

FIGURE 15.27 Geologic map of part of the Observation Peak quadrangle, Wyoming.

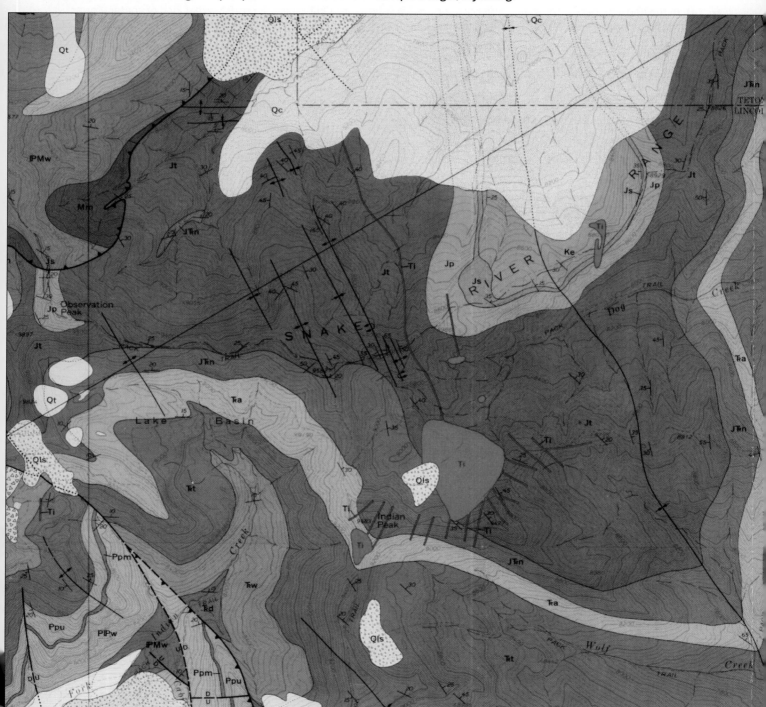

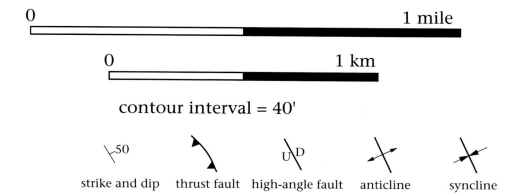

contour interval = 40'

strike and dip   thrust fault   high-angle fault   anticline   syncline

Geologic Map of the
Observation Peak Quadrangle, Wyoming
by
Howard F. Albee
1973

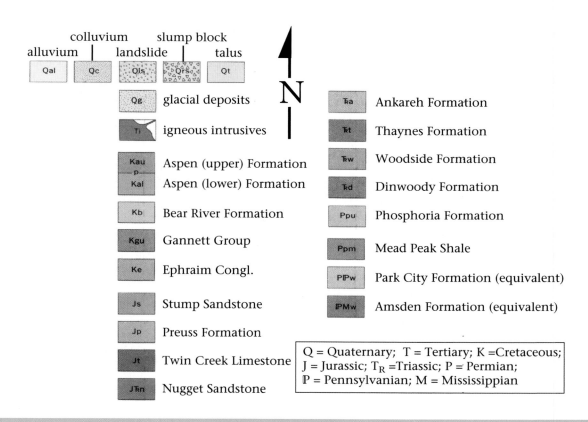

alluvium   colluvium   landslide   slump block   talus

Qal   Qc   Qls   Qrs   Qt

Qg   glacial deposits   N

Ti   igneous intrusives

Kau / Kal   Aspen (upper) Formation / Aspen (lower) Formation

Kb   Bear River Formation

Kgu   Gannett Group

Ke   Ephraim Congl.

Js   Stump Sandstone

Jp   Preuss Formation

Jt   Twin Creek Limestone

JTRn   Nugget Sandstone

TRa   Ankareh Formation

TRt   Thaynes Formation

TRw   Woodside Formation

TRd   Dinwoody Formation

Ppu   Phosphoria Formation

Ppm   Mead Peak Shale

PIPw   Park City Formation (equivalent)

IPMw   Amsden Formation (equivalent)

Q = Quaternary; T = Tertiary; K =Cretaceous;
J = Jurassic; T$_R$ =Triassic; P = Permian;
IP = Pennsylvanian; M = Mississippian

15.6.2 Geologic Maps of State- to Continent-Sized Areas

North America boasts a long geologic history. The continent is the product of multiple cycles of mountain building in which plate collisions formed supercontinents and rifting created smaller continents and numerous ocean basins.

Outcrop patterns resulting from these tectonic cycles identify areas with different deformation histories, even at the scale of the entire United States (Figure 15.25). Without a map legend, you can't tell *when* any of the folding or faulting took place, but you can certainly tell from the map patterns that some areas have been folded and others faulted. And some haven't experienced much deformation at all.

EXERCISE 15.18 **Regional Structural Provinces of the United States**

Name:	Section:
Course:	Date:

The geologic map of the United States (**Figure 15.28**), at first glance, may look like abstract art. But look a little closer, and in the bright splash of colors, you can begin to see patterns that have geologic meaning. Specifically, you can recognize distinct geologic provinces. The stratigraphic column at the right uses colors to distinguish rocks of different age, with the youngest rocks on top, and the oldest on the bottom. Keeping in mind the patterns displayed by different structures, answer the following questions. (*Note:* On this map, the colors representing geologic units have been placed over a DEM of the landscape.)

With a marking pen, trace out boundaries between the following provinces, and label them in the margins of the map. Your instructor will provide a full-scale version of this map, and a comparable DEM without the geology, for you to examine to get a clearer picture.

(1) The coastal plain, a region along the Atlantic coast and the Gulf coast, where the subdued topography is underlain by subhorizontal sheets of Cretaceous and Tertiary strata.

(2) The Appalachian Mountains, a sinuous (snake-like) mountain belt in which rocks have been folded into generally north to northeast trending folds.

(3) The Basin and Range Province, a rift, in which the crust has been broken by normal faults into roughly north-south trending ridges of bedrock, separated by valleys filled with young sedimentary strata.

(4) The Midcontinent Platform, where strata have been warped into regional-scale (i.e., hundreds of kilometers across) basins and domes. Within this province, label two of the basins with 'B' and two of the domes with 'D'.

(5) The North American Cordillera, the entire region of mountainous landscape that occupies the western third of the country.

(6) The Rocky Mountains, an array of individual mountain ranges within which Precambrian rocks have been uplifted and exposed.

(7) Note that the Rocky Mountains, in geologic jargon, occur only on the eastern side of the overall mountainous area of the western third of the country—the entire region of mountainous area in the western third is called the North American Cordillera.

(8) The Colorado Plateau, a roughly rectangular area bounded by the basin and range on the west, and the southern Rocky Mountains on the east.

(9) The "gangplank" region of the Rocky Mountains underlain by nearly flat-lying strata

(10) The Sierra Nevada, a mountain range in California bounded on the east by the Basin and Range, and on the west by the Great Valley (a basin filled with young strata), in which Mesozoic igneous rocks (granites) are exposed.

FIGURE 15.28 Bedrock geologic map of the United States.

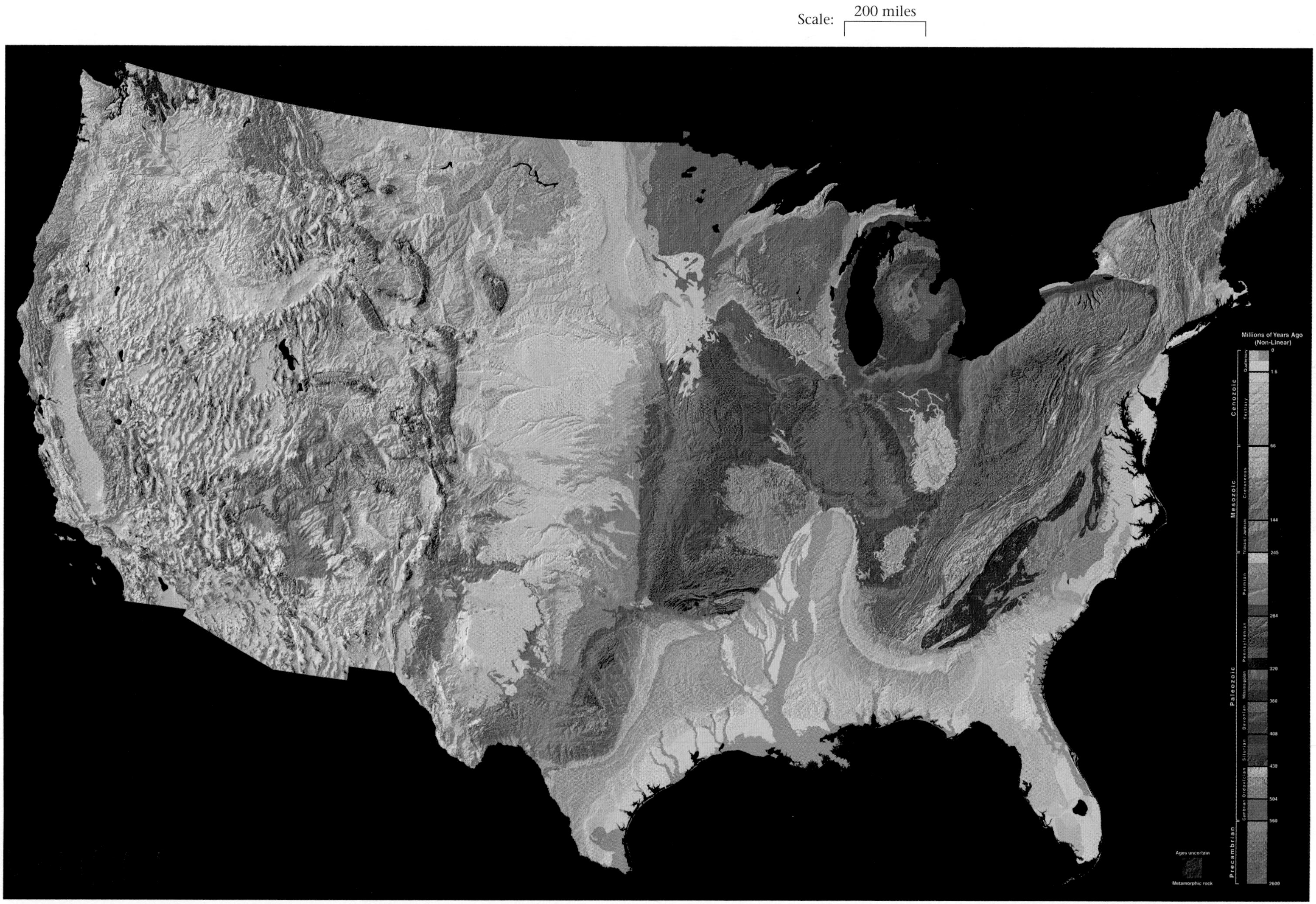

Scale: 200 miles
Millions of Years Ago
(Non-Linear)
Cenozoic
Quaternary
Tertiary
Mesozoic
Cretaceous
Triassic / Jurassic
Paleozoic
Permian
Pennsylvanian
Mississippian
Devonian
Silurian
Ordovician
Cambrian
Precambrian
0
1.6
66
144
245
284
320
360
408
438
504
560
2600
Ages uncertain
Metamorphic rock

15.6.3 Walk a Mile in Our Field Boots

We finish this chapter by talking about how geologists make geologic maps in the first place. It isn't easy! Students who want to learn the skill generally attend a summer geology field camp, where they practice the art of mapping for several weeks and gradually develop an eye for identifying rock types, contacts, folds, and faults. Typically, in arid regions, not much soil forms, so bedrock may be abundantly exposed; in such areas, geologists may actually see contacts, and can walk out the traces of contacts. Commonly, however, soil and vegetation cover much of the rocks, so outcrops are discontinuous and separated from one another by "covered intervals." In such cases, geologists must extrapolate contacts, using common sense and an understanding of geologic structures. Exercise 15.19 provides the opportunity for you to construct a map, in an area where limited outcrop data are available. The map shown is called an "outcrop map" because individual outcrops of rock are outlined. To complete the map, you need to extrapolate contacts.

EXERCISE 15.19   **Making a Geologic Map from Outcrop Data**

Name: _____   Section: _____
Course: _____   Date: _____

Figure 15.29 is an outcrop map of an area showing several different rock layers and their attitudes. Using the structural information available on the map and the stratigraphic column, draw a geologic map that shows the contacts between the rock units present. Note that in a few outcrop areas, a contact is visible.

FIGURE 15.29 An outcrop map for structural interpretation.

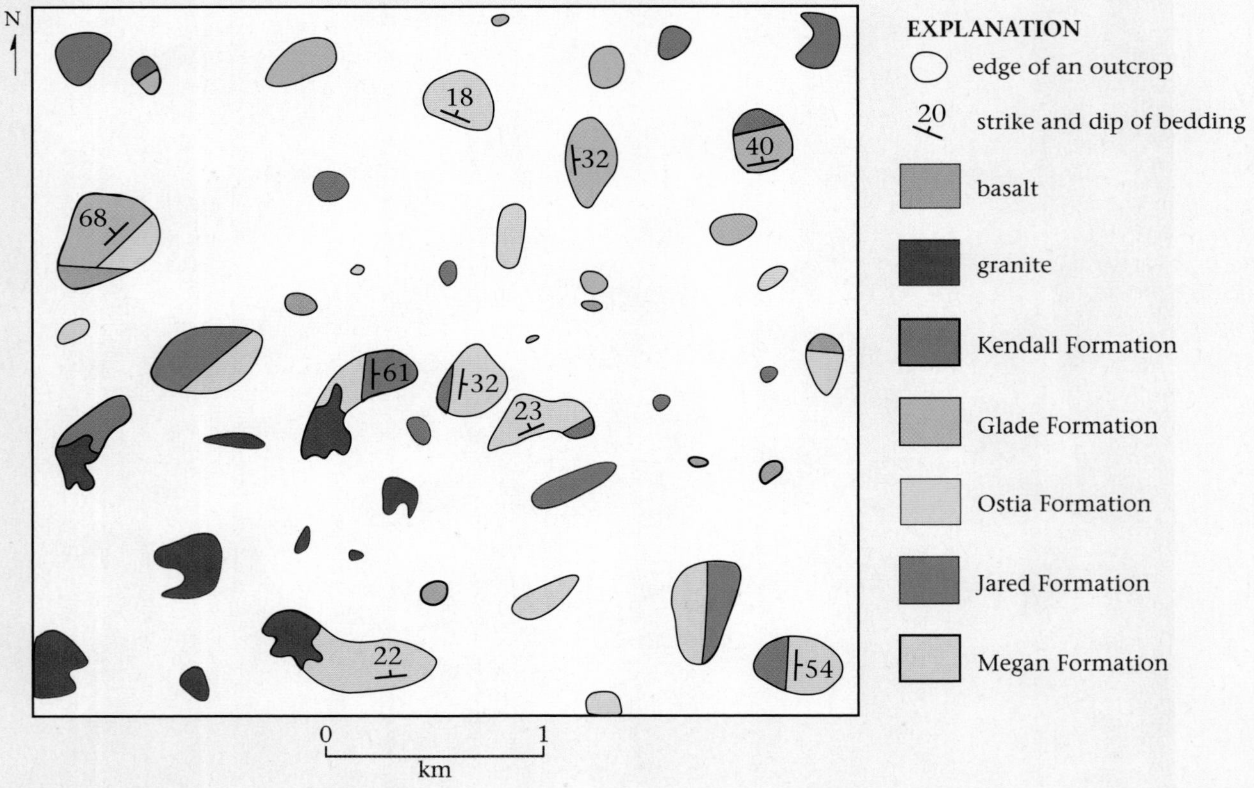

continued

Name: _____   **Section:** _____

Course: _____   **Date:** _____

Completing an Outcrop Map

A geologist has just finished mapping a few square kilometers. She has drawn the approximate shapes of outcrops, has plotted strike and dip measurements, and has located contacts where they were exposed. Complete the map by extrapolating contacts across the covered areas and by adding appropriate symbols for folds, if they are present. When you have completed the map describe the basic structure of the map area, in words in the space provided.

Structural Description: _____

_____

_____

BONUS EXERCISE 15.20   **A Practical Application of Structural Geology**

Name: _____   **Section:** _____

Course: _____   **Date:** _____

Knowledge of an area's deformation history helps avoid potentially calamitous situations, such as building a school directly on an active fault. Structural information can also pay off—big time—in our search for energy and mineral resources. For example, geologists have learned that oil and natural gas are often trapped in the crests of anticlines. Since it costs millions of dollars to drill an exploratory well, knowing where the anticlines are (or aren't) can mean the difference between a fortune and bankruptcy.

Imagine that a company has dug a shaft for an underground mine into a coal bed (**Fig. 15.30**). The coal bed and all beds above and below it dip 10°W. A drill hole through the coal layer has provided the stratigraphic sequence and the thickness of the layers, as represented in the column on the left (drawn to scale). Mapping at the ground surface reveals two faults and the map traces of two distinctive beds. The miners have found that the coal layer terminates at the faults, as shown in the cross-section face. It is not economical to mine at a depth greater than the red line. Based on the data available, determine whether the mining company should buy the mineral rights beneath Region A (west of the western fault) or Region B (east of the eastern fault). Complete the cross section face to help you visualize the answer.

continued

Name: _____   **Section:** _____

Course: _____   **Date:** _____

FIGURE 15.30 Practical application problem.

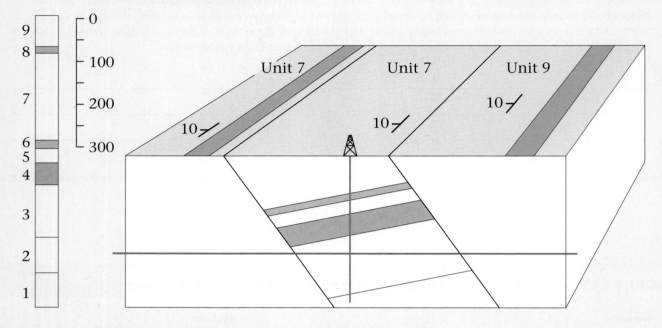

Which area (A or B) will likely provide more coal worth mining? Explain your answer.

_____

_____

_____

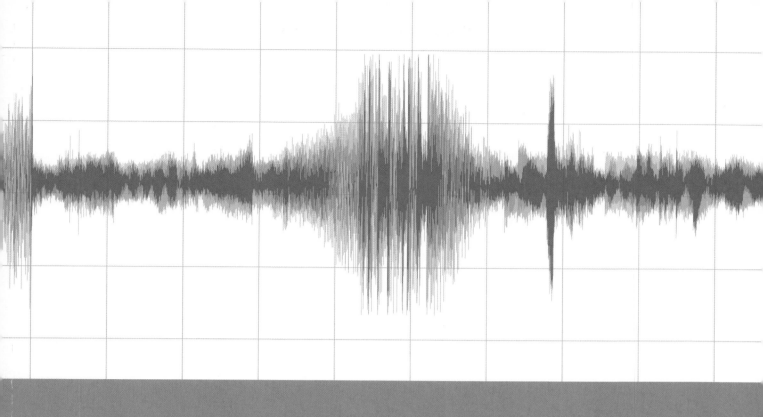

CHAPTER 16

EARTHQUAKES AND SEISMOLOGY

PURPOSE

- Understand how faulting causes the ground to move during earthquakes.
- Recognize how earthquakes cause damage to buildings and other structures.
- Learn how to locate an earthquake epicenter and determine its magnitude and when it occurred.

MATERIALS NEEDED

- Sharp pencil
- Ruler with divisions in millimeters
- Draftsman's compass (or piece of string)

16.1 Introduction

Few things are as fearsome as a major earthquake. Unpredictable and enormously powerful, a great earthquake destroys more than buildings and other structures. It shakes our sense of safety and stability as it shakes the solid rock beneath our feet. But it is not just the shaking that is dangerous. The devastating tsunamis of 2004 reminded us that oceanic earthquakes can ravage coastlines thousands of miles from the earthquake origin; landslides and mudslides triggered by earthquakes can engulf towns and villages; and the loss of water when rigid pipes break beneath city streets can cause health problems and make it difficult to fight fires.

In our attempt to understand earthquakes, we have developed tools that reveal Earth's internal structure, define the boundaries between tectonic plates, prove that the asthenosphere exists, and track the movement of plates as they are subducted into the mantle. In this chapter, we look at what an earthquake is and why it causes so much damage, and learn how seismologists locate earthquakes and estimate the amount of energy they release. You will learn to read a seismogram and use it to locate an earthquake, determine when it happened, and measure its strength. First, let's review some basic facts about the causes and nature of earthquakes that are discussed in detail in your textbook.

16.2 Causes of Earthquakes: Seismic Waves

• Earthquakes occur when rocks in a fault zone break, releasing energy. The energy is brought to the surface by two kinds of seismic waves called **body waves** because they travel through the body of the Earth. It is this energy that causes the ground to shake.

• The point beneath the surface where the energy is released is called the **focus** (or hypocenter) of the earthquake. The point on the surface directly above the focus is called the **epicenter** (**Fig. 16.1**). In most cases, the epicenter, being closest to the focus, is the site of greatest ground motion and damage.

FIGURE 16.1 The focus and epicenter of an earthquake.

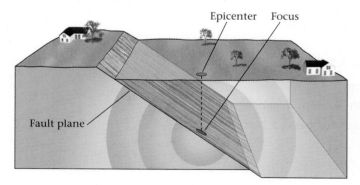

• There are two kinds of body waves, distinguished by how particles vibrate as the waves pass through rocks. **P-waves** (**Fig. 16.2a**) are a form of longitudinal wave in which particles vibrate back and forth *in the direction in which the wave is traveling.* **S-waves** (**Fig. 16.2b**) are a form of transverse wave in which particle motion *is perpendicular to the direction of energy transport.* P-waves reach the surface first because they are faster than S-waves.

- When P- or S-waves reach the surface, some of their energy is converted to two **surface waves**: the **Love wave**, a transverse wave in which particles vibrate horizontally (**Fig. 16.2c**), and the **Rayleigh wave**, a unique wave in which particles move in a circular pattern opposite the direction in which the wave is traveling (**Fig. 16.2d**).

FIGURE 16.2 Particle vibration during passage of the four seismic waves.

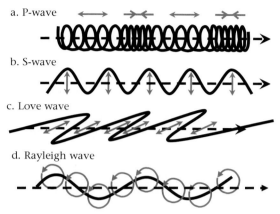

a. P-wave

b. S-wave

c. Love wave

d. Rayleigh wave

Dashed black arrows indicate wave travel direction. Red arrows indicate particle vibration.

- Seismic waves are detected with instruments called **seismometers**. These are anchored in bedrock to measure the amount of ground movement associated with each type of wave. Seismic stations use separate seismometers to measure vertical and horizontal motion (**Fig. 16.3**). The printed or digital record of ground motion is a **seismogram**.

- The strength of an earthquake is measured either by the Richter magnitude scale, based on the amount of ground motion and energy released, or by the Mercalli intensity scale, based on the amount of damage to structures.

FIGURE 16.3 Vertical and horizontal seismometers.

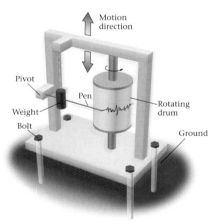

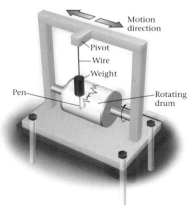

a. Vertical seismometer: The pivot and spring let the pen record only vertical motion.

b. Horizontal seismometer: The pivot allows the pen to record only horizontal motion.

As the ground vibrates, the inertia of the weight keeps the pen in position. The rotating drum moves, causing the pen to trace ground movement on the seismogram.

Name: _____   Section: _____

Course: _____   Date: _____

When P- and S-waves reach the surface, the ground vibrates and anything built on it is shaken in directions that depend on which wave is involved and the distance from the epicenter. **Figure 16.4** shows the arrival of P-, S-, Love, and Rayleigh waves at a skyscraper.

FIGURE 16.4 Ground shaking caused by seismic waves.

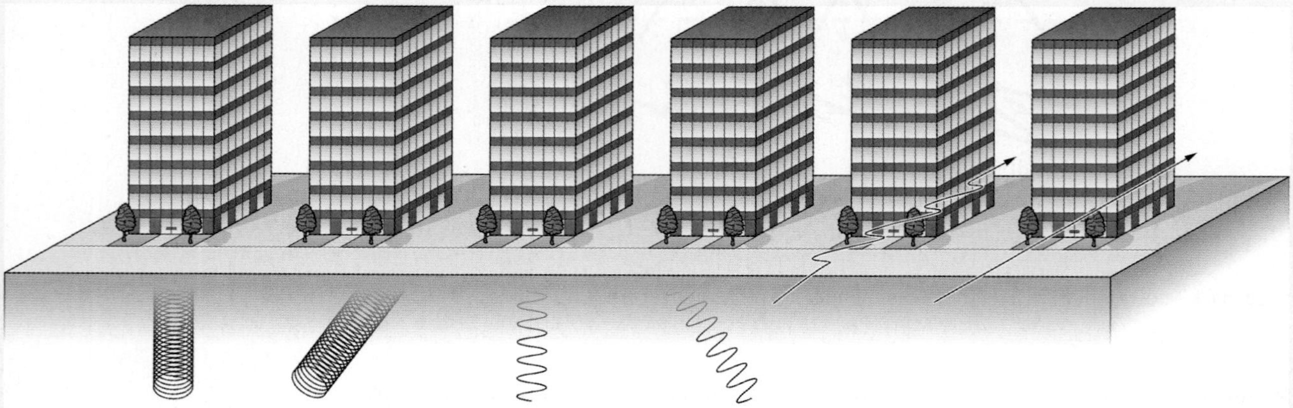

| P-wave at epicenter | P-wave far from epicenter | S-wave at epicenter | S-wave far from epicenter | Love wave | Rayleigh wave |

(a) Draw arrows to indicate how each of the buildings will move in response to the different type of waves.

(b) Why do the P- and S-waves at an earthquake epicenter make the ground shake differently from the ground in an area far from the epicenter?

_____

_____

_____

16.3 Locating Earthquakes and Measuring Magnitude

Locating earthquakes helps us to understand what causes them and to predict if an area will experience more in the future. Most earthquakes occur in linear belts caused by faulting at the three kinds of plate boundaries (Fig. 16.5) and are used to define those boundaries. But intraplate earthquakes also occur, and their causes are less well understood. How can we locate earthquake epicenters, especially when they are in remote areas? Seismologists triangulate epicenter locations by using sophisticated mathematical analysis of the arrival times of the different seismic waves from many seismic recording stations.

FIGURE 16.5 Worldwide distribution of earthquakes.

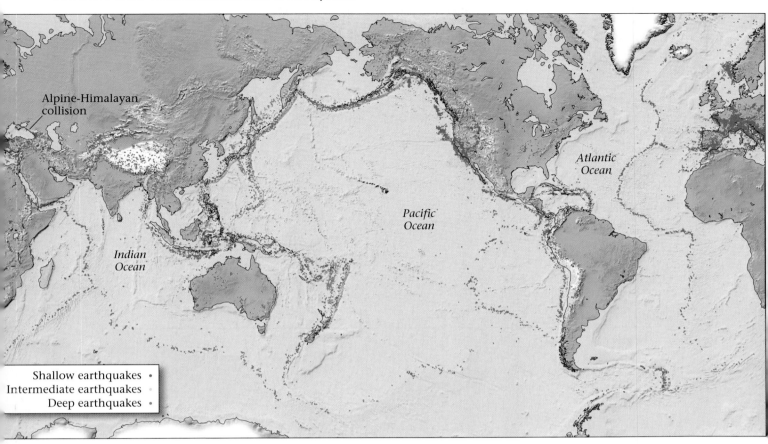

Shallow earthquakes ·
Intermediate earthquakes ·
Deep earthquakes ·

In this book, we use a much simpler method to locate an epicenter from only three seismic stations and determine the precise time at which the earthquake occurred. The basis for this method is the fact that the four types of seismic waves travel at different velocities. Figure 16.6 shows how the reasoning can be applied to an everyday, nonseismic situation. Analogous reasoning permits us to locate epicenters and determine the precise time of the faulting that causes individual earthquakes. Let's look at location first, then timing.

EXERCISE 16.2   **Distance to Epicenter**

Name: _____   Section: _____
Course: _____   Date: _____

This exercise shows the reasoning involved in measuring the distance from a seismic station to an earthquake epicenter. Two cars start along a road at exactly the same time. Both are using cruise control set at 1 mile per minute (60 miles per hour) but the controls are not exactly the same. Car 1 covers each mile in precisely 60 seconds, but Car 1 is a little slower, taking 1 second longer per mile. Car 1 therefore arrives at the first mile marker, followed by Car 2 a second later. The delay after 2 miles is 2 seconds, after 3 miles 3 seconds, and so on. With earthquakes, the delay is between arrivals of P-, S-, Love, and Rayleigh waves, but the reasoning is the same.

(a) Indicate on the blank clocks in **Figure 16.6** the amount of time between the arrival of the two cars.

Now work backwards. Car 1 drives by your window followed a short time later by Car 2. You don't know where they came from, but you do know (1) their velocities (described above), (2) they left at exactly the same time, and (3) the precise delay between their arrivals.

continued

Name: _____   Section: _____
Course: _____   Date: _____

(b) Using that knowledge, how far did the two cars travel if Car 2 arrived
    (i) 25 seconds after Car 1? _____ miles
    (ii) 45 seconds after Car 1? _____ miles

FIGURE 16.6 The basis for locating earthquake epicenters based on the arrival times of different seismic waves.

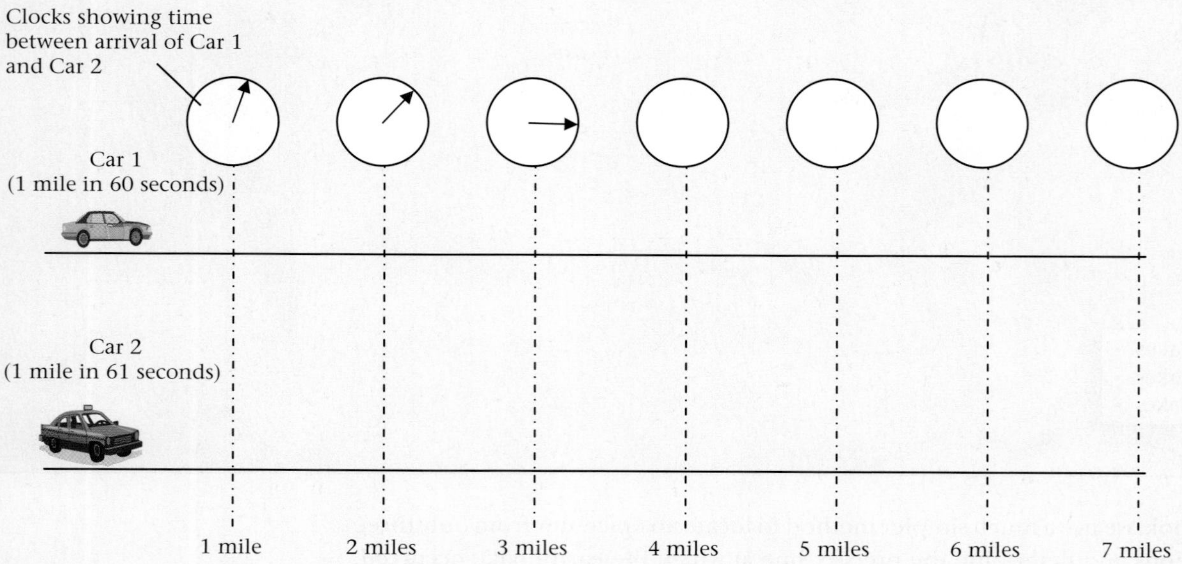

Clocks showing time between arrival of Car 1 and Car 2

Car 1 (1 mile in 60 seconds)

Car 2 (1 mile in 61 seconds)

1 mile   2 miles   3 miles   4 miles   5 miles   6 miles   7 miles

Time of Faulting: You now have enough information to deduce the time at which the two cars started their trip. Knowing that Car 2 lags behind Car 1 by 1 second for every mile they travel, you can calculate how many miles they traveled from their starting point.

(c) If Car 1 arrived at precisely 12:25:00, at what time did it start if it arrived 25 seconds before Car 2? _____
45 seconds before Car 2? _____ (*Hint:* Determine how far the cars traveled and use the velocity of Car 1 to estimate the time it needed to travel that distance.)

Now put the two pieces of reasoning together.

(d) If Car 1 arrives at precisely 1:45:22, followed at 1:46:37 by Car 2, how far did they travel? _____ miles

(e) At what time did they leave? _____ : _____ : _____ (hour:minute:second)

This simple exercise illustrates the reasoning used to determine the distance from a seismometer to an earthquake epicenter. Seismologists have measured velocities of P-, S-, Love, and Rayleigh waves in different rock types and know that P-waves average approximately 6.3 kilometers per *second* at the surface and S-waves about 3.6 km/s. A simple graphical method called a **travel-time diagram** is used to calculate the distance to an epicenter (**Fig. 16.7**).

The red curves in Figure 16.7 show how much time it takes (vertical axis) the four types of seismic waves to travel the distances shown on the horizontal axis. To find the time it takes a P-wave to reach a point 4,000 km from an epicenter, find the intersection of the 4,000 km vertical line with the P-wave travel-time curve. Draw a horizontal line from the intersection to the time axis on the left and read the time required—7 minutes in this example.

FIGURE 16.7 Travel-time curves show the relationship between distance traveled and relative seismic wave velocity.

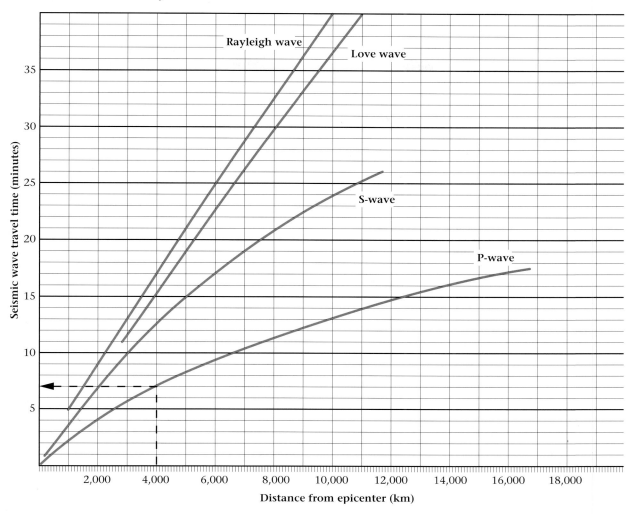

Name: _____   **Section:** _____

Course: _____   **Date:** _____

Refer to the travel-time curve in Figure 16.7.

(a) How long does it take a P-wave to travel 11,000 km? _____ minutes

(b) How long does it take an S-wave to travel 4,000 km? _____ minutes

(c) How long does it take a Love wave to travel 4,000 km? _____ minutes

(d) How long does it take a Rayleigh wave to travel 4,000 km? _____ minutes

(e) Based on these answers, which type of seismic wave is most efficient at transmitting energy? The least efficient? Explain your reasoning.

_____

_____

_____

Name: _____   Section: _____
Course: _____   Date: _____

Park the cars and let's tackle an earthquake. Use the following series of steps to get all the information you need to identify the location of the earthquake and pinpoint when it occurred:

1. Identify the four different seismic waves on seismograms from three stations.

2. Determine the arrival times and measure the delays between different waves.

3. Use this data and the travel-time diagram to estimate each station's distance from the epicenter.

4. Use triangulation to locate the epicenter.

5. Determine the time of faulting with the travel-time diagram.

Step 1: Reading a Seismogram

You need to know what P-, S-, Love, and Rayleigh waves look like on a seismogram to identify them correctly. Each has a unique combination of wavelength and amplitude (**Fig. 16.8**) that makes identification easy.

FIGURE 16.8 Characteristics of seismic waves.

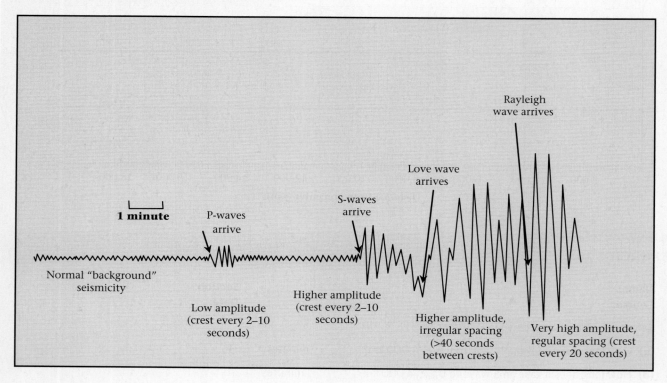

Measuring Time: To determine arrival times, you have to know how to read time on a seismogram and how to recognize the arrival of a new wave (**Fig. 16.9**). The squiggly horizontal lines are the traces of ground motion. Most of the record is normal Earth activity, essentially background noise. The dashed vertical lines are time markers, one minute apart. Time moves forward from left to right along each horizontal row. At the end of a row, time continues on the left side of the next row down (see top two rows in Fig. 16.9). Seismic waves are rarely considerate enough to arrive precisely on a minute marker, so you have to estimate the number of seconds before or after the nearest minute. A millimeter ruler helps, but you may find it useful to prepare your own scale as shown by the green lines in the enlarged inset that divide each minute

continued

Name: _____   **Section:** _____

Course: _____   **Date:** _____

FIGURE 16.9 Recognizing seismic wave arrivals and measuring time on a seismogram.

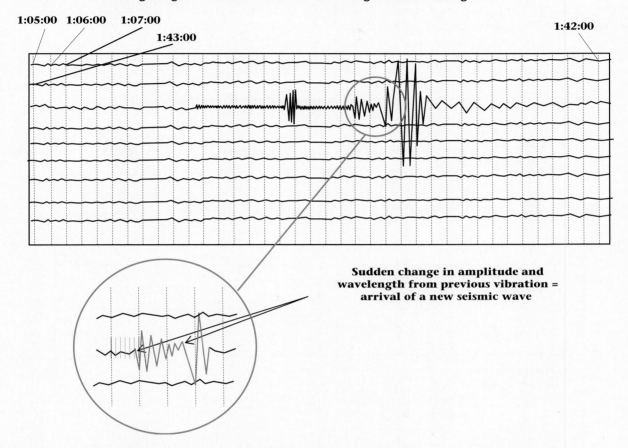

Sudden change in amplitude and wavelength from previous vibration = arrival of a new seismic wave

into 10-second intervals. (*Note:* To avoid confusion involving time zones and changes to and from daylight savings time, seismogram times are standardized to Greenwich mean time (GMT).)

Recognizing Seismic Wave Arrivals: Each seismic wave has a distinctive combination of wave amplitude and frequency. The key to successful epicenter location is determining the arrival times of each of the different waves accurately. Arrival of a new wave is indicated by a sudden change in the amplitude and frequency, as shown by the arrows in the inset in Figure 16.9.

To practice your seismogram reading skills, determine the time of arrival of the

- first wave indicated in the inset. _____:_____:_____
- second wave in the inset. _____:_____:_____
- first wave associated with this earthquake. _____:_____:_____
- last wave associated with this earthquake. _____:_____:_____

continued

Name: _____   **Section:** _____

Course: _____   **Date:** _____

Step 2: Measuring the Delay between Arrival of Different Waves

Figure 16.10 shows seismograms from three stations. Identify each of the waves and record their arrival times in Table 16.1. Then calculate the times between arrivals by subtracting the P-wave arrival time from that of the S-wave, the S-wave from the Love wave, and the Love from the Rayleigh.

Table 16.1 Arrival times and delays between seismic waves (from Fig. 16.10).

Seismic wave arrival times	Urbana	Los Angeles	Austin
P-wave			
S-wave			
Love wave			
Rayleigh wave			
Delays between seismic waves			
S – P			
Love – S			
Rayleigh – Love			

Step 3: Estimating Distance from the Epicenter to Each Station Using the Travel-Time Diagram

Figure 16.11 shows how to use seismic wave delay data to locate the distance from each station to the epicenter. Start with one of the stations and then repeat for the others. Draw an arrow to represent the P-wave arrival anywhere near the bottom on a station worksheet (**Appendix 16.1**) or a graph with exactly the same vertical (time) scale as the travel-time curve (as shown in Fig. 16.11). Then indicate the arrivals of the S-, Love, and Rayleigh waves using the appropriate time delays you recorded in Table 16.1. Slide the station worksheet across the travel-time diagram until all four arrows coincide with the appropriate travel-time curves as shown in the example.

continued

Name: _____   Section: _____

Course: _____   Date: _____

FIGURE 16.10 Seismograms for Exercise 16.4.

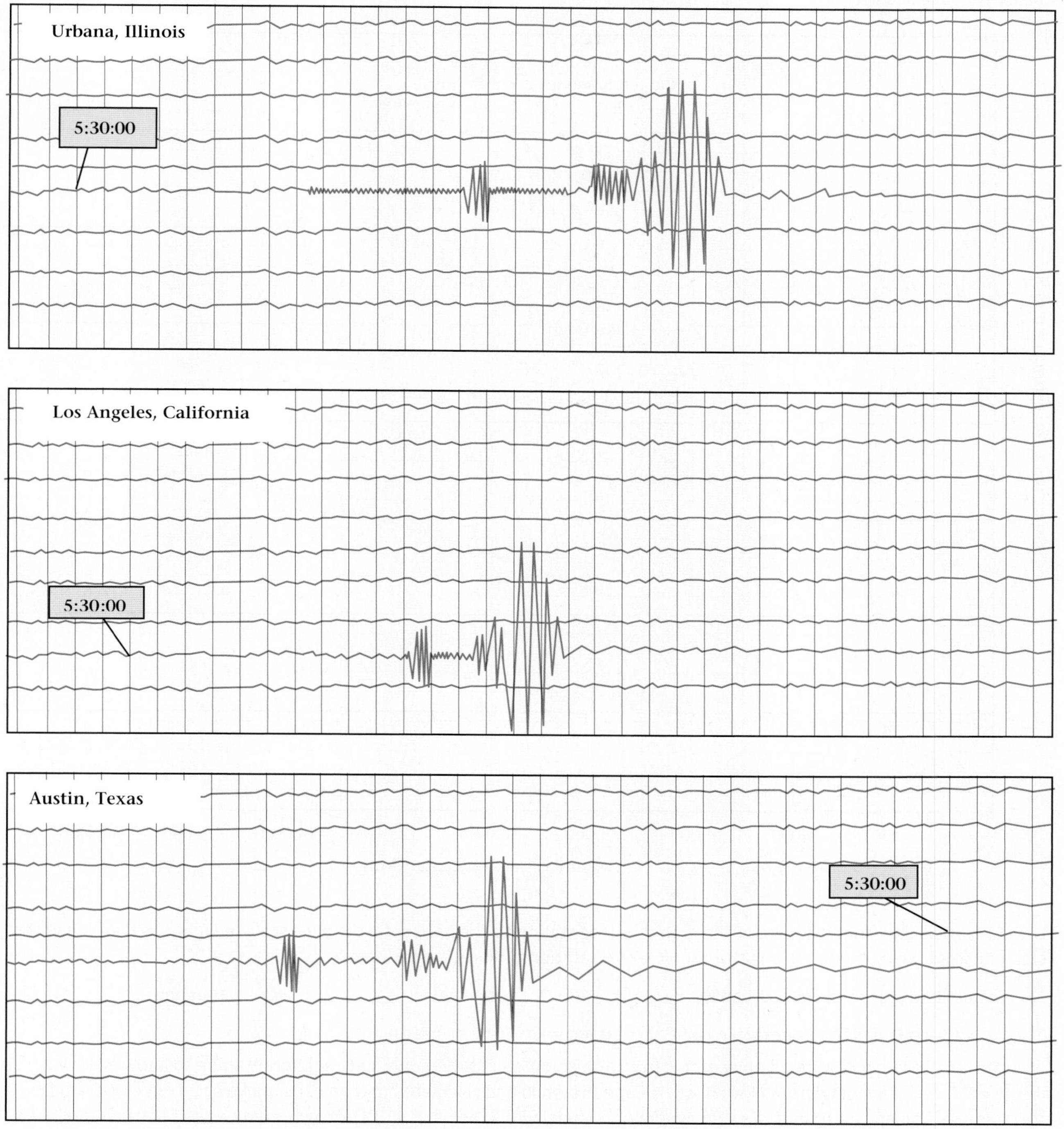

continued

FIGURE 16.11 Using a travel-time diagram to determine distance to an earthquake epicenter and time the earthquake occurred.

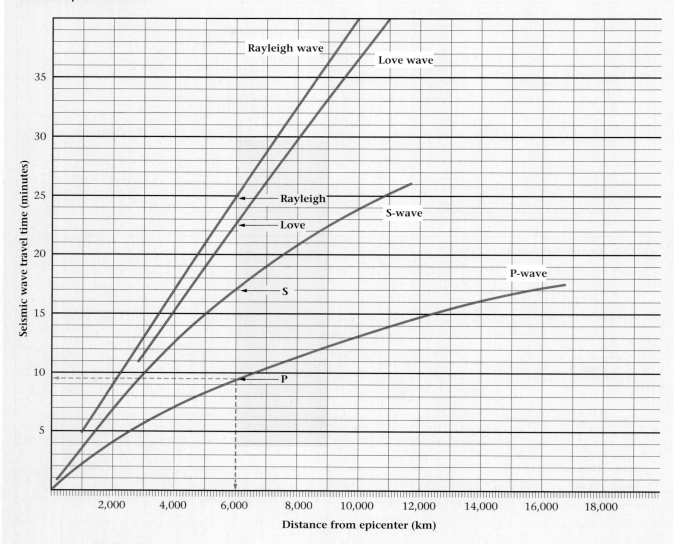

Draw a vertical line from the edge of the graph for that station to the horizontal distance scale in Figure 16.11 and read the distance from the epicenter. Repeat for the other two stations and record the data.

Distance to epicenter from Urbana: _____ km   Los Angeles: _____ km   Austin: _____ km

Step 4: Determining the Location of an Earthquake with Triangulation

It's one thing to know *how far* an epicenter is from a seismometer station, but quite a different thing to know exactly *where* it is (**Fig. 16.12**). Seismologists in Nova Scotia's Cape Breton Island calculated that an earthquake occurred 4,000 km from their station. The epicenter must lie somewhere on a circle with a radius of 4,000 km centered on their station—but that could be in the Atlantic Ocean, Hudson Bay, Mexico's Yucatán peninsula, or the front ranges of the Rocky Mountains.

continued

Name: _____   Section: _____

Course: _____   Date: _____

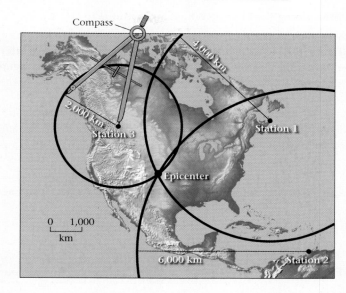

FIGURE 16.12 Locating an earthquake.

A second seismometer station in Caracas, Venezuela, estimates a distance of 6,000 km to the epicenter—somewhere on a circle with a radius of 6,000 km centered on Caracas. The two circles cross in two places, one in the Rockies, the other somewhere in the eastern Atlantic Ocean outside Figure 16.12. The epicenter must be at one of these intersections, but both locations are equally possible. A third station is needed to settle the question, in this case pinpointing the epicenter in the front ranges of the Rocky Mountains. *Data from at least three seismic stations is needed to locate any epicenter, and the process is called* **triangulation**. The more stations used, the more accurate the location.

To locate the epicenter (at last!), use a draftsman's compass or piece of string scaled to the appropriate distance in **Figure 16.13**. Draw an arc representing the distance for the first station and repeat until all three (ideally) intersect at the epicenter.

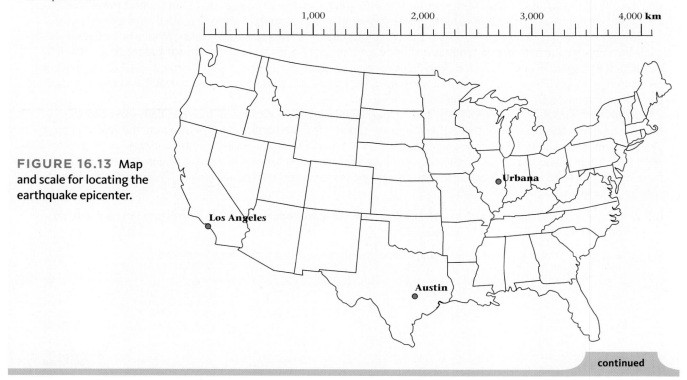

FIGURE 16.13 Map and scale for locating the earthquake epicenter.

continued

Name: _____   **Section:** _____
Course: _____   **Date:** _____

Step 5: Determining the Time at Which an Earthquake Occurred

When, exactly, did the earthquake occur at the epicenter? Align each data graph from Appendix 16.1 with the travel-time curves in Figure 16.7. For each, draw a horizontal line from the **P-wave intercept** on your graph to the time scale on the left side of the travel-time diagram, as shown by the blue dashed arrow in Figure 16.11. Read directly how many minutes the P-wave took to travel to the station, using a millimeter ruler to estimate the number of seconds—in this case, 9 minutes and 35 seconds.

Subtract that amount of time from the P-wave arrival time in Table 16.1 to get the time at which the earthquake occurred. Repeat for the other two stations—the answer should be the same for each.

Time of the earthquake based on: Urbana _____ : _____ : _____   Los Angeles _____ : _____ : _____
Austin _____ : _____ : _____

EXERCISE 16.5   Determining the Magnitude of an Earthquake

Name: _____   **Section:** _____
Course: _____   **Date:** _____

Congratulations! All that remains is to determine the strength of the event. There are two ways to describe the strength of an earthquake. The Mercalli *intensity* scale is based on the amount of damage sustained by buildings. Because damage depends on factors unrelated to the energy released by an earthquake—such as the quality and nature of construction and type of ground beneath buildings—this method is not very useful in our study of the Earth. In 1935, Charles Richter designed the first widely accepted method for estimating energy released in an earthquake—the Richter *magnitude* scale, which is based on the amount of energy released during faulting and calculated from the amount of actual bedrock motion. Each level of magnitude indicates an earthquake with ground motion *10 times greater* than the next lower level. Thus, a magnitude 4 earthquake has 10 times more ground motion than a magnitude 3, and one-tenth that of a magnitude 5.

We now know that Richter's method is accurate only for local, shallow-focus earthquakes. Modern estimates of earthquake strength require different methods using body waves, surface waves, and a *moment magnitude* scale to accurately calculate the magnitude of shallow- and deep-focus, local and distant, and large and small earthquakes. They also depend on complex analyses of the rock type that broke in the fault, how much offset took place at the fault, and other factors that can't be determined from seismic records alone.

In this simplified exercise, you will estimate m_b, the magnitude based on the amplitude of a body wave (P-wave). Because ground motion decreases the farther a seísmic station is from an earthquake, distance from the epicenter must also be taken into account. This is done graphically (**Fig. 16.14**). To determine m_b, mark the left-hand scale at the appropriate S-P delay for one of your stations to account for distance from the epicenter. Measure the maximum P-wave amplitude and mark it on the right-hand scale. Now draw a line connecting these two points. The value for m_b is where the line intersects the center magnitude scale.

(a) What is m_b for an earthquake with exactly the same **P-wave amplitude** as the example in Figure 16.14 but with an S-P delay of 50 seconds? _____ 10 seconds? _____

(b) What is the relationship between wave amplitude, magnitude, and distance from the epicenter?

_____

_____

_____

continued

(c) What is m_b for an earthquake with exactly the same **S-P delay time** as the example in Figure 16.14 but with a P-wave amplitude of 2 mm? _____ 100 mm? _____

(d) What is m_b for the Exercise 16.4 earthquake as recorded at Los Angeles? _____

Note: The seismograms in Figure 16.10 are artificial. If they were from a real earthquake, the values for m_b calculated from each should be nearly identical.

FIGURE 16.14 Determining the body wave magnitude of an earthquake.

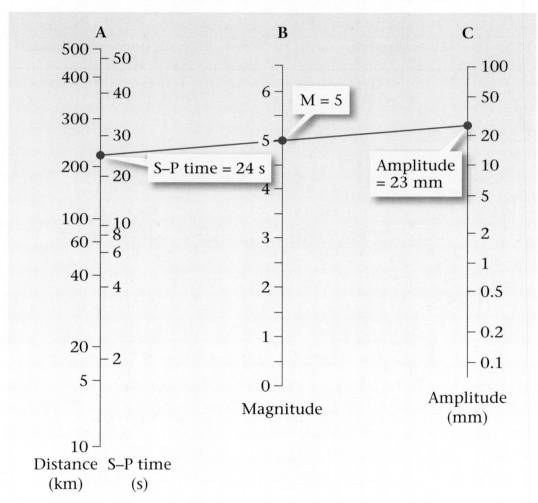

Name: _____   Section: _____

Course: _____   Date: _____

With the modern worldwide network of seismic stations, any earthquake epicenter can be located quickly and accurately. But what about earthquakes that took place *before* seismometers were invented? How can we define Earth's zones of seismic activity if we can't include earthquakes that occurred as (geologically) recently as 100 or 200 years ago? If there are records of damage associated with those events, geologists can estimate epicenter locations by making *isoseismal maps* based on the modified Mercalli intensity scale that show the geographic distribution of damage. First, historic reports of damage are analyzed for each location and given an approximate intensity value. These data are plotted on a map and the map is contoured to show the variability of damage. Ideally, the epicenter is located within the area of greatest damage.

    Figure 16.15 shows Mercalli intensity values for an 1872 earthquake that shook much of the Pacific Northwest. Contour the map to show areas of equal damage (isoseismal areas) and suggest a location for the epicenter of this earthquake.

FIGURE 16.15 Modified Mercalli intensity values for the December 1872 Pacific Northwest earthquake.

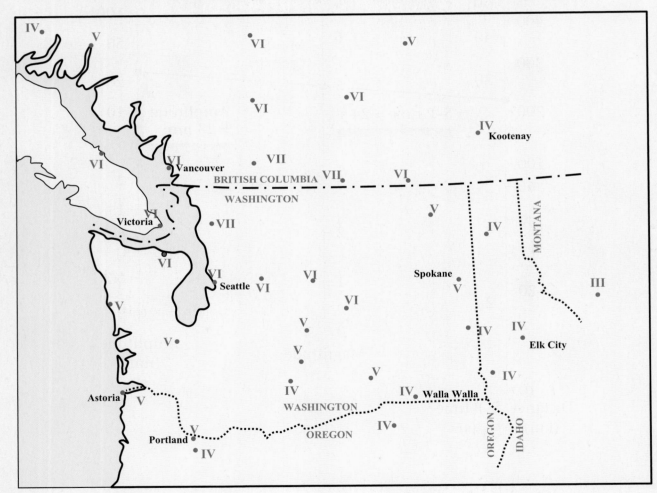

16.4 Predicting Earthquake Hazards: Liquefaction

One of the hazards associated with earthquakes is **liquefaction**, a process by which seismic vibration causes friction between sand grains in waterlogged sediment to be reduced so the sediment loses its ability to support overlying weight and flows like a liquid. At a small scale, this can be annoying (**Fig. 16.16a**), but on a larger scale it can be catastrophic (**Fig. 16.16b**).

Liquefaction requires several key conditions: unconsolidated sediment, pores saturated or nearly saturated with water, and ground vibration. The sediment may be natural, such as rapidly deposited deltaic or other coastal sediments, or sandy fill used to create new land for shoreline development.

Landfill is increasingly used in crowded cities for new housing and business districts, but these could be severely damaged by liquefaction during an earthquake. Emergency planners must prepare for such an event, and they begin by identifying areas in which conditions for liquefaction are likely. **Figure 16.17** shows data from

FIGURE 16.16 Results of liquefaction.

(b) Apartment houses in Niigata, Japan rotated and sank when an earthquake caused liquefaction of the ground beneath them.

(a) Upwelling liquefied sand in San Francisco broke through the sidewalk and curbs, causing minor disruption.

a study of liquefaction potential for San Francisco County. It separates areas where bedrock is exposed at the surface from those underlain by unconsolidated sediment and shows the depth to the water table in the sediment. Examine the map carefully to locate places where liquefaction has occurred in the past and for clues to why it happened in those locations.

Name: _____   Section: _____
Course: _____   Date: _____

FIGURE 16.17 Liquefaction history in San Francisco related to depth to bedrock and water table height.

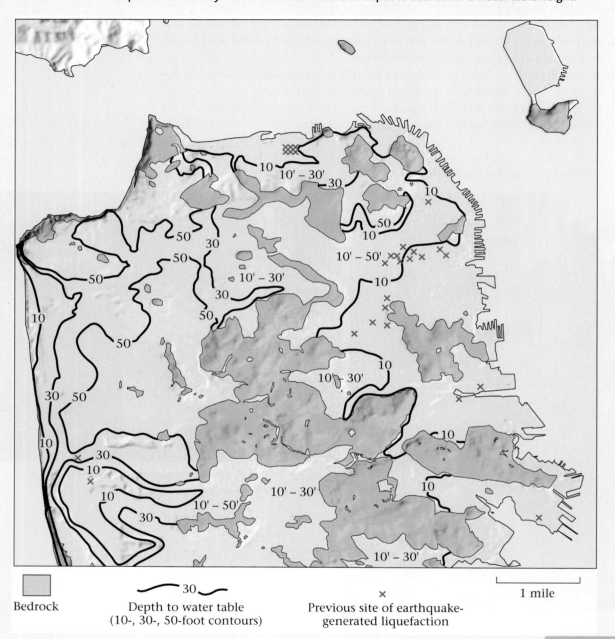

Bedrock

Depth to water table
(10-, 30-, 50-foot contours)

×   Previous site of earthquake-
generated liquefaction

1 mile

continued

Name: _____   **Section:** _____

Course: _____   **Date:** _____

(a) Why is the location of the water table important in a liquefaction-potential study?

_____

_____

_____

_____

(b) Compare the sites of previous liquefaction events with the water table information. What range of water table depths is associated with those events? _____ Shade areas with a colored pencil to show where liquefaction is likely to take place in the future.

(c) Compare San Francisco today with the city in 1898 (**Figure 16.18**). Locate new land areas created by landfill. Color them on the satellite image.

(d) What major structures in these areas might be affected by liquefaction?

_____

_____

_____

_____

continued

FIGURE 16.18 Changes in San Francisco, California.

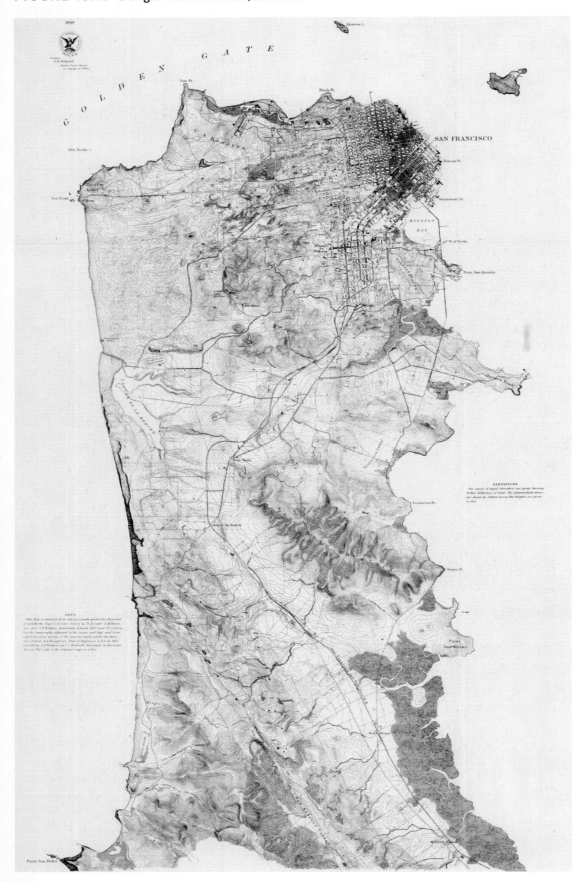

(a) Topographic map, 1898.

FIGURE 16.18 continued.

(b) Satellite image, 2005.

SEISMIC ANALYSIS WORKSHEETS

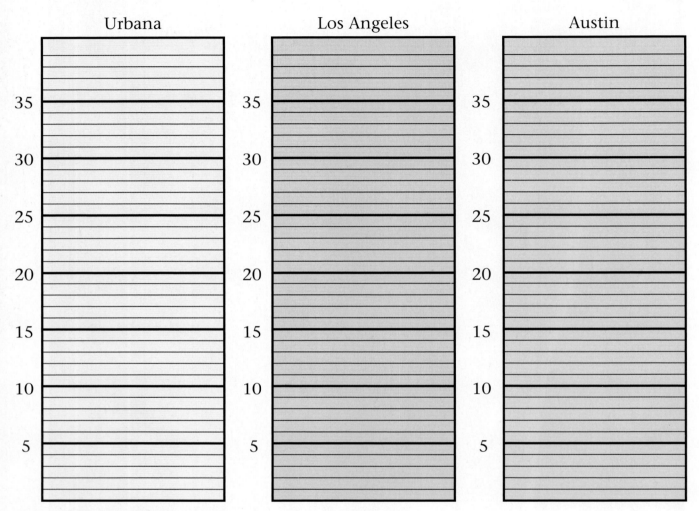

Cut out and use separately for the appropriate seismograph station on Figure 16.11.

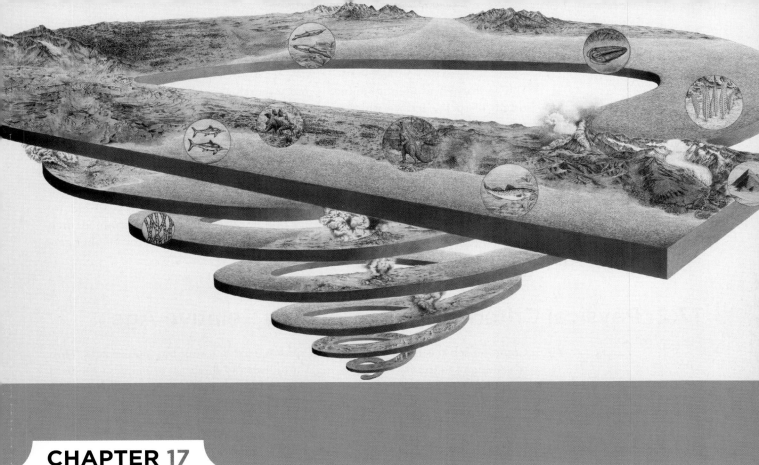

CHAPTER 17

INTERPRETING GEOLOGIC HISTORY: WHAT HAPPENED AND WHEN DID IT HAPPEN?

PURPOSE

- Determine the relative ages of rocks and geologic processes and use these methods to interpret complex geological histories.
- Learn how numeric (absolute) ages of rocks are calculated and apply them to dating geologic materials and events.
- Understand how geologists piece together Earth history from widely separated areas.

MATERIALS NEEDED

- Pen, pencils, calculator
- Draftsman's compass

17.1 Introduction

You've learned to identify minerals, use mineralogy and texture to interpret the origin of rocks, deduce which agents of erosion have affected a given area, and recognize evidence of tectonic events. With these skills you can construct a three-dimensional picture of Earth, using topography and surface map patterns to infer underground relationships.

This chapter adds the fourth dimension—time: the ages of rocks and processes. Geologists ask two different questions about age: "Is a rock or process older or younger than another?" (their **relative** ages) and "Exactly how many years old are they?" (their **numeric** ages). We look first at how relative ages are determined, then at methods for calculating numeric age, and finally combine them to decipher geologic histories of varying complexity.

17.2 Physical Criteria for Determining Relative Age

Common sense is the most important resource for determining relative ages. Most reasoning used in relative age dating is intuitive, and the basic principles were used for hundreds of years before we could measure numeric ages. Geologists use two types of information to determine relative age: **physical methods** based on features in rocks and relationships between them, and **biological methods** that use fossils. We focus first on the physical methods and return to fossils later.

17.2.1 Principles of Original Horizontality and Superposition

The **principle of original horizontality** states that *most* sedimentary rocks are deposited in horizontal beds (there are exceptions, such as inclined sedimentation in alluvial fans, dunes, and deltas). Deposition as horizontal beds makes it easy to determine relative ages in a sequence of sedimentary beds *if the rocks are still in their original horizontal position.*

EXERCISE 17.1   **Relative Ages of Horizontal Rocks**

Name: _____   Section: _____
Course: _____   Date: _____

(a) **Figure 17.1** shows sedimentary rocks in Painted Desert National Park. Assuming original horizontality, label the oldest and the youngest rocks. Where will the oldest rocks be in any sequence of undeformed, horizontal sedimentary rocks? Explain.

_____

_____

_____

(b) Would your answers be different if the rocks were volcanic ash deposits or lava flows? Explain.

_____

_____

_____

continued

FIGURE 17.1 Horizontal strata in Painted Desert National Park, Arizona.

This bit of common sense was first applied to sedimentary rocks by Nils Stensen (Nicolaus Steno in Latin) almost 400 years ago and is called the **principle of superposition**. It cannot be used if rocks have been tilted or folded because, in some cases, the rocks may have been completely overturned so that the oldest is on top.

17.2.2 Principle of Cross-Cutting Relationships

Figure 17.2 shows cross bedding in Zion National Park in Utah. The person in the photograph is pointing to a contact that cuts across several inclined beds.

FIGURE 17.2 Cross-bedded sandstones, Zion National Park, Utah.

Name: _____   **Section:** _____

Course: _____   **Date:** _____

(a) Which is younger? The inclined beds in Figure 17.2 or the layer that cuts them?

_____

(b) Why can't the principle of superposition be used here?

_____

_____

(c) The thick red bed near the top of Figure 17.1 seems to cut into the gray bed below. How could this relationship help determine the relative ages of the layers even if the rocks were turned completely upside down?

_____

_____

_____

Congratulations! Your common sense got these answers right, too. This **principle of cross-cutting relationships** is useful for many types of geologic materials and processes, such as those in **Figure 17.3**.

FIGURE 17.3 Applying the principle of cross-cutting relationships.

(a) Dikes intruding granite of the Sierra Nevada batholith in Yosemite National Park, California.

(b) Vertical fault offsetting volcanic ash deposits in Kingman, Arizona.

(c) Folded sedimentary rocks in Utah.

continued

Name: _____   **Section:** _____

Course: _____   **Date:** _____

(d) An intrusive igneous rock that cuts across other rock units (Fig. 17.3a) must be _____ than the units it cuts across. Using this principle, label the order of intrusion of the granite and dikes in Figure 17.3a.

(e) A fault (Fig. 17.3b) must be _____ than the rocks it offsets. While you're at it, use the principle of original horizontality to label the oldest and youngest (horizontal) volcanic ash layers that the fault cuts.

(f) The process that folded the sedimentary rocks in Figure 17.3c must be _____ than the rocks. Why can't you use the principle of original horizontality to determine the relative ages of the sedimentary layers in this photograph?

_____

_____

Interpret "cross-cutting" broadly and use it not only for features that physically cut others, but also for processes that have *affected* materials, such as in the folding in Figure 17.3c. Contact metamorphism is an example of a process that affects rocks without physically cutting across them, as a lava flow bakes the rock or sediment it flows across. A sill intruded between two sedimentary beds doesn't cut across them, but its contact metamorphism changes their mineralogy and texture. In both cases, the metamorphism, and therefore the igneous rock that caused it, must be younger than the rocks it affects. Consider a horizontal layer of basalt that has been found between two horizontal beds of sandstone.

(g) Using your knowledge of the cross-cutting nature of contact metamorphism, how could you tell whether the basalt was a lava flow or a sill?

_____

_____

_____

17.2.3 Principle of Inclusions

The reasoning here applies to any material found within another rock or sediment, whether the rock or inclusion is igneous, sedimentary, or metamorphic. For example, examine the rock in **Figure 17.4a**, a granitic intrusive containing inclusions

FIGURE 17.4 Inclusions in igneous and sedimentary rocks.

(a) Gabbro inclusions in the Baring Granite in Maine.

(b) Fragments of rhyolite tuff in conglomerate from Maine.

Name: _____   **Section:** _____

Course: _____   **Date:** _____

(a) Which is older, the granite in Figure 17.4a or the xenoliths? Explain.

_____

_____

_____

(b) Clasts in sedimentary rock must be _____ than the sedimentary rock in which they are included. Explain.

_____

_____

_____

of medium-grained gabbro. Inclusions of different rock types in an igneous rock are called **xenoliths** (from the Greek *xenos*, meaning stranger, and *lith*, meaning rock).

Now examine the rock in **Figure 17.4b**. The clasts in this sedimentary conglomerate are fragments of rhyolitic tuff eroded from an Ordovician island arc.

17.2.4 Sedimentary Structures

Some sedimentary features indicate where the top or bottom of a bed was at the time the rock formed. These "top-and-bottom" structures are extremely helpful when sedimentary rocks have been deformed and the principle of superposition can no longer be used. **Figure 17.5** shows some of the features that help determine the relative ages of sedimentary rocks.

FIGURE 17.5 Sedimentary "top-and-bottom" indicators of relative age.

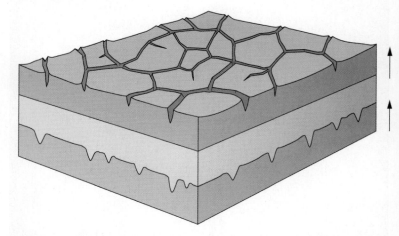

(a) Mud cracks: Mud cracks are widest at the top and narrow downward. The diagram at right is therefore right-side up, indicating that the bottom-most bed is older than the blue-gray bed above it.

(b) Graded beds: The coarsest grains settle first and lie at the bottom of the bed. They are followed by progressively smaller grains, producing a size gradation. Arrows in the diagram at right show how a geologist would interpret the upright nature of the graded beds.

(c) Symmetrical ripple marks: The sharp points of the ripple marks point toward the top of the bed.

(d) Impressions: Features such as dinosaur footprints (left) and raindrop impressions (right) formed when something (here, a dinosaur and raindrops) sank into soft sediment.

Name: _____   Section: _____
Course: _____   Date: _____

(a) Sketch diagrams showing how the three sedimentary structures indicated below would appear in beds that had been turned upside-down.

(i) Mud cracks   (ii) Graded beds   (iii) Symmetrical ripple marks

(b) Are the sedimentary features in Figure 17.5d right-side up or upside-down as shown? Explain your reasoning.

_____

_____

_____

(c) **Figure 17.6** shows the value of sedimentary structures in relative age dating. A cliff face exposing horizontal rock (left side of Fig. 17.6) might be incorrectly interpreted as undeformed horizontal strata without the top-and-bottom information that prove the layers have been folded and some must have been overturned. The right side of Figure 17.6 shows what the geologist might have seen had the adjacent rocks not been eroded away or covered by glacial deposits.

(d) Number the layers, with 1 representing the oldest rock. Which is the youngest?

_____

FIGURE 17.6 Reversals of top-and-bottom features reveal folding.

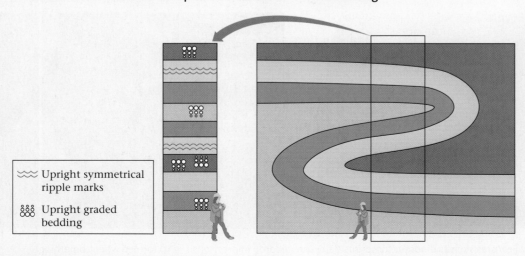

~~~ Upright symmetrical ripple marks

ooo Upright graded bedding

17.3 Unconformities: Evidence for a Gap in the Geologic Record

When rocks are deposited continuously in a basin without interruption by tectonic activity, uplift, or erosion, the result is a stack of parallel beds. Beds in a continuous sequence are said to be **conformable** because each conforms to, or has the same shape and orientation of, the others, as in Figure 17.1.

Tilting, folding, and uplift leading to erosion interrupt this simple history and break the continuity of deposition. In these cases, younger layers are not parallel to the older folded or tilted beds, and erosion may remove large parts of the rock record, leaving a gap in an area's history. We may recognize that deposition was interrupted but can't always tell how long the interruption lasted or what happened during it. A contact indicating a gap in the geologic record is called an **unconformity** because the layer above it is not conformable with those below.

There are three kinds of unconformity (**Fig. 17.7**). A **disconformity** separates parallel beds when some older rocks below the disconformity were removed by erosion. A **nonconformity** is an erosion surface separating older igneous rocks from sedimentary rocks deposited after the pluton was eroded. When rocks above an unconformity cut across folded or tilted rocks below it, the contact is called an **angular unconformity**.

Figure 17.8 shows an angular unconformity in the Grand Canyon that cuts across vertically layered Precambrian metamorphic rocks and light-colored dikes that intrude them. Rocks above the unconformity are nearly horizontal, unmetamorphosed Cambrian sedimentary rocks. This angular unconformity separates rocks that had very different geologic histories, and the gap it represents marks a period of history for which the rock record in this area is missing. Without numerical age dating, we could not know how much geologic history is missing.

In this case, nearly 500,000,000 years (half a billion!) elapsed between the Precambrian metamorphism and dike intrusion and the Cambrian deposition. To put this in perspective, consider that in the 500,000,000 years since the Cambrian rocks were deposited, a supercontinent broke into several plates separated by ocean basins; the ocean basins were subducted, forming a new supercontinent and creating huge mountains; the mountains were worn down by millions of years of erosion; and the second supercontinent also broke into several plates separated by today's oceans and continents.

FIGURE 17.7 Origin of unconformities.

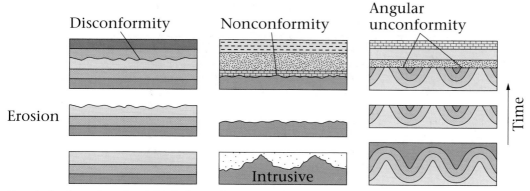

Steps in the formation of each kind of unconformity are shown, with the first at the bottom and the last at the top.

FIGURE 17.8 Angular unconformity in the Grand Canyon.

FIGURE 17.8 Angular unconformity in the Grand Canyon.

The red line marks a profound unconformity separating vertical Precambrian metamorphic rocks (below) and unmetamorphosed, nearly horizontal Cambrian sedimentary rocks (above).

EXERCISE 17.5   **Applying Physical Principles of Relative Age Dating**

Name: _____   Section: _____

Course: _____   Date: _____

Now apply the principles you've just learned to interpreting geologic histories of various degrees of complexity. **Figures 17.9, 17.10, 17.11,** and **17.12** simulate four Grand Canyon–scale cliff exposures in which rock units have been labeled for easy identification. The labels have no meaning with respect to relative age. Arrange them from oldest to youngest and explain your reasoning based on the principles outlined above. You should have no doubt about some relative ages, but others are not so clear cut. Indicate the uncertainties and why they exist. Note that rock units and some contacts are labeled, but the *geologic events* that produced the relationships are not. *Include those events in your histories.*

FIGURE 17.9 Geologic cross section 1.

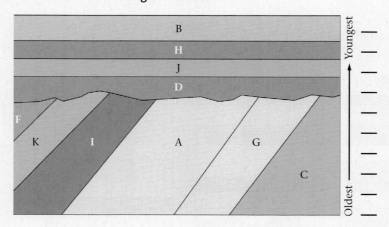

continued

Name: _____   Section: _____

Course: _____   Date: _____

FIGURE 17.10 Geologic cross section 2.

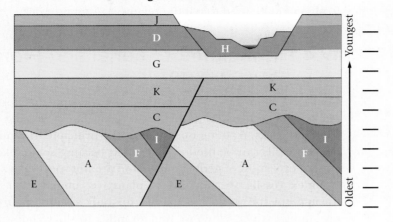

FIGURE 17.11 Geologic cross section 3.

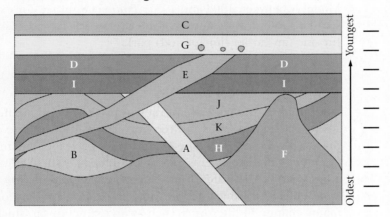

FIGURE 17.12 Geologic cross section 4.

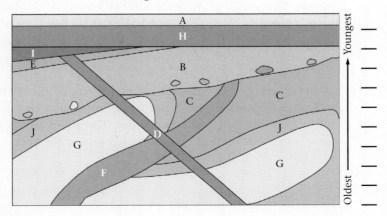

17.4 Biological Methods for Relative Age Dating and Correlation

In 1793, British canal builder and amateur naturalist William Smith noticed different fossils in the layers he was excavating. He discovered wherever he worked that some fossils were always found in rocks that lay above others, and he suggested that fossils could be used to tell the relative ages of the rocks.

17.4.1 Principle of Faunal and Floral Succession

Geologists confirmed his hypothesis by showing that fossil animals (fauna) and plants (flora) throughout the world record an increasing complexity of life forms from old rocks to young ones. This is the **principle of faunal and floral succession**. Not all fossils can be used to date rocks because some, like blue-green algae, have existed over most of geologic time and are not specific to a narrow span.

Index fossils are remains of plants or animals that were distributed widely throughout the world but existed for only a short span of geologic time before becoming extinct. When we find an index fossil, we know that the rock in which it is found dates from that unique segment of time. This is like knowing that a Ford Edsel could only have been made in the three years between 1957 and 1960, or the Model A between 1903 and 1931. *Tyrannosaurus rex*, for example, lived only in the span of geologic time known as the Cretaceous Period; it should not have been in a Jurassic (an earlier time period) park.

17.4.2 The Geologic Time Scale

Determining the sequence of rock units by physical methods and using index fossils to place those units in their correct position in time resulted in the geologic time scale (**Fig. 17.13**). The chart divides geologic time into progressively smaller segments called eons, eras, periods, and epochs (not shown). Era names reveal the complexity of their life forms: Paleozoic, meaning *ancient* life; Mesozoic, meaning *middle* life; and Cenozoic, meaning *recent* life. The end of an era is marked by a major change in life forms, such as an extinction in which most life forms disappear and others fill their ecologic niches. For example, nearly 90% of fossil genera became extinct at the end of the Paleozoic Era, making room for the dinosaurs, and the extinction of the dinosaurs at the end of the Mesozoic Era made room for us mammals. Several period names come from areas where rocks of that particular age were best documented: Devonian from Devonshire in England; Permian from the Perm Basin in Russia; and the Mississippian and Pennsylvanian from U.S. states.

The geologic time scale was originally based entirely on *relative age* dating. We knew that Ordovician rocks and fossils were older than Silurian rocks and fossils, but had no way to tell *how much* older. The ability to calculate the numerical ages came nearly 100 years after the original time scale. Section 17.6 will explore methods of numerical age dating.

17.4.3 Fossil Age Ranges

Some index fossil ages are very specific. The trilobites *Elrathia* and *Redlichia* lived only during the Middle and Early Cambrian, respectively. Others lived over a longer span, like *Cybele* (Ordovician and Silurian) and the brachiopod *Leptaena* (Middle Ordovician to Mississippian). But even index fossils with broad ranges can yield specific information if they occur with other index fossils whose overlap in time limits the possible age of the rock. This can be seen even in the broad fossil groups shown in Figure 17.13. Trilobite, fish, and reptile fossils each span several periods of geologic time, but if specimens of all three are found together, the rock that contains them could only have been formed during the Pennsylvanian or Permian.

FIGURE 17.13 The geologic time scale.

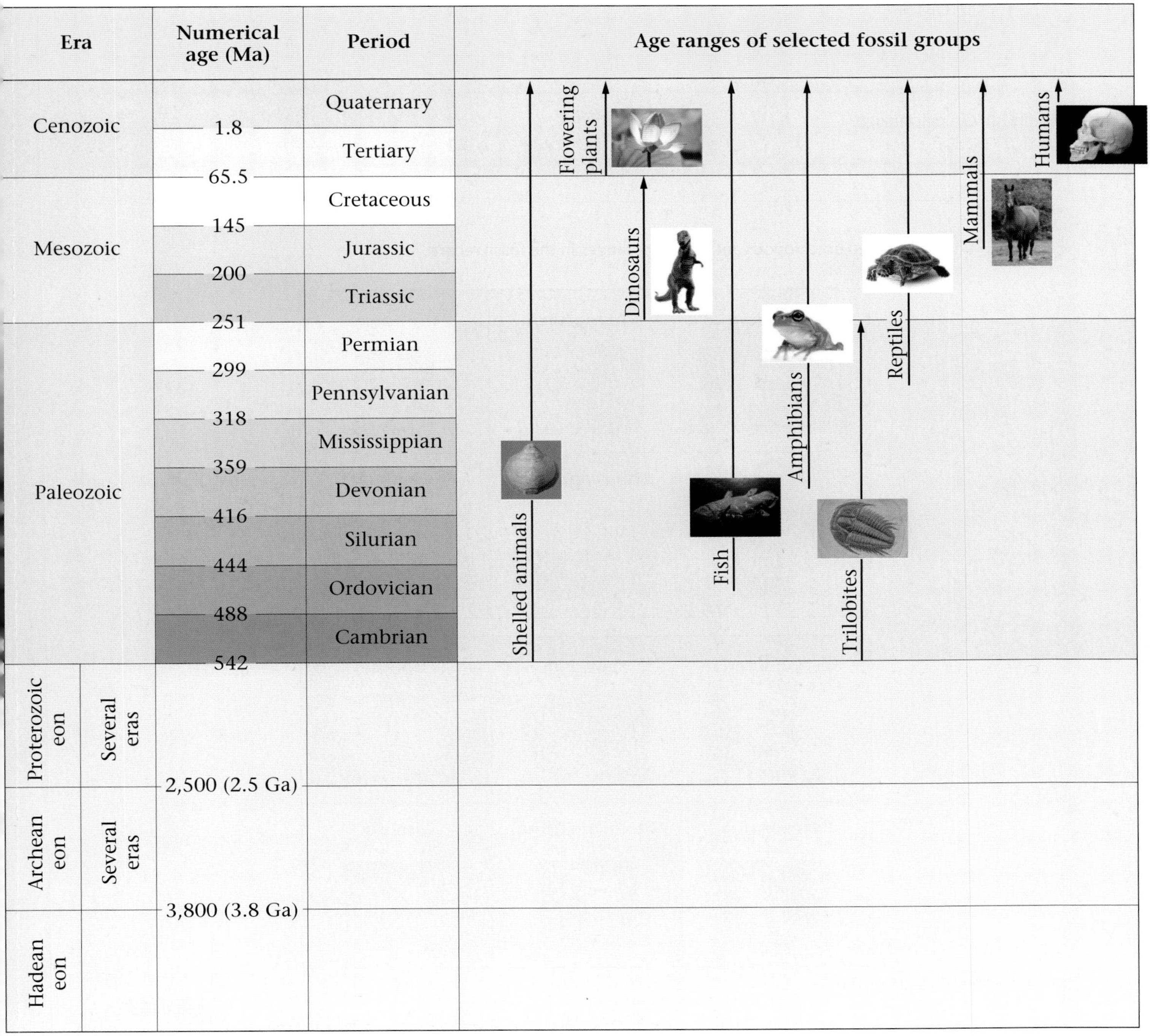

Name: _____ Section: _____
Course: _____ Date: _____

Figure 17.14 shows selected Paleozoic brachiopod species and graphs their ranges within the geologic record.

(a) Based on the overlaps in their ranges, what brachiopod fossil assemblage would indicate

(i) a Permian age?

_____

_____

(ii) a Silurian age?

_____

_____

(iii) an Ordovician age?

_____

_____

FIGURE 17.14 Selected brachiopods and their age ranges in the fossil record.

(a) Selected brachiopod species.

continued

EXERCISE 17.6 continued

Name: ______________________ Section: ____________
Course: ______________________ Date: ____________

FIGURE 17.14 Selected brachiopods and their age ranges in the fossil record. (*con't.*)

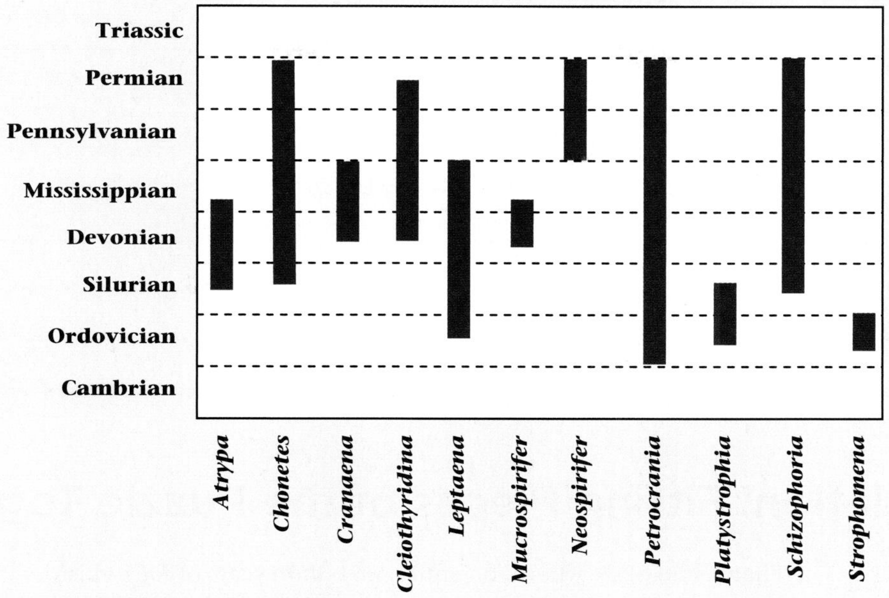

(b) Age ranges of brachiopod species

(b) Now apply overlapping ranges to the cross section in Figure 17.9.

(i) If *Neospirifer* is found in Unit D, *Platystrophia* in F, and *Strophomena* in A, suggest an age for C. Explain your reasoning.

____

____

____

(ii) What is the extent of the gap in the geologic record represented by the angular unconformity (E)?

____

____

____

(c) In Figure 17.10, *Strophomena* is found in E, *Platystrophia* and *Petrocrania* in F, and *Petrocrania*, *Chonetes*, and *Neospirifer* in C.

(i) When were units E, A, F, and I tilted? Explain.

____

____

____

continued

Name: _____   **Section:** _____
Course: _____   **Date:** _____

(ii) What is the extent of the gap in the geologic record represented by the contact between D and the tilted rocks beneath it? Explain.

_____

_____

_____

(iii) When in geologic time did the fault cutting units K, C, I, F, A, and E occur? Explain.

_____

_____

_____

17.5 Correlation: Fitting Pieces of the Puzzle Together

There is no place where the entire 4.6 billion years of Earth history is revealed—not even in the Grand Canyon. To work out a history (and time chart) for the entire planet, we must decipher the records of local areas and combine them to build the big picture. But it is hard to know which sandstone in Kansas was deposited at the same time as sandstones in Pennsylvania—much less in Japan, Kenya, or France—especially if none have index fossils. Geologists use physical and biological methods to **correlate** units from different areas, showing that they formed at the same time.

Name: _____   **Section:** _____
Course: _____   **Date:** _____

Five years ago, geologists determined the relative ages of rocks in two parts of the midcontinent, but the sequences were not the same (**Fig. 17.15**). Unfortunately, similar rocks appear at several places in each section, making it difficult to know exactly which limestone, for example, in the western section correlates with a particular limestone in the east. It is better to compare *sequences* of units based on similar sequences of depositional environments rather than similarities in single rock units.

(a) Examine the two stratigraphic columns below for similar sequences of rock types and environments. Were environments indicated by the sedimentary rocks the same in both places? Was deposition continuous (conformable) in both areas? If not, explain what might have caused the differences and unconformities.

_____

_____

_____

_____

continued

Name: _____   **Section:** _____

Course: _____   **Date:** _____

(b) Suggest correlations between units of the two sequences by connecting the tops and bottoms of units in the western section to those of what you infer are their correlatives in the eastern section. Explain your reasoning.

_____

_____

_____

_____

FIGURE 17.15 Correlation of lithologic sequences.

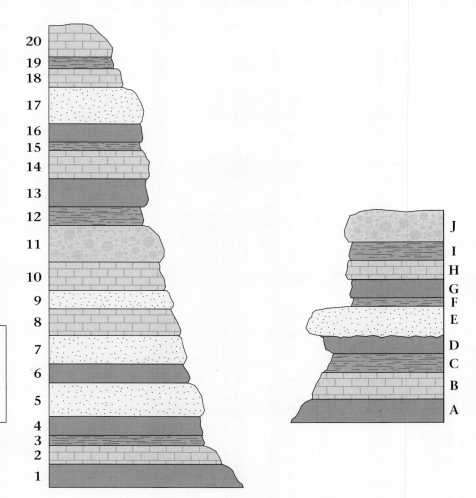

Western sequence ◄——— 250 miles ———► Eastern sequence

Legend:
- Limestone
- Siltstone
- Shale
- Sandstone
- Conglomerate

continued

Name: _____   Section: _____

Course: _____   Date: _____

During field mapping last year, trilobites were found in limestone units in both areas (**Fig. 17.16**). Paleontologists reported that these were identical index fossils from a narrow span of Middle Cambrian time.

(c) Using this additional information, draw lines indicating correlative units in Figure 17.16.

(d) Why does the presence of index fossils make correlation more accurate than correlation based on lithologic and sequence similarities alone?

_____

_____

_____

FIGURE 17.16 Correlation assisted by index fossils.

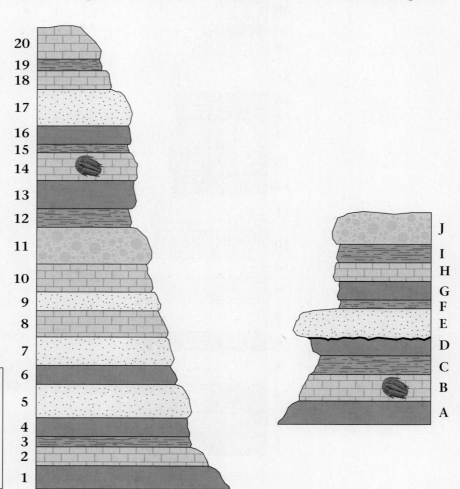

17.6 Numerical Age Dating

The geologic time chart (Fig. 17.13) was used for relative age dating for about 100 years before numerical ages could be added. Numerical age dating is based on the fact that the nuclei of some atoms of elements found in minerals (**parent elements**) decay to form atoms of new elements (**daughter elements**) at a fixed rate *regardless of conditions*.

The decay "clock" begins when a mineral containing the parent element crystallizes during igneous and metamorphic processes. With time, the amount of the parent decreases and the amount of the daughter increases (**Fig. 17.17**). The amount of time the process takes depends on the *decay rates* of the different isotopes. The **half-life** is the amount of time it takes for half of the parent atoms in a mineral sample to decay to an equal number of daughter atoms (the center hourglass).

Table 17.1 lists the isotopes used commonly in numerical age dating, their half-lives, and the minerals used in dating. In general, an isotope can date ages as old as ten of its half-lives; any older than that and there wouldn't be enough parent atoms to measure accurately. Isotopes with short half-lives are therefore best used for relatively recent ages. Conversely, isotopes with very long half-lives, like ^{147}samarium, decay so slowly that they can only be used to date very old rocks.

To calculate the numerical age of a rock, geologists crush it to separate minerals containing the desired isotope. A mass spectrometer determines the parent:daughter ratio and then a logarithmic equation is solved to calculate the age. Your calculation is easier: once you know the parent:daughter ratio, you can use Table 17.2 to calculate age through simple multiplication.

FIGURE 17.17 Parent:daughter ratios during radioactive decay.

Percent parent	100	75	50	25	0
Percent daughter	0	25	50	75	100
Parent:daughter ratio	—	3:1	1:1	1:3	—

TABLE 17.1 Geologically important radioactive decay schemes.

Parent isotope	Daughter decay product	Half-life (years)	Useful dating range (years)	Dateable materials
^{147}Samarium	^{143}Neodymium	106 billion	>10,000,000	Garnets, micas
^{87}Rubidium	^{87}Strontium	48.8 billion	>10,000,000	Potassium-bearing minerals (mica, feldspar, hornblende)
^{238}Uranium	^{206}Lead	4.5 billion	>10,000,000	Uranium-bearing minerals (zircon, apatite, uraninite)
^{235}Uranium	^{207}Lead	713 million	>10,000,000	Uranium-bearing minerals (zircon, apatite, uraninite)
^{40}Potassium	^{40}Argon	1.3 billion	>10,000	Potassium-bearing minerals (mica, feldspar, hornblende)
^{14}Carbon	^{14}Nitrogen	5,370	100 to 70,000	Organic materials

TABLE 17.2 Calculating the numeric age of a rock from decay scheme half-lives.

Percent of parent atoms remaining	Parent: daughter ratio	Number of half-lives elapsed	Multiply half-life by ___ to determine age	Percent of parent atoms remaining	Parent: daughter ratio	Number of half-lives elapsed	Multiply half-life by ___ to determine age
100	—	0	0	35.4	0.547	1½	1.500
98.9	89.90	1/64	0.016	25	0.333	2	2.000
97.9	46.62	1/32	0.031	12.5	0.143	3	3.000
95.8	22.81	1/16	0.062	6.2	0.066	4	4.000
91.7	11.05	1/8	0.125				
84.1	5.289	1/4	0.250				
70.7	2.413	1/2	0.500	0.05		11	Don't bother! There are too few parent atoms to measure accurately enough.
50	1.000	1	1.000	0.025		12	

EXERCISE 17.8   Combining Numerical and Relative Age Dating

Name: _____   Section: _____
Course: _____   Date: _____

(a) First, get practice calculating ages using Tables 17.1 and 17.2. How old is a rock if it contains:

  (i) a ^{235}uranium:^{207}lead ratio of 46.62? _____ years

  (ii) a ^{87}rubidium:^{87}strontium ratio of 89.9? _____ years

  (iii) 6.2% of its original ^{14}carbon? _____ years

  (iv) 97.9% of its original ^{40}potassium? _____ years

Now add more detail to the cross sections in Exercise 17.6.

(b) In Figure 17.11, Dike E has zircon with a ^{235}uranium:^{207}lead ratio of 11.05; Dike A has zircon with a ^{238}uranium:^{206}lead ratio of 22.81; and Pluton F has hornblende with 84.1% of its parent ^{40}potassium.

  (i) How old is E? _____ A? _____ F? _____

  (ii) How old (in years and using period names) are Layers B, H, K, and J?

_____

_____

continued

Name: _____   Section: _____
Course: _____   Date: _____

(iii) When were these layers folded?

_____

_____

(iv) When did the unconformity separating Layer I from the underlying rocks form?

_____

_____

(v) How old are Layers I and D?

_____

_____

(c) In Figure 17.12, Layer H has a thin volcanic ash bed at its base and zircon with a ^{235}uranium:^{207}lead ratio of 46.62; Layer D has zircon with a ^{235}uranium:^{207}lead ratio of 2.413; and Dike F has hornblende with 50% of its parent ^{40}potassium.

 (i) How old is H? _____ D? _____ F? _____

 (ii) When (in years and using period names) were Layers C, G, and J folded?

_____

_____

 (iii) How large a gap in the geologic record is represented by the unconformity below Layer B?

_____

_____

 (iv) How old is Layer I? _____

 (v) If Layer I contains the brachiopods *Neospirifer*, *Chonetes*, and *Schizophoria*, how does this help narrow its possible age?

_____

_____

 (vi) In that case, how large a gap in the geologic record is represented by the unconformity below Layer H?

_____

_____

CREDITS

Chapter 1

Chapter Opener 1: Gary Whitton/Dreamstime.com; **1.2:** NASA; **1.6a:** G.R. Roberts, NSIL; **1.6b:** Belgian Federal Science Policy Office; **1.9a:** Dr. Stephen Hughes, courtesy of Coastal Hydraulic Laboratories; **1.9b:** US Army Corps of Engineers/Steven Hughes; **1.10a:** Courtesy of Stephen Marshak; **1.10b:** Courtesy of Stephen Marshak; **1.11a:** Courtesy of Allan Ludman; **1.11b:** Courtesy of Stephen Marshak.

Chapter 2

Chapter Opener 2: Bertrandb/Dreamstime.com; **2.3:** NGDC/NOAA; **2.11a:** Marine Geoscience Data System/Natural Science Foundation; **2.11b:** Marine Geoscience Data System/Natural Science Foundation; **2.15:** NGDC/NOAA; **2.16:** NGDC/NOAA.

Chapter 3

Chapter Opener 3: Javier Trueba/MSF/Photo Researchers, Inc.; **3.1a:** Courtesy of Allan Ludman; **3.1b:** Courtesy of Allan Ludman; **3.1c:** Courtesy of Allan Ludman; **3.1d:** Courtesy of Allan Ludman; **3.2a:** Courtesy of Allan Ludman; **3.2b:** Courtesy of Allan Ludman; **3.4a:** Richard P. Jacobs/JLM Visuals; **3.4b:** Scientifica/Visuals Unlimited, Inc.; **3.4c:** Courtesy of Stephen Marshak; **3.4d:** Courtesy of Allan Ludman; **3.4e:** Courtesy of Allan Ludman; **3.4f:** Courtesy of Allan Ludman; **3.4g:** 1995-1998 by Amethyst Galleries, Inc. http://mineral.galleries.com; **3.4h:** 1993 Jeff Scovil; **3.5:** Courtesy of Allan Ludman; **3.6 top left:** Richard P. Jacobs/JLM Visuals; **3.6 top right:** Courtesy of Allan Ludman; **3.6 bottom left:** Courtesy of Allan Ludman; **3.6 bottom right:** Courtesy of Allan Ludman.

Chapter 4

Chapter Opener 4 left: Spbphoto/Dreamstime.com; **Chapter Opener 4 center:** Scientifica/Visuals Unlimited, Inc.; **Chapter Opener 4 right:** Ralph Hutchings/Visuals Unlimited, Inc.; **4.2b:** Courtesy of Stephen Marshak; **4.2a:** Courtesy of Stephen Marshak; **4.2c:** Courtesy of Stephen Marshak; **4.2c:** Joel Arem/Photo Researchers, Inc.; **4.2d:** Courtesy of Stephen Marshak; **4.3a:** Dr. John D. Cunningham/Visuals Unlimited, Inc.; **4.3b:** Courtesy of Allan Ludman; **4.3c:** Marli Miller/Visuals Unlimited, Inc.; **4.3d:** Ken Lucas/Visuals Unlimited, Inc.; **4.4a:** Courtesy of Stephen Marshak; **4.4b:** Courtesy of Stephen Marshak; **4.4c:** Rafael Laguillo/Dreamstime.com; **4.4d:** Courtesy of Allan Ludman.

Chapter 5

Chapter Opener 5: Patricia Marroquin/Dreamstime.com; **5.2a:** Courtesy of Allan Ludman; **5.2b:** Albert Copley/Visuals Unlimited, Inc.; **5.2c:** Courtesy of Allan Ludman; **5.3a:** Courtesy of Allan Ludman; **5.3b:** Courtesy of Allan Ludman; **5.3c:** Courtesy of Stephen Marshak; **5.3d:** Marli Miller/Visuals Unlimited, Inc.; **5.4:** Doug Sokell/Visuals Unlimited; **5.5a:** Courtesy of Stephen Marshak; **5.5b:** Courtesy of Stephen Marshak; **5.6a:** Courtesy of Douglas W Rankin; **5.6b:** Courtesy of Allan Ludman; **5.6c:** Courtesy of Allan Ludman; **5.9a:** Marli Miller/Visuals Unlimited; **5.9b:** Courtesy of Allan Ludman; **5.11:** Courtesy of Allan Ludman.

Chapter 6

Chapter Opener 6: Paul Moore/Dreamstime.com; **6.1a:** Courtesy of Allan Ludman; **6.1b:** Courtesy of Allan Ludman; **6.1c:** Courtesy of Allan Ludman; **6.2a:** Courtesy of Allan Ludman; **6.2b:** Courtesy of Allan Ludman; **6.4a:** Courtesy of Allan Ludman; **6.4b:** Courtesy of Allan Ludman; **6.6a:** Courtesy of Allan Ludman; **6.6b:** Courtesy of Allan Ludman; **6.7a:** Courtesy of Allan Ludman; **6.7b:** Courtesy of Allan Ludman; **6.8a:** Courtesy of Allan Ludman; **6.8b:** Courtesy of Allan Ludman; **6.9:** Courtesy of Allan Ludman; **6.10a:** Courtesy of Allan Ludman; **6.10b:** Courtesy of Allan Ludman; **6.11:** Dr. Marli Miller/Visuals Unlimited, Inc.; **6.12a:** Dr. Marli Miller/Visuals Unlimited, Inc.; **6.12b:** Dr. Marli Miller/Visuals Unlimited, Inc.; **6.13a:** Courtesy of Stephen Marshak; **6.13b:** Omikron/Photo Researchers, Inc.; **6.14a top left:** Kandi Traxel/Dreamstime.com; **6.14a top right:** Dirk Wiersma/SPL/Photo Researchers, Inc.; **6.14a center right:** Imv/Dreamstime.com; **6.14a bottom left:** Dr. John D. Cunningham/Visuals Unlimited, Inc.; **6.14a bottom right:** Bloopiers/Dreamstime.com; **6.14b top left:** American Museum of Natural History; **6.14b center right:** Xiao Bin Lin/Dreamstime.com; **6.14b bottom left:** Sam Pierson/Photo Researchers, Inc.; **6.15a:** Courtesy of Allan Ludman; **6.15b:** Courtesy of Stephen Marshak.

Chapter 7

Chapter Opener 7: Andreas Strauss/Getty Images; **7.1a:** Courtesy of Allan Ludman; **7.1b:** Courtesy of Allan Ludman; **7.4a:** Courtesy of Allan Ludman; **7.5a:** Courtesy of Allan Ludman; **7.5b:** Courtesy of Allan Ludman; **7.5c:** Scientifica/Visuals Unlimited, Inc.; **7.5d:** Courtesy of Allan Ludman; **7.6a:** Courtesy of Allan Ludman; **7.6b:** Courtesy of Allan Ludman; **7.6c:** Courtesy of Allan Ludman; **7.6d:** Courtesy of Allan Ludman; **7.7a:** Courtesy of Allan Ludman; **7.7b:** University of Vermont Digital Repository; **7.7c:** Scientifica/Visuals Unlimited, Inc.; **7.7d:** Dirk Wiersma/SPL/Photo Researchers, Inc.; **7.8:** Courtesy of Allan Ludman; **7.9a:** Courtesy of Allan Ludman; **7.9b:** Courtesy of Allan Ludman; **7.9c:** Courtesy of Allan Ludman;

Chapter 8

Chapter Opener 8: Bandit/Dreamstime.com; **8.1a:** TOPO! (c)2009 National Geographic; **8.1b:** TOPO! (c)2009 National Geographic; **8.1c:** TOPO! (c)2009 National Geographic; **8.1d:** TOPO! (c)2009 National Geographic; **8.7a:** USGS; **8.7b:** USGS; **8.8:** USGS; **8.10a:** TOPO! (c)2009 National Geographic; **8.10b:** USGS; **8.10c:** TOPO! (c)2009 National Geographic; **8.12:** TOPO! (c)2009 National Geographic.

Chapter 9

Chapter Opener 9: TOPO! (c)2009 National Geographic; **9.5:** TOPO! (c)2009 National Geographic; **9.6:** TOPO! (c)2009 National Geographic; **9.10:** Willtu/Dreamstime.com; **9.10:** Outdoorsman/Dreamstime.com; **9.10:** Mikhail Lavrenov/Dreamstime.com; **9.10:** Mike Tan/Dreamstime.com; **9.10:** Harris Shiffman/Dreamstime.com; **9.10:** Gavril Margittai/Dreamstime.com;

9.14a: TOPO! (c)2009 National Geographic; 9.16a: Jim Wark/Visuals Unlimited, Inc.; 9.16b: Gene Blevins/LA Daily News/Corbis; 9.17: TOPO! (c)2009 National Geographic; 9.18: TOPO! (c)2009 National Geographic; 9.19: TOPO! (c)2009 National Geographic; A9.1: USGS.

Chapter 10
Chapter Opener 10: Blake Darche/Dreamstime.com; Chapter Opener 10: Jay McCain/Dreamstime.com;Chapter Opener 10: Pierre Jean Durieu/Dreamstime.com; 10.1a: Courtesy of Allan Ludman; 10.1b: Courtesy of Allan Ludman; 10.4: TOPO! (c)2009 National Geographic; 10.5: TOPO! (c)2009 National Geographic; 10.6: TOPO! (c)2009 National Geographic; 10.6 inset: TOPO! (c)2009 National Geographic; 10.9a: TOPO! (c)2009 National Geographic; 10.9b: TOPO! (c)2009 National Geographic; 10.9c: TOPO! (c)2009 National Geographic; 10.11: TOPO! (c)2009 National Geographic; 10.14: TOPO! (c)2009 National Geographic; 10.15: TOPO! (c)2009 National Geographic; 10.16: TOPO! (c)2009 National Geographic; 10.17: TOPO! (c)2009 National Geographic; 10.18: TOPO! (c)2009 National Geographic; 10.20a: USGS/Perry Rahn; 10.20b: USGS/Perry Rahn; 10.20c: USGS/Perry Rahn; 10.20d: USGS/Perry Rahn; 10.21: TOPO! (c)2009 National Geographic.

Chapter 11
Chapter Opener 11: Jkey/Dreamstime.com; 11.1: Ben Renard-wiart/Dreamstime.com; 11.2a: Dr. Marli Miller/Visuals Unlimited, Inc.; 11.2b: Ashely Cooper/Visuals Unlimited; Inc.; 11.3a: Courtesy of Stephen Marshak; 11.3b: Courtesy Allan Ludman; 11.4a: Courtesy of Allan Ludman; 11.4b: Courtesy of Allan Ludamn; 11.5: Gail Johnson/Dreamstime.com; 11.6: TOPO! (c)2009 National Geographic; 11.7: TOPO! (c)2009 National Geographic; 11.9: Marine Geoscience Data System/Natural Science Foundation; 11.10: TOPO! (c)2009 National Geographic; 11.11a: TOPO! (c)2009 National Geographic; 11.11b: TOPO! (c)2009 National Geographic; 11.12a: TOPO! (c)2009 National Geographic; 11.12b: TOPO! (c)2009 National Geographic; 11.16: Courtesy of Allan Ludman; 11.17: TOPO! (c)2009 National Geographic; 11.18 : TOPO! (c)2009 National Geographic; 11.19: TOPO! (c)2009 National Geographic.

Chapter 12
Chapter Opener 12: Shariff Che' Lah/Dreamstime.com; 12.1a: Courtesy of Allan Ludman; 12.1b: Courtesy of Allan Ludman; 12.5a: Courtesy of Stephen Marshak; 12.6: TOPO! (c)2009 National Geographic; 12.7: TOPO! (c)2009 National Geographic; 12.9: TOPO! (c)2009 National Geographic.

Chapter 13
Chapter Opener 13: Rollie Rodriguez/Alamy; 13.1a: Courtesy of Stephen Marshal; 13.1b: Courtesy of Allan Ludman; 13.1c: Magda Moiola/Dreamstime.com; 13.2a: USGS EROS Data Center; 13.2b: Pixonnet.com/Alamy; 13.2c: Ron Chapple Studios/Dreamstime.com; 13.3a: Courtesy of Allan Ludman; 13.3b: Courtesy of Allan Ludman; 13.3c: ClassicStock/Alamy; 13.3d: Courtesy of Stephen Marshak; 13.4e: Duncan Gilbert/Dreamstime.com; 13.5: TOPO! (c)2009 National Geographic; 13.6: TOPO! (c)2009 National Geographic; 13.7: TOPO! (c)2009 National Geographic; 13.8: TOPO! (c)2009 National Geographic; 13.9: TOPO! (c)2009 National Geographic; 13.11a: Sinclair Stammers/SPL/Photo Researchers; 13.12: TOPO! (c)2009 National Geographic; 13.13: TOPO! (c)2009 National Geographic.

Chapter 14
Chapter Opener 14: John Kershner/Dreamstime.com; Chapter Opener 14: Fever Pitch Productions/Dreamstime.com; Chapter Opener 14: Jonathlee/Dreamstime.com; 14.2a: Courtesy of Allan Ludman; 14.2b: Courtesy of Allan Ludman; 14.3a: Courtesy of Allan Ludman; 14.3b: Courtesy of Allan Ludman; 14.3c: Courtesy of Allan Ludman; 14.3d: Courtesy of Allan Ludman; 14.3e: Courtesy of Stephen Marshak; 14.3f : Courtesy of Stephen Marshak; 14.3g: Courtesy of Stephen Marshak; 14.4: Courtesy of Allan Ludman; 14.5: TOPO! (c)2009 National Geographic; 14.6: TOPO! (c)2009 National Geographic; 14.7: TOPO! (c)2009 National Geographic; 14.8: TOPO! (c)2009 National Geographic; 14.8 inset: TOPO! (c)2009 National Geographic; 14.9a: G.R. Roberts (c) NSIL; 14.9b: Courtesy of Stephen Marshak; 14.9e: Keith Livingston/Dreamstime.com; 14.9f: Courtesy of Allan Ludman; 14.10: TOPO! (c)2009 National Geographic; 14.11: AP Photo/Phil Coale; 14.12: TOPO! (c)2009 National Geographic; 14.13a: TOPO! (c)2009 National Geographic; 14.13b: TOPO! (c)2009 National Geographic; 14.14a: TOPO! (c)2009 National Geographic; 14.14b: TOPO! (c)2009 National Geographic; 14.15: GOES 12 Satellite, NASA, NOAA; 14.16: AP Photo/Dave Martin; 14.17a: AP Photo/John David Mercer, Pool; 14.17b: AP Photo/Rogelio Solis; 14.17c: AP Photo/David J. Phillip; 14.17d: AP Photo/Vincent Laforet, Pool; 14.18a: TOPO! (c)2009 National Geographic; 14.18b: TOPO! (c)2009 National Geographic; 14.21: TOPO! (c)2009 National Geographic.

Chapter 15
Chapter Opener 15: Kevin Walsh/Dreamstime.com; 15.2a: John S. Shelton; 15.13: Maine Geological Survey; 15.24b: NED/EROS/USGS; 15.22a: TOPO! (c)2009 National Geographic; 15.22b: TOPO! (c)2009 National Geographic; 15.23: (c) 1980 by the Grand Canyon Natural History Association; 15.24: USGS; 15.25: José F. Vigil, Richard J. Pike, and David G. Howell/USGS.

Chapter 16
Chapter Opener 16: Oriontrail/Dreamstime.com; 16.16b: NGDC; 16.18a: David Rumsey Map Collection, www.davidrumsey.com; 16.18b: NASA.

Chapter 17
Chapter Opener 17: The Natural History Museum/Alamy; 17.1: Suchan/Dreamstime.com; 17.2: Courtesy of Allan Ludman; 17.3a: Courtesy of Allan Ludman; 17.3b: Courtesy of Allan Ludman; 17.3c: Courtesy of Allan Ludman; 17.4a: Courtesy of Allan Ludman; 17.4b: Courtesy of Allan Ludman; 17.5a: Courtesy of Stephen Marshak; 17.5b: Courtesy of Allan Ludman; 17.5c: Courtesy of Allan Ludman; 17.5d: Dr. Marli Miller/Visuals Unlimited, Inc.; 17.5d: Omikron/Photo Researchers, Inc.; 17.8: Courtesy of Allan Ludman 17.13: Mishoo/Dreamstime.com; 17.13: Stef Bennett/Dreamstime.com; 17.13: Warren Rosenberg/Dreamstime.com; 17.13: Pawe Szpytma/Dreamstime.com; 17.13: Martin Valigursky/Dreamstime.com; 17.13: José Alonso/Dreamstime.com; 17.13: American Museum of Natural History; 17.13: David Fleetham/Visuals Unlimited, Inc.; 17.13: Scott P. Orr/istockphoto.com; 17.14a: American Museum of Natural History.

Cut along solid red lines

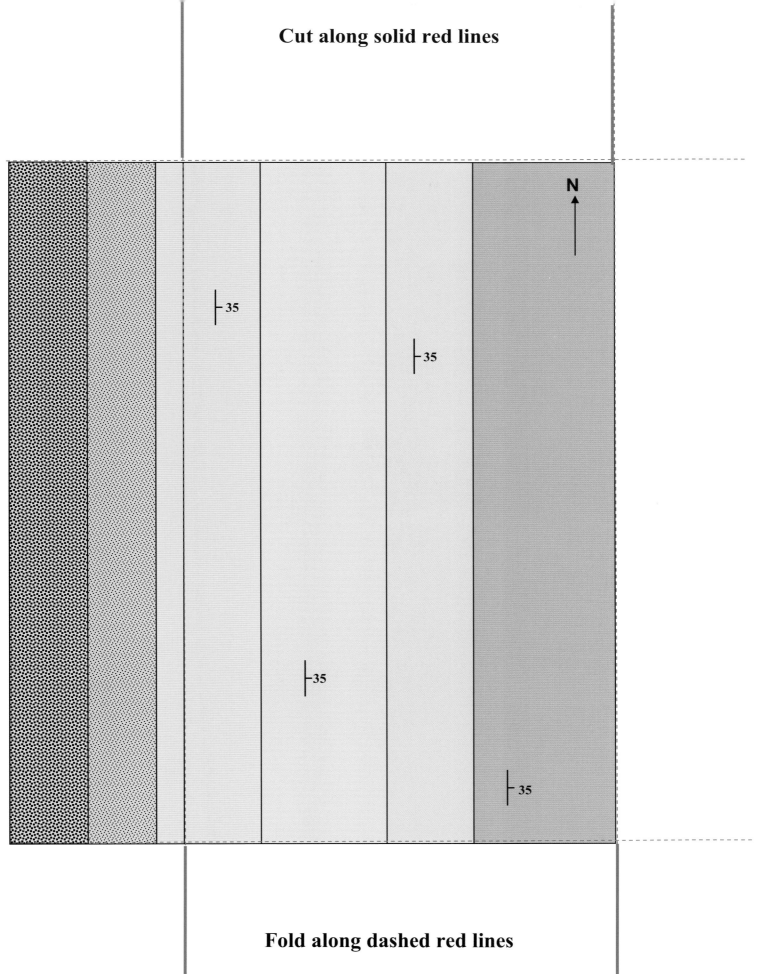

Fold along dashed red lines

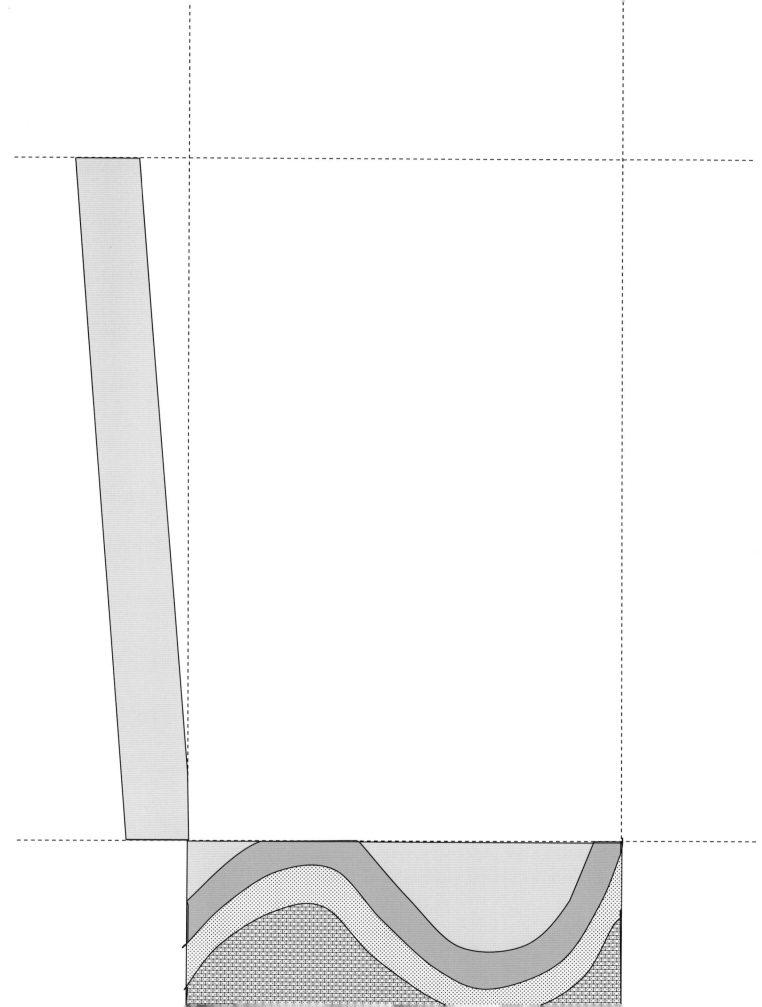

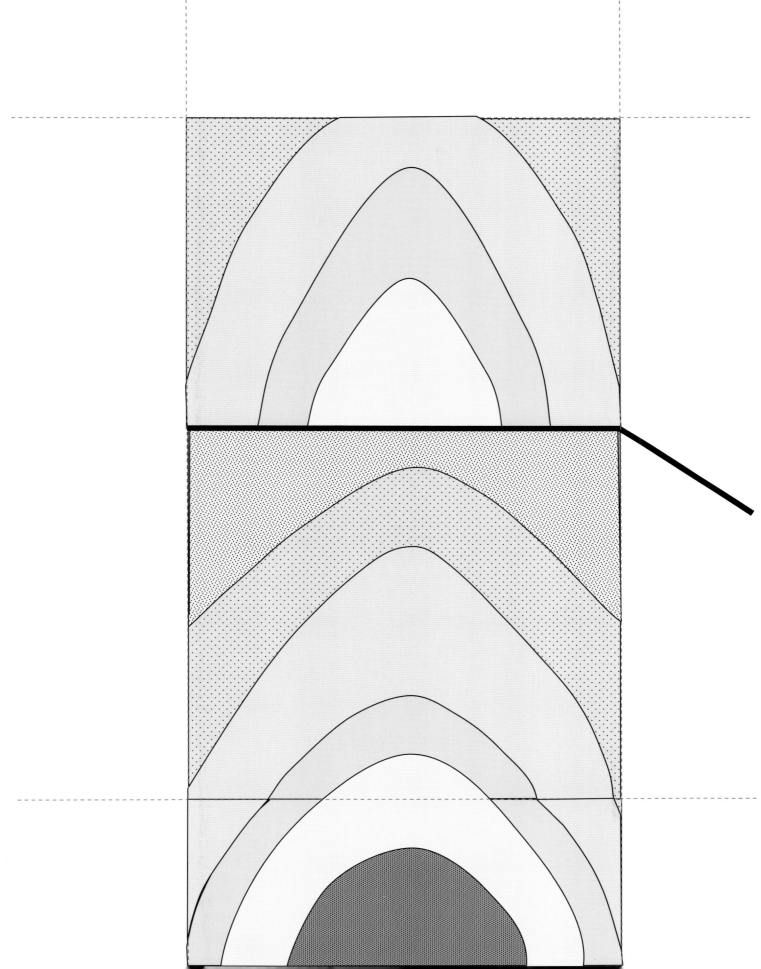